An Introduction to Statistical Inference and Its Applications with R

CHAPMAN & HALL/CRC
Texts in Statistical Science Series

Analysis of Failure and Survival Data
P. J. Smith

The Analysis of Time Series — An Introduction, Sixth Edition
C. Chatfield

Applied Bayesian Forecasting and Time Series Analysis
A. Pole, M. West and J. Harrison

Applied Nonparametric Statistical Methods, Fourth Edition
P. Sprent and N.C. Smeeton

Applied Statistics — Handbook of GENSTAT Analysis
E.J. Snell and H. Simpson

Applied Statistics — Principles and Examples
D.R. Cox and E.J. Snell

Applied Stochastic Modelling, Second Edition
B.J.T. Morgan

Bayesian Data Analysis, Second Edition
A. Gelman, J.B. Carlin, H.S. Stern and D.B. Rubin

Bayesian Methods for Data Analysis, Third Edition
B.P. Carlin and T.A. Louis

Beyond ANOVA — Basics of Applied Statistics
R.G. Miller, Jr.

Computer-Aided Multivariate Analysis, Fourth Edition
A.A. Afifi and V.A. Clark

A Course in Categorical Data Analysis
T. Leonard

A Course in Large Sample Theory
T.S. Ferguson

Data Driven Statistical Methods
P. Sprent

Decision Analysis — A Bayesian Approach
J.Q. Smith

Elementary Applications of Probability Theory, Second Edition
H.C. Tuckwell

Elements of Simulation
B.J.T. Morgan

Epidemiology — Study Design and Data Analysis, Second Edition
M. Woodward

Essential Statistics, Fourth Edition
D.A.G. Rees

Extending the Linear Model with R — Generalized Linear, Mixed Effects and Nonparametric Regression Models
J.J. Faraway

A First Course in Linear Model Theory
N. Ravishanker and D.K. Dey

Generalized Additive Models: An Introduction with R
S. Wood

Interpreting Data — A First Course in Statistics
A.J.B. Anderson

An Introduction to Generalized Linear Models, Third Edition
A.J. Dobson and A.G. Barnett

Introduction to Multivariate Analysis
C. Chatfield and A.J. Collins

Introduction to Optimization Methods and Their Applications in Statistics
B.S. Everitt

Introduction to Probability with R
K. Baclawski

Introduction to Randomized Controlled Clinical Trials, Second Edition
J.N.S. Matthews

Introduction to Statistical Inference and Its Applications with R
M.W. Trosset

Introduction to Statistical Methods for Clinical Trials
T.D. Cook and D.L. DeMets

Large Sample Methods in Statistics
P.K. Sen and J. da Motta Singer

Linear Models with R
J.J. Faraway

Logistic Regression Models
J.M. Hilbe

Markov Chain Monte Carlo — Stochastic Simulation for Bayesian Inference, Second Edition
D. Gamerman and H.F. Lopes

Mathematical Statistics
K. Knight

Modeling and Analysis of Stochastic Systems
V. Kulkarni

Modelling Binary Data, Second Edition
D. Collett

Modelling Survival Data in Medical Research, Second Edition
D. Collett

Multivariate Analysis of Variance and Repeated Measures — A Practical Approach for Behavioural Scientists
D.J. Hand and C.C. Taylor

Multivariate Statistics — A Practical Approach
B. Flury and H. Riedwyl

Pólya Urn Models
H. Mahmoud

Practical Data Analysis for Designed Experiments
B.S. Yandell

Practical Longitudinal Data Analysis
D.J. Hand and M. Crowder

Practical Statistics for Medical Research
D.G. Altman

A Primer on Linear Models
J.F. Monahan

Probability — Methods and Measurement
A. O'Hagan

Problem Solving — A Statistician's Guide, Second Edition
C. Chatfield

Randomization, Bootstrap and Monte Carlo Methods in Biology, Third Edition
B.F.J. Manly

Readings in Decision Analysis
S. French

Sampling Methodologies with Applications
P.S.R.S. Rao

Statistical Analysis of Reliability Data
M.J. Crowder, A.C. Kimber, T.J. Sweeting, and R.L. Smith

Statistical Methods for Spatial Data Analysis
O. Schabenberger and C.A. Gotway

Statistical Methods for SPC and TQM
D. Bissell

Statistical Methods in Agriculture and Experimental Biology, Second Edition
R. Mead, R.N. Curnow, and A.M. Hasted

Statistical Process Control — Theory and Practice, Third Edition
G.B. Wetherill and D.W. Brown

Statistical Theory, Fourth Edition
B.W. Lindgren

Statistics for Accountants
S. Letchford

Statistics for Epidemiology
N.P. Jewell

Statistics for Technology — A Course in Applied Statistics, Third Edition
C. Chatfield

Statistics in Engineering — A Practical Approach
A.V. Metcalfe

Statistics in Research and Development, Second Edition
R. Caulcutt

Survival Analysis Using S — Analysis of Time-to-Event Data
M. Tableman and J.S. Kim

The Theory of Linear Models
B. Jørgensen

Time Series Analysis
H. Madsen

Texts in Statistical Science

An Introduction to Statistical Inference and Its Applications with R

Michael W. Trosset

Indiana University

Bloomington, Indiana, U.S.A.

CRC Press is an imprint of the
Taylor & Francis Group an informa business
A CHAPMAN & HALL BOOK

Chapman & Hall/CRC
Taylor & Francis Group
6000 Broken Sound Parkway NW, Suite 300
Boca Raton, FL 33487-2742

Printed in the United States of America on acid-free paper
10 9 8 7 6 5 4 3 2 1

International Standard Book Number: 978-1-58488-947-2 (Hardback)

Library of Congress Cataloging-in-Publication Data

Trosset, Michael W.
An introduction to statistical inference and its applications with R / Michael W. Trosset.
p. cm.
Includes bibliographical references and index.
ISBN 978-1-58488-947-2 (hard back : alk. paper)
1. Mathematical statistics. 2. Probabilities. 3. R (Computer program language) I. Title.

QA276.T756 2009
519.5'4--dc22 2009015981

Visit the Taylor & Francis Web site at
http://www.taylorandfrancis.com

and the CRC Press Web site at
http://www.crcpress.com

I of dice possess the science
and in numbers thus am skilled.

From *The Story of Nala*, the third book of the Indian epic *Mahábarata*.[1]

This book is dedicated to

Richard A. Tapia,

my teacher, mentor, collaborator, and friend.

[1]Translated by H. Milman in 1860. Quoted by Ian Hacking in *The Emergence of Probability*, Cambridge University Press, 1975, page 7.

Contents

List of Figures

List of Tables

Preface

To paraphrase Tuco (Eli Wallach) in *The Good, the Bad, and the Ugly*, there are two kinds of textbooks: those that sketch essential ideas, on which an instructor elaborates in class, and those that include all of the details for which an instructor rarely has time. This book is of the second kind. I began writing for the simple reason that I didn't have enough time in class to tell my students all of the things that I wanted to share with them.

This book has a long history. In the spring semesters of 1994 and 1995, I taught a second semester of statistics for undergraduate psychology majors at the University of Arizona. That course was an elective, recommended for strong students contemplating graduate study. I originally planned to teach nonparametric methods, then expanded the course and began developing *Lecture Notes on Univariate Location Problems*, which eventually evolved into Chapters 10–12 of the present book.

From Spring 1994 to Spring 1998, I regularly taught a service course at the University of Arizona, an introduction to statistics for graduate students in various quantitative disciplines. By 1998 my preferred text[1] was no longer in print; hence, when I moved to the College of William & Mary and created a comparable course for undergraduates (Math 351: Applied Statistics), I was obliged to find another. It was only then, and with considerable trepidation, that I decided to expand my own lecture notes and use them as my primary text.

Most of the material that I covered in Math 351 had been completed by 2006. When I moved to Indiana University, I continued to teach the same material as Stat S320: Introduction to Statistics. Presented with the opportunity to help build a new department, I found that at last I was ready to finish my manuscript and move on to other challenges.

[1] L. H. Koopmans (1987). *Introduction to Contemporary Statistical Methods*, Second Edition. Duxbury Press, Boston.

Who Should Read This Book? This book is for students who like to think and who want to understand. It is a self-contained introduction to the methods of statistical inference for students who are comfortable with mathematical concepts and who like to be challenged. Students who are not mathematically inclined, who are intimidated by mathematical notation and formulas, should seek a more gentle approach. There are many ways to begin one's study of statistics and no one approach is ideal for everyone.

At the College of William & Mary, my primary constituency was a subset of undergraduate math majors, those who were more interested in how mathematics is used than in mathematical theory. Math 351 had to contain enough mathematics to be a math course, but it was always conceived as an applied course. A second constituency was a subset of students from other disciplines, often students contemplating graduate study, who wanted more than a superficial survey of various statistical recipes. I particularly remember a biology major, Emilie Snell-Rood, who told me that she wanted to understand what she was doing in her lab when she entered data into a computer program and drew scientific conclusions based on the numbers that it returned.

The prerequisite for Math 351 was two semesters of univariate calculus. This requirement provides a crude way of deciding who is prepared to read this book, but it is potentially misleading. Although the book alludes to several basic concepts from calculus, the student is never required to use the methods of calculus. For example, the probability that a continuous random variable X produces a value in the interval $[a, b]$ is described as an area under a probability density function, but these areas are computed by elementary geometry or by computer software, never by integration. What a student needs to read this book is a certain degree of mathematical maturity.

Over the years, I have taught this material to many students who were taking a first course in statistics and quite a few who were taking a second. The ratio of the latter to the former is increasing as the popularity of AP statistics courses grows. Although this book does not suppose previous study of statistics, I can't recall a single student who felt that she/he had already learned the material in this book in a previous course.

Content This book looks intimidating. It contains a great deal of mathematical notation and it is not until one immerses oneself in the material that the benefits of that notation become apparent. I have adopted the convention of stating mathematical truths, i.e., facts that are true of logical necessity, as theorems. I have even included a small number of proofs![2] To students who point to these theorems and complain that the book is too theoretical, my standard retort is that the theorems describe the conditions under which the statistical procedures can be used, and what's more useful than that? Although my decisions about how much theory to include are idiosyncratic, I hope that anyone who examines the examples, the case studies, and the exercises will be persuaded that this book is first and foremost about *applied* statistical inference.

The organization of this book is also slightly idiosyncratic. I place greater emphasis on probability models than have most authors of recent introductory texts. Furthermore, I develop the tools of probability before progressing to the study of statistics. Although Chapter 1 motivates the study of probability, many pages intervene before Chapter 7 begins the transition from probability models to statistical inference. Chapter 7 introduces a variety of summary statistics and diagnostic techniques that are used to extract information from random samples. Many introductions to statistics begin with this information, usually represented as *descriptive statistics*. The potential problem with such an arrangement is that it may be difficult to communicate to the student the sense that sample quantities are calculated for the purpose of drawing inferences about population quantities. This difficulty is avoided by deferring discussion of samples until the probabilistic tools for describing populations have been introduced. One can then introduce a single unifying concept—the plug-in principle—and derive a variety of sample quantities by applying the tools of discrete probability to the empirical distribution.

I have attempted to organize the material on statistical inference so as to emphasize the importance of identifying the structure of the experiment and making appropriate assumptions about it. Thus, Chapter 10 on 1-sample location problems and Chapter 11 on 2-sample location problems each include material on both parametric and nonparametric procedures. I had originally intended to continue this program in subsequent chapters, but eventually it became clear that doing so would make the book far too long.

Computing plays a special role in this book. When I began teaching, either one did not use the computer or one relied on statistical packages like

---

[2]Some proofs were included to illustrate the nature of mathematical deduction. Others use elementary arguments. My hope is that certain students, possibly the statisticians of tomorrow, will be excited by the discovery that they already possess the tools to follow these arguments. All of the proofs are optional and I never cover them when I teach this material.

Minitab to perform entire analyses. Believing that students should not rely on computers before understanding the procedures themselves, I preferred the former approach. Naturally, this approach limits one to very small data sets, and even then the calculations can be extremely tedious.

The rise of S-Plus and R created a third possibility: a course that frees students from tedious calculations and obsolete tables, but still requires them to understand the procedures that they use. This book relies heavily on R, but in a way that privileges statistics over computation. It does not use advanced functions to perform entire analyses; rather, it explains how to use elementary functions to perform the individual steps of a procedure. For example, a student who performs a 1-sample t-test must still compute a sample mean and variance, insert these quantities into a test statistic, then compute a significance probability. Using R renders the calculations painless, but the procedure cannot be implemented without thought.

The examples and exercises that appear in this book were constructed with great care. They include idealized experiments that allow one to focus on the concept at hand, imagined experiments that were inspired by real-world considerations, sanitized experiments that are simplified versions of actual experiments, and actual experiments that produced actual data. I have drawn extensively from my own experiences, but I have also made liberal use of data sets included in *A Handbook of Small Data Sets*.[3] Instructors may find it convenient to acquire a copy of this extremely useful resource.

Web Resources I have created a web page that I will use to disseminate materials that supplement the text. These materials include the R functions that accompany the text (see Appendix R), many of the data sets used in the text, and the inevitable list of errata. Here is the URL:

http://mypage.iu.edu/~mtrosset/StatInfeR.html

Acknowledgments As this work progressed, I frequently found myself reflecting on how much I owe to so many people. My thinking about statistics has been shaped by many fine teachers and scholars, but C. Ralph Buncher, James R. Thompson, Erich L. Lehmann, and Peter J. Bickel especially influenced my early development. My unusual career trajectory (from academia to consulting to academia) has served me well, but I might not be

[3] D. J. Hand, E. Daly, A. D. Lunn, K. J. McConway, and E. Ostrowski (1994). *A Handbook of Small Data Sets*. Chapman & Hall, London.

in academia today without the support of David W. Scott, Yashaswini D. Mittal, and especially Richard A. Tapia, to whom this book is dedicated.

A great many people have contributed to this project. I am grateful for the feedback (sometimes unintentional!) of all the students on whom I tested material. I retain a special fondness for my students at the College of William & Mary, where most of this book was written. Many colleagues offered suggestions. Eva Czabarka, Larry Leemis, and Sebastian Schreiber taught courses from early drafts and offered invaluable criticism. When my creativity waned, I drew inspiration from the interests and exploits of friends and acquaintances. If Arlen Fuhrman appears too frequently in this book, it is because he is so extraordinarily entertaining.

I did not set out to write a book, only notes that would supplement my lectures. David Scott encouraged me to be more ambitious, but in December 2005 I wrote the following:

> Occasionally I toy with the notion of finding a publisher and finishing the project of Math 351, but always I conclude that this is a project that should continue to evolve. One of the challenges of teaching introductory courses lies in finding new and better ways to explain difficult concepts without trivializing them. Even after years of teaching the same material, I continue to discover new examples and improve my exposition of familiar concepts.

Bob Stern approached me at precisely the right moment and persuaded me to seek closure. Nevertheless, I took longer than expected to complete this project and I am profoundly grateful to Bob for his forbearance.

Finally, a special thanks to two very dear friends. This project began in Arizona and, 15 years later, it concludes in Indiana. Susan White always believed in me and bravely supported my decision to leave Tucson. Dana Fielding welcomed me to Bloomington and reminded me that life is an adventure.

Michael W. Trosset
Bloomington, IN

Chapter 1

Experiments

Statistical methods have proven enormously valuable in helping scientists interpret the results of their experiments—and in helping them design experiments that will produce interpretable results. In a quite general sense, the purpose of statistical analysis is to organize a data set in ways that reveal its structure. Sometimes this is so easy that one does not think that one is doing "statistics;" sometimes it is so difficult that one seeks the assistance of a professional statistician.

This is a book about how statisticians draw conclusions from experimental data. Its primary goal is to introduce the reader to an important type of reasoning that statisticians call "statistical inference." Rather than provide a superficial introduction to a wide variety of inferential methods, we will concentrate on fundamental concepts and study a few methods in depth.

Although statistics can be studied at many levels with varying degrees of sophistication, there is no escaping the simple fact that statistics is a mathematical discipline. Statistical inference rests on the mathematical foundation of probability. The better one desires to understand statistical inference, the more that one needs to know about probability. Accordingly, we will devote several chapters to probability before we begin our study of statistics. To motivate the reader to embark on this program of study, the present chapter describes the important role that probability plays in scientific investigation.

1.1 Examples

This section describes several scientific experiments. Each involves chance variation in a different way. The common theme is that chance variation cannot be avoided in scientific experimentation.

1.1.1 Spinning a Penny

In August 1994, while attending the 15th International Symposium on Mathematical Programming in Ann Arbor, MI, I read an article in which the author asserted that spinning (as opposed to tossing/flipping) a typical penny is not fair, i.e., that `Heads` and `Tails` are not equally likely to result. Specifically, the author asserted that the chance of obtaining `Heads` by spinning a penny is about 30%.[1]

I was one of several people in a delegation from Rice University. That evening, we ended up at a local Subway restaurant for dinner and talk turned to whether or not spinning pennies is fair. Before long we were each spinning pennies and counting `Heads`. At first it seemed that about 70% of the spins were `Heads`, but this proved to be a temporary anomaly. By the time that we tired of our informal experiment, our results seemed to confirm the plausibility of the author's assertion.

I subsequently used penny-spinning as an example in introductory statistics courses, each time asserting that the chance of obtaining `Heads` by spinning a penny is about 30%. Students found this to be an interesting bit of trivia, but no one bothered to check it—until 2001. In the spring of 2001, three students at the College of William & Mary spun pennies, counted `Heads`, and obtained some intriguing results.

For example, Matt, James, and Sarah selected one penny that had been minted in the year 2000 and spun it 300 times, observing 145 `Heads`. This is very nearly 50% and the discrepancy might easily be explained by chance variation—perhaps spinning their penny is fair! They tried different pennies

---

[1] Years later, I have been unable to discover what I read or who wrote it. It seems to be widely believed that the chance is less than 50%. The most extreme assertion that I have discovered is by R. L. Graham, D. E. Knuth, and O. Patashnik (*Concrete Mathematics, Second Edition*, Addison-Wesley, 1994, page 401), who claimed that the chance is approximately 10% "when you spin a newly minted U.S. penny on a smooth table." A fairly comprehensive discussion of "Flipping, spinning, and tilting coins" can be found at

`http://www.dartmouth.edu/~chance/chance_news/recent_news/`
`chance_news_11.02.html#item2`,

in which various individuals emphasize that the chance of `Heads` depends on such factors as the year in which the penny was minted, the surface on which the penny is spun, and the quality of the spin. For pennies minted in the 1960s, one individual reported 1878 `Heads` in 5520 spins, about 34%.

and obtained different percentages. Perhaps all pennies are not alike! (Pennies minted before 1982 are 95% copper and 5% zinc; pennies minted after 1982 are 97.5% zinc and 2.5% copper.) Or perhaps the differences were due to chance variation.

Were one to undertake a scientific study of penny spinning, there are many questions that one might ask. Here are several:

- Choose a penny. What is the chance of obtaining `Heads` by spinning that penny? (This question is the basis for Exercise 1 at the end of this chapter.)

- Choose two pennies. Are they equally likely to produce `Heads` when spun?

- Choose several pennies minted before 1982 and several pennies minted after 1982. As groups, are pre-1982 pennies and post-1982 pennies equally likely to produce `Heads` when spun?

1.1.2 The Speed of Light

According to Albert Einstein's special theory of relativity, the speed of light through a vacuum is a universal constant c. Since 1974, that speed has been given as $c = 299,792.458$ kilometers per second.[2] Long before Einstein, however, philosophers had debated whether or not light is transmitted instantaneously and, if not, at what speed it moved. In this section, we consider Albert Abraham Michelson's famous 1879 experiment to determine the speed of light.[3]

Aristotle believed that light "is not a movement" and therefore has no speed. Francis Bacon, Johannes Kepler, and René Descartes believed that light moved with infinite speed, whereas Galileo Galilei thought that its speed was finite. In 1638 Galileo proposed a terrestrial experiment to resolve the dispute, but two centuries would pass before this experiment became technologically practicable. Instead, early determinations of the speed of light were derived from astronomical data.

---

[2]Actually, a second is defined to be $9,192,631,770$ periods of radiation from cesium-133 and a kilometer is defined to be the distance travelled by light through a vacuum in 1/299792458 seconds!

[3]A. A. Michelson (1880). Experimental determination of the velocity of light made at the U.S. Naval Academy, Annapolis. *Astronomical Papers*, 1:109–145. The material in this section is taken from R. J. MacKay and R. W. Oldford (2000), Scientific method, statistical method and the speed of light, *Statistical Science*, 15:254–278, with the permission of the Institute of Mathematical Statistics.

The first empirical evidence that light is not transmitted instantaneously was presented by the Danish astronomer Ole Römer, who studied a series of eclipses of Io, Jupiter's largest moon. In September 1676, Römer correctly predicted a 10-minute discrepancy in the time of an impending eclipse. He argued that this discrepancy was due to the finite speed of light, which he estimated to be about 214,000 kilometers per second. In 1729, James Bradley discovered an annual variation in stellar positions that could be explained by the earth's motion *if* the speed of light was finite. Bradley estimated that light from the sun took 8 minutes and 12 seconds to reach the earth and that the speed of light was 301,000 kilometers per second. In 1809, Jean-Baptiste Joseph Delambre used 150 years of data on eclipses of Jupiter's moons to estimate that light travels from sun to earth in 8 minutes and 13.2 seconds, at a speed of 300,267.64 kilometers per second.

In 1849, Hippolyte Fizeau became the first scientist to estimate the speed of light from a terrestrial experiment, a refinement of the one proposed by Galileo. An accurately machined toothed wheel was spun in front of a light source, automatically covering and uncovering it. The light emitted in the gaps between the teeth travelled 8633 meters to a fixed flat mirror, which reflected the light back to its source. The returning light struck either a tooth or a gap, depending on the wheel's speed of rotation. By varying the speed of rotation and observing the resulting image from reflected light beams, Fizeau was able to measure the speed of light.

In 1851, Leon Foucault further refined Galileo's experiment, replacing Fizeau's toothed wheel with a rotating mirror. Michelson further refined Foucault's experimental setup. A precise account of the experiment is beyond the scope of this book, but MacKay's and Oldford's account of how Michelson produced each of his 100 measurements of the speed of light provides some sense of what was involved. More importantly, their account (Section 3.4, pp. 262–263) reveals the multiple ways in which Michelson's measurements were subject to error.

1. The distance $|RM|$ from the rotating mirror to the fixed mirror was measured five times, each time allowing for temperature, and the average used as the "true distance" between the mirrors for all determinations.
2. The fire for the pump was started about a half hour before measurement began. After this time, there was sufficient pressure to begin the determinations.
3. The fixed mirror M was adjusted... and the heliostat placed and adjusted so that the Sun's image was directed at the slit.

4. The revolving mirror was adjusted on two different axes....
5. The distance $|SR|$ from the revolving mirror to the crosshair of the eyepiece was measured using the steel tape.
6. The vertical crosshair of the eyepiece of the micrometer was centred on the slit and its position recorded in terms of the position of the screw.
7. The electric tuning fork was started. The frequency of the fork was measured two or three times for each set of observations.
8. The temperature was recorded.
9. The revolving mirror was started. The eyepiece was set approximately to capture the displaced image. If the image did not appear in the eyepiece, the mirror was inclined forward or back until it came into sight.
10. The speed of rotation of the mirror was adjusted until the image of the revolving mirror came to rest.
11. The micrometer eyepiece was moved by turning the screw until its vertical crosshair was centred on the return image of the slit. The number of turns of the screw was recorded. The displacement is the difference in the two positions. To express this as the distance $|IS|$ in millimetres the measured number of turns was multiplied by the calibrated number of millimetres per turn of the screw.
12. Steps 10 and 11 were repeated until 10 measurements of the displacement $|IS|$ were made.
13. The rotating mirror was stopped, the temperature noted and the frequency of the electric fork was determined again.

Michelson used the procedure described above to obtain 100 measurements of the speed of light in air. Each measurement was computed using the average of the 10 measured displacements in Step 12. These measurements, reported in Table 1.1, subsequently were adjusted for temperature and corrected by a factor based on the refractive index of air. Michelson reported the speed of light in a vacuum as $299,944 \pm 51$ kilometers per second.

1.1.3 Termite Foraging Behavior

In the mid-1980s, Susan Jones was a USDA entomologist and a graduate student in the Department of Entomology at the University of Arizona. Her dissertation research concerned the foraging ecology of subterranean

50	−60	100	270	130	50	150	180	180	80
200	180	130	−150	−40	10	200	200	160	160
160	140	160	140	80	0	50	80	100	40
30	−10	10	80	80	30	0	−10	−40	0
80	80	80	60	−80	−80	−180	60	170	150
80	110	50	70	40	40	50	40	40	40
90	10	10	20	0	−30	−40	−60	−50	−40
110	120	90	60	80	−80	40	50	50	−20
90	40	−20	10	−40	10	−10	10	10	50
70	70	10	−60	10	140	150	0	10	70

Table 1.1: Michelson's 100 unadjusted measurements of the speed of light in air. Add $299,800$ to obtain measurements in units of kilometers per second.

termites in the Sonoran Desert.[4] Her field studies were conducted on the Santa Rita Experimental Range, about 40 kilometers south of Tucson, AZ:

> The foraging activity of *H. aureus*[5] was studied in 30 plots, each consisting of a grid (6 by 6 m) of 25 toilet-paper rolls which served as baits... Plots were selected on the basis of two criteria: the presence of *H. aureus* foragers in dead wood, and separation by at least 12 m from any other plot. A 6-by-6-m area was then marked off within the vicinity of infested wood, and toilet-paper rolls were aligned in five rows and five columns and spaced at 1.5-m intervals. The rolls were positioned on the soil surface and each was held in place with a heavy wire stake. All pieces of wood ca. 15 cm long and longer were removed from each plot and ca. 3 m around the periphery to minimize the availability of natural wood as an alternative food source. Before infested wood was removed from the site, termites were allowed to retreat into their galleries in the soil to avoid depleting the numbers of surface foragers. All plots were established within a 1-wk period during late June 1984.

[4]The material in this section and Section 16.1 was taken from S. C. Jones, M. W. Trosset, and W. L. Nutting (1987), Biotic and abiotic influences on foraging of *Heterotermes aureus* (Snyder) (Isoptera: Rhinotermitidae), *Environmental Entomology* 16:791–795, with the permission of the Entomological Society of America.

[5]*Heterotermes aureus* (Snyder) is the most common subterranean termite species in the Sonoran Desert. M. I. Haverty, W. L. Nutting, and J. P. LaFage (Density of colonies and spatial distribution of foraging territories of the desert subterranean termite, *Heterotermes aureues* (Snyder), *Environmental Entomology*, 4:105–109, 1975) estimated the population density of this species in the Santa Rita Experimental Range at 4.31×10^6 termites per hectare.

> Plots were examined once a week during the first 5 wk after establishment, and then at least once monthly thereafter until August 1985.[6]

An important objective of the above study was

> ...to investigate the relationship between food-source distance (on a scale 6 by 6 m) and foraging behavior. This was accomplished by analyzing the order in which different toilet-paper rolls in the same plot were attacked.... Specifically, a statistical methodology was developed to test the null hypothesis that any previously unattacked roll was equally likely to be the next roll attacked (random foraging). Alternative hypotheses supposed that the likelihood that a previously unattacked roll would be the next attacked roll decreased with increasing distance from previously attacked rolls (systematic foraging).[7]

The order in which the toilet-paper rolls in Plot 20 were attacked is displayed in Figure 1.1. The unattacked rolls are denoted by ○, the initially attacked rolls are denoted by ●, and the subsequently attacked rolls are denoted (in order of attack) by 1, 2, 3, 4, and 5. Notice that these numbers do not specify a unique order of attack:

> ...because the plots were not observed continuously, a number of rolls seemed to have been attacked simultaneously. Therefore, it was not always possible to determine the exact order in which they were attacked. Accordingly, all permutations consistent with the observed ties in order were considered...[8]

In Section 16.1 we will return to the question of whether or not *H. aureus* forages randomly and describe the statistical methodology that was developed to answer it. Along the way, we will develop rigorous interpretations of the phrases that appear in the above passages, e.g., *permutations*, *equally likely*, *null hypothesis*, *alternative hypotheses*, etc.

1.2 Randomization

This section illustrates an important principle in the design of experiments. We begin by describing two famous studies that produced embarrassing

[6] S. C. Jones, M. W. Trosset, and W. L. Nutting (1987). Biotic and abiotic influences on foraging of *Heterotermes aureus* (Snyder) (Isoptera: Rhinotermitidae), *Environmental Entomology*, 16:791–795, pp. 791–792.

[7] Ibid., p. 792.

[8] Ibid., p. 792.

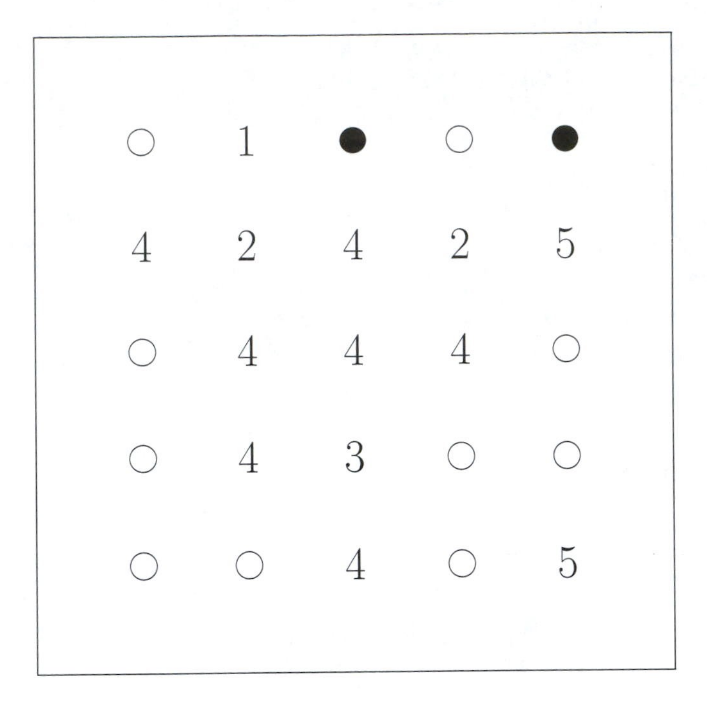

Figure 1.1: Order of *H. aureus* attack in Plot 20.

results because they failed to respect this principle.

The Lanarkshire Milk Experiment A 1930 experiment in the schools of Lanarkshire attempted to ascertain the effect of milk supplements on Scottish children. For four months, 5000 children received a daily supplement of 3/4 pint of raw milk, 5000 children received a daily supplement of 3/4 pint of pasteurized milk, and 10,000 children received no daily milk supplement. Each child was weighed (while wearing indoor clothing) and measured for height before the study commenced (in February) and after it ended (in June). The final observations of the control group exceeded the final observations of the treatment groups by average amounts equivalent to 3 months growth in weight and 4 months growth in weight, thereby suggesting that the milk supplements actually retarded growth! What went wrong?

To explain the results of the Lanarkshire milk experiment, one must examine how the 20,000 children enrolled in the study were assigned to the study groups. An initial division into treatment versus control groups was made arbitrarily, e.g., using the alphabet. However, if the initial division appeared to produce groups with unbalanced numbers of well-fed or ill-nourished children, then teachers were allowed to swap children between the two groups in order to obtain (apparently) better balanced groups. It is thought that well-meaning teachers, concerned about the plight of ill-nourished children and "knowing" that milk supplements would be beneficial, consciously or subconsciously availed themselves of the opportunity to swap ill-nourished children into the treatment group. This resulted in a treatment group that was lighter and shorter than the control group. Furthermore, it is likely that differences in weight gains were confounded with a tendency for well-fed children to come from families that could afford warm (heavier) winter clothing, as opposed to a tendency for ill-nourished children to come from poor families that provided shabbier (lighter) clothing.[9]

The Pre-Election Polls of 1948 The 1948 presidential election pitted Harry Truman, the Democratic incumbent who had succeeded to the presidency when Franklin Roosevelt died in office, against Thomas Dewey, the Republican governor of New York.[10] Each of the three major polling organizations that covered the campaign predicted that Dewey would win: the Crossley poll predicted 50% of the popular vote for Dewey and 45% for Truman, the Gallup poll predicted 50% for Dewey and 44% for Truman, and the Roper poll predicted 53% for Dewey and 38% for Truman. Dewey's election was considered a foregone conclusion until the votes were actually counted: in one of the great upsets in American politics, Truman received slightly less than 50% of the popular vote and Dewey received slightly more than 45%. What happened?

The failure of the pre-election polls to predict Truman's victory caused considerable controversy. In response, just eight days after the election, the Social Science Research Council appointed a Committee on Analysis of Pre-election Polls and Forecasts to investigate the causes of forecasting error in 1948 and to recommend improved polling methods for future use. Their

---

[9]For additional details and commentary, see Student (1931), The Lanarkshire milk experiment, *Biometrika*, 23:398–406, and Section 5.4 (Justification of Randomization) of D. R. Cox (1958), *Planning of Experiments*, John Wiley & Sons, New York.

[10]As a crusading district attorney in New York City, Dewey was a national hero in the 1930s. In late 1938, two Hollywood films attempted to capitalize on his popularity, RKO's *Smashing the Rackets* and Warner Brothers' *Racket Busters*. The special prosecutor in the latter film was played by Walter Abel, who bore a strong physical resemblance to Dewey.

report was released to the press on December 27, 1948; a more comprehensive report prepared by the Committee's technical staff was published in 1949.[11]

Poll predictions are based on data collected from a sample of prospective voters. To assess the quality of a prediction, one must examine how the sample was obtained. In 1948, all three polling organizations used a method called *quota sampling* that attempts to hand-pick a sample that is representative of the entire population. First, one attempts to identify several important characteristics that may be associated with different voting patterns, e.g., place of residence, sex, age, race, etc. Second, one attempts to obtain a sample that resembles the entire population with respect to those characteristics. For example, in a national quota sample survey conducted by the National Opinion Research Center in December 1947,[12] an interviewer in St. Louis was assigned to interview 15 people. Of these 15 people, 7 were to be men and 8 were to be women. Of the 7 men, 3 were to be under 40 years of age; of the 8 women, 4 were to be under 40 years of age. Also, 1 of the 7 men and 1 of the 8 women was to be black. Monthly rent categories were specified for the 6 white men and the 7 white women. Finally, a total of 6 persons were to be from the suburbs.

Unfortunately, quota sampling does not work especially well. The reason is that quota sampling does not specify how to choose the sample *within* the quotas—these choices are left to the discretion of the interviewer. Human choice is often subject to bias, which means that samples obtained by quota sampling may be unrepresentative of the population. For example, Table 7.8 of Stephan and McCarthy (1958) compares the education distribution for two national quota samples to U.S. Bureau of the Census estimates: 20.4% of the two quota samples had some college education, versus just 12% of the population.[13] Similarly, in the pre-election polls of 1948, "the Gallup and Roper samples show considerable underrepresentation of the less highly educated classes, the underrepresentation being much greater in Roper's

---

[11]F. Mosteller, H. Hyman, P. J. McCarthy, E. S. Marks, and D. B. Truman (1949). *The Pre-election Polls of 1948: Report to the Committee on Analysis of Pre-election Polls and Forecasts.* Bulletin 60, Social Science Research Council, New York. Appendix A contains the Committee's original report of December 27, 1948.

[12]F. F. Stephan and P. J. McCarthy (1958). *Sampling Opinions: An Analysis of Survey Procedure.* John Wiley & Sons, New York. See Chapter 12 (An Analysis of the Field Operations of a National Quota Sample Survey).

[13]Ibid. See Sections 7.3A and 8.2B. The surveys were made in June and August of 1946; the Census estimate was for April 1947.

samples than in Gallup's."[14] Because education is related to socio-economic status, and because voters of high socio-economic status are more likely to vote Republican than voters of low socio-economic status, the tendency of quota sampling interviewers to choose more educated prospective voters created an unintentional bias toward Republicans. This bias had distorted previous polls without dramatic effect; in 1948, the election was close enough that the polls picked the wrong candidate.

Beyond the systematic bias that typically plagues quota samples, "the very nature of a quota sample makes it impossible to assess its adequacy, the entire process being too dependent upon the unknown behavior of human agents, namely, the interviewers."[15] Stephan and McCarthy (1958, p. 191) concluded that "it seems impossible to place quota sampling on a sound theoretical basis." One consequence of Truman's upset victory in 1948 is that the practice of quota sampling has fallen from favor.

In both the Lanarkshire milk experiment and the pre-election polls of 1948, subjective attempts to hand-pick representative samples resulted in embarrassing failures. Let us now exploit our knowledge of what *not* to do and design a simple experiment. An instructor—let's call him Ishmael—of one section of Math 106 (Elementary Statistics) has prepared two versions of a final exam. Ishmael hopes that the two versions are equivalent, but he recognizes that this will have to be determined experimentally. He therefore decides to divide his class of 40 students into two groups, each of which will receive a different version of the final. How should he proceed?

Ishmael recognizes that he requires two comparable groups if he hopes to draw conclusions about his two exams. For example, suppose that he administers one exam to the students who attained an `A` average on the midterms and the other exam to the other students. If the average score on exam `A` is 20 points higher than the average score on exam `B`, then what can he conclude? It might be that exam `A` is 20 points easier than exam `B`. Or it might be that the two exams are equally difficult, but that the `A` students are 20 points more capable than the `B` students. Or it might be that exam `A` is actually 10 points more difficult than exam `B`, but that the `A` students are 30 points more capable than the `B` students. There is no

[14] F. Mosteller, H. Hyman, P. J. McCarthy, E. S. Marks, and D. B. Truman (1949). *The Pre-election Polls of 1948: Report to the Committee on Analysis of Pre-election Polls and Forecasts.* Bulletin 60, Social Science Research Council, New York, p. 114. "The educational distribution of the Crossley samples could not be checked since his ballot did not contain a question on education."

[15] Ibid, p. 115.

way to decide—exam version and student capability are *confounded* in this experiment.

The lesson of the Lanarkshire milk experiment and the pre-election polls of 1948 is that it is difficult to hand-pick representative samples. Accordingly, Ishmael decides to randomly assign the exams, relying on chance variation to produce balanced groups. This can be done in various ways, but a common principle prevails: each student is equally likely to receive exam `A` or `B`. Here are two possibilities:

1. Ishmael creates 40 identical slips of paper. He writes the name of each student on one slip, mixes the slips in a large jar, then draws 20 slips. (After each draw, the selected slip is set aside and the next draw uses only those slips that remain in the jar, i.e., sampling occurs *without replacement.*) The 20 students selected receive exam `A`; the remaining 20 students receive exam `B`. This is called *simple random sampling.*

2. Ishmael notices that his class comprises 30 freshmen and 10 nonfreshmen. Believing that it is essential to have 3/4 freshmen in each group, he assigns freshmen and nonfreshmen separately. Again, Ishmael creates 40 identical slips of paper and writes the name of each student on one slip. This time he separates the 30 freshman slips from the 10 nonfreshman slips. To assign the freshmen, he mixes the 30 freshman slips and draws 15 slips. The 15 freshmen selected receive exam `A`; the remaining 15 freshmen receive exam `B`. To assign the nonfreshmen, he mixes the 10 nonfreshman slips and draws 5 slips. The 5 nonfreshmen selected receive exam `A`; the remaining 5 nonfreshmen receive exam `B`. This is called *stratified random sampling.*

1.3 The Importance of Probability

Each of the experiments described in Sections 1.1 and 1.2 reveals something about the role of chance variation in scientific experimentation.

It is beyond our ability to predict with certainty if a spinning penny will come to rest with `Heads` facing up. Even if we believe that the outcome is completely determined, we cannot measure all the relevant variables with sufficient precision, nor can we perform the necessary calculations, to know what it will be. We express our inability to predict `Heads` versus `Tails` in the language of probability, e.g., "there is a 30% chance that `Heads` will result." (Section 3.1 discusses how such statements may be interpreted.)

Thus, *even when studying allegedly deterministic phenomena,* probability models may be of enormous value.

When measuring the speed of light, it is not the phenomenon itself but the experiment that admits chance variation. Despite his excruciating precautions, Michelson was unable to remove chance variation from his experiment—his measurements differ. Adjusting the measurements for temperature removes one source of variation, but it is impossible to remove them all. Later experiments with more sophisticated equipment produced better measurements, but did not succeed in completely removing all sources of variation. Experiments are never perfect,[16] and probability models may be of enormous value in modelling errors that the experimenter is unable to remove or control.

Probability plays another, more subtle role in statistical inference. When studying termites, it is not clear whether or not one is observing a systematic foraging strategy. Probability was introduced as a hypothetical benchmark: *what if* termites forage randomly? Even if termites actually do forage deterministically, understanding how they would behave if they foraged randomly provides insights that inform our judgments about their behavior.

Thus, probability helps us answer questions that naturally arise when analyzing experimental data. Another example arose when we remarked that Matt, James, and Sarah observed *nearly* 50% `Heads`, specifically 145 `Heads` in 300 spins. What do we mean by "nearly"? Is this an important discrepancy or can chance variation account for it? To find out, we might study the behavior of penny spinning under the mathematical assumption that it is fair. If we learn that 300 spins of a fair penny rarely produce a discrepancy of 5 (or more) `Heads`, then we might conclude that penny spinning is not fair. If we learn that discrepancies of this magnitude are common, then we would be reluctant to draw this conclusion.

The ability to use the tools of probability to understand the behavior of inferential procedures is so powerful that good experiments are designed with this in mind. Besides avoiding the pitfalls of subjective methods, randomization allows us to answer questions about how well our methods work. For example, Ishmael might ask "How likely is simple random sampling to result in exactly 5 nonfreshmen receiving exam `A`?" Such questions derive meaning from the use of probability methods.

---

[16] Another example is described by D. Freedman, R. Pisani, and R. Purves in Section 6.2 (Chance Error) of *Statistics* (Third Edition, W. W. Norton & Company, New York, 1998). The National Bureau of Standards repeatedly weighs the national prototype kilogram under carefully controlled conditions. The measurements are extremely precise, but nevertheless subject to small variations.

When a scientist performs an experiment, s/he observes a *sample* of possible experimental values. The set of all values that might have been observed is a *population*. Probability helps us describe the population and understand the data generating process that produced the sample. It also helps us understand the behavior of the statistical procedures used to analyze experimental data, e.g., averaging 100 measurements to produce an estimate. This linkage, of sample to population through probability, is the foundation on which statistical inference is based. Statistical inference is relatively new, but the linkage that we have described is wonderfully encapsulated in a remarkable passage from *The Story of Nala*, the third book of the ancient Indian epic *Mahábarata*.[17] Rtuparna examines a single twig of a spreading tree and accurately estimates the number of fruit on two great branches. Nala marvels at this ability, and Rtuparna rejoins:

> I of dice possess the science
> and in numbers thus am skilled.

1.4 Games of Chance

In *The Story of Nala*, Rtuparna's skill in estimation is connected with his prowess at dicing. Throughout history, probabilistic concepts have invariably been illustrated using simple games of chance. There are excellent reasons for us to embrace this pedagogical cliché. First, many fundamental probabilistic concepts were invented for the purpose of understanding certain games of chance; it is pleasant to incorporate a bit of this fascinating, centuries-old history into a modern program of study. Second, games of chance serve as idealized experiments that effectively reveal essential issues without the distraction of the many complicated nuances associated with most scientific experiments. Third, as idealized experiments, games of chance provide canonical examples of various recurring experimental structures. For example, tossing a coin is a useful abstraction of such diverse experiments as observing whether a baby is male or female, observing whether an Alzheimer's patient does or does not know the day of the week, or observing whether a pond is or is not inhabited by geese. A scientist who is familiar with these idealized experiments will find it easier to diagnose the mathematical structure of an actual scientific experiment.

Many of the examples and exercises in subsequent chapters will refer to simple games of chance. The present section collects some facts and trivia

[17]This passage is summarized in Ian Hacking's *The Emergence of Probability*, Cambridge University Press, 1975, pp. 6–7, which quotes H. H. Milman's 1860 translation.

about several of the most common.

Coins According to the *Encylopædia Britannica*,

> Early cast-bronze animal shapes of known and readily identifiable weight, provided for the beam-balance scales of the Middle Eastern civilizations of the 7th millennium BC, are evidence of the first attempts to provide a medium of exchange. ... The first true coins, that is, cast disks of standard weight and value specifically designed as a medium of exchange, were probably produced by the Lydians of Anatolia in about 640 BC from a natural alloy of gold containing 20 to 35 percent silver.[18]

Despite (or perhaps because of) the simplicity of tossing a coin and observing which side (canonically identified as `Heads` or `Tails`) comes to lie facing up, it appears that coins did not play an important role in the early history of probability. Nevertheless, the use of coin tosses (or their equivalents) as randomizing agents is ubiquitous in modern times. In football, an official tosses a coin and a representative of one team calls `Heads` or `Tails`. If his call matches the outcome of the toss, then his team may choose whether to kick or receive (or, which goal to defend); otherwise, the opposing team chooses. A similar practice is popular in tennis, except that one player spins a racquet instead of tossing a coin. In each of these practices, it is presumed that the "coin" is *balanced* or *fair*, i.e., that each side is equally likely to turn up; see Section 1.1.1 for a discussion of whether or not spinning a penny is fair.

Dice The noun *dice* is the plural form of the noun *die*.[19] A die is a small cube, marked on each of its six faces with a number of pips (spots, dots). To generate a random outcome, the die is cast (tossed, thrown, rolled) on a smooth surface and the number of pips on the uppermost face is observed. If each face is equally likely to be uppermost, then the die is *balanced* or *fair*; otherwise, it is *unbalanced* or *loaded*.

The casting of dice is an ancient practice. According to F. N. David,

---

[18] Coins and coinage. *The New Encyclopædia Britannica in 30 Volumes*, Macropædia, Volume 4, 1974, pp. 821–822.

[19] In *The Devil's Dictionary*, Ambrose Bierce defined *die* as the singular of *dice*, remarking that "we seldom hear the word, because there is a prohibitory proverb, 'Never say die.' "

> The earliest dice so far found are described as being of well-fired buff pottery and date from the beginning of the third millenium. ... consecutive order of the pips must have continued for some time. It is still to be seen in dice of the late XVIIIth Dynasty (Egypt *c.* 1370 B.C.), but about that time, or soon after, the arrangement must have settled into the 2-partitions of 7 familiar to us at the present time. Out of some fifty dice of the classical period which I have seen, forty had the 'modern' arrangement of the pips.[20]

Today, pure dice games include craps, in which two dice are cast, and Yahtzee™, in which five dice are cast. More commonly, the casting of dice is used as a randomizing agent in a variety of board games, e.g., backgammon and Monopoly™. Typically, two dice are cast and the outcome is defined to be the total number of pips on the two uppermost faces.

Astragali Even more ancient than dice are *astragali*, the singular form of which is *astragalus*. The astragalus is a bone in the heel of many vertebrate animals; it lies directly above the talus, and is roughly symmetrical in hooved mammals, e.g., deer. Such astragali have been found in abundance in excavations of prehistoric man, who may have used them for counting. They were used for board games at least as early as the First Dynasty in Egypt (*c.* 3500 B.C.) and were the principal randomizing agent in classical Greece and Rome. According to F. N. David,

> The astragalus has only four sides on which it will rest, since the other two are rounded... A favourite research of the scholars of the Italian Renaissance was to try to deduce the scoring used. It was generally agreed from a close study of the writings of classical times that the upper side of the bone, broad and slightly convex, counted 4; the opposite side, broad and slightly concave, 3; the lateral side, flat and narrow, scored 1, and the opposite narrow lateral side, which is slightly hollow, 6. The numbers 2 and 5 were omitted.[21]

Accordingly, we can think of an astragalus as a 4-sided die with possible outcomes 1, 3, 4, and 6. An astragalus is not balanced. From tossing a modern sheep's astragalus, David estimated the chances of throwing a 1 or a 6 at roughly 10 percent each and the chances of throwing a 3 or a 4 at roughly 40 percent each.

---

[20] F. N. David (1962). *Games, Gods and Gambling: A History of Probability and Statistical Ideas.* Dover Publications, New York, p. 10.

[21] Ibid., p. 7.

The Greeks and Romans invariably cast four astragali. The most desirable result, the *venus*, occurred when the four uppermost sides were all different; the *dog*, which occurred when each uppermost side was a 1, was undesirable. In Asia Minor, five astragali were cast and different results were identified with the names of different gods, e.g., the throw of Saviour Zeus (one with 1, two with 3, and two with 4), the throw of child-eating Cronos (three with 4 and two with 6), etc. In addition to their use in gaming, astragali were cast for the purpose of divination, i.e., to ascertain if the gods favored a proposed undertaking.

In 1962, David reported that "it is not uncommon to see children in France and Italy playing games with them [astragali] today;"[22] for the most part, however, unbalanced astragali have given way to balanced dice. A whimsical contemporary example of unbalanced dice that evoke astragali are the pig dice used in Pass the Pigs™ (formerly Pigmania™).

Cards David estimated that playing cards "were not invented until *c.* A.D. 1350, but once in use, they slowly began to displace dice both as instruments of play and for fortune-telling."[23] By a *standard deck of playing cards*, we shall mean the familiar deck of 52 cards, organized into four *suits* (clubs, diamonds, hearts, spades) of thirteen *ranks* or *denominations* (2–10, jack, queen, king, ace). The diamonds and hearts are red; the clubs and spades are black. When we say that a deck has been shuffled, we mean that the order of the cards in the deck has been randomized. When we say that cards are dealt, we mean that they are removed from a shuffled deck in sequence, beginning with the top card. The cards received by a player constitute that player's *hand*. The quality of a hand depends on the game being played; however, unless otherwise specified, the order in which the player received the cards in her hand is irrelevant.

Poker involves hands of five cards. The following types of hands are arranged in order of decreasing value. An ace is counted as either the highest or the lowest rank, whichever results in the more valuable hand. Thus, every possible hand is of exactly one type.

1. A *straight flush* contains five cards of the same suit and of consecutive ranks.

2. A hand with *4 of a kind* contains cards of exactly two ranks, four cards of one rank and one of the other rank.

---

[22] Ibid., p. 3.
[23] Ibid., p. 20.

3. A *full house* contains cards of exactly two ranks, three cards of one rank and two cards of the other rank.

4. A *flush* contains five cards of the same suit, not of consecutive rank.

5. A *straight* contains five cards of consecutive rank, not all of the same suit.

6. A hand with *3 of a kind* contains cards of exactly three ranks, three cards of one rank and one card of each of the other two ranks.

7. A hand with *two pairs* contains cards of exactly three ranks, two cards of one rank, two cards of a second rank, and one card of a third rank.

8. A hand with *one pair* contains cards of exactly four ranks, two cards of one rank and one card each of a second, third, and fourth rank.

9. Any other hand contains *no pair.*

Urns For the purposes of this book, an urn is a container from which objects are drawn, e.g., a box of raffle tickets or a jar of marbles. Modern lotteries often select winning numbers by using air pressure to draw numbered ping pong balls from a clear plastic container. When an object is drawn from an urn, it is presumed that each object in the urn is equally likely to be selected.

That urn models have enormous explanatory power was first recognized by J. Bernoulli (1654–1705), who used them in *Ars Conjectandi*, his brilliant treatise on probability. It is not difficult to devise urn models that are equivalent to other randomizing agents considered in this section.

Example 1.1: Urn Model for Tossing a Fair Coin Imagine an urn that contains one red marble and one black marble. A marble is drawn from this urn. If it is red, then the outcome is `Heads`; if it is black, then the outcome is `Tails`. This is equivalent to tossing a fair coin *once.*

Example 1.2: Urn Model for Throwing a Fair Die Imagine an urn that contains six tickets, labelled 1 through 6. Drawing one ticket from this urn is equivalent to throwing a fair die *once.* If we want to throw the die a second time, then we return the selected ticket to the urn and repeat the procedure. This is an example of drawing *with replacement.*

Example 1.3: Urn Model for Throwing an Astragalus Imagine an urn that contains ten tickets, one labelled 1, four labelled 3, four labelled 4, and one labelled 6. Drawing one ticket from this urn is equivalent to throwing an astragalus *once*. If we want to throw four astragali, then we repeat this procedure four times, each time returning the selected ticket to the urn. This is another example of drawing *with replacement*.

Example 1.4: Urn Model for Drawing a Poker Hand Place a standard deck of playing cards in an urn. Draw one card, then a second, then a third, then a fourth, then a fifth. Because each card in the deck can only be dealt once, we do not return a card to the urn after drawing it. This is an example of drawing *without replacement*.

In the preceding examples, the statements about the equivalence of the urn model and another randomizing agent were intended to appeal to your intuition. Subsequent chapters will introduce mathematical tools that will allow us to validate these assertions.

1.5 Exercises

1. Select a penny minted in any year other than 1982. Find a smooth surface on which to spin it. Practice spinning the penny until you are able to do so in a reasonably consistent manner. Develop an experimental protocol that specifies precisely how you spin your penny. Spin your penny 100 times in accordance with this protocol. Record the outcome of each spin, including aberrant events (e.g., the penny spun off the table and therefore neither `Heads` nor `Tails` was recorded). Your report of this experiment should include the following:

    - The penny itself, taped to your report. Note any features of the penny that seem relevant, e.g., the year and city in which it was minted, its condition, etc.
    - A description of the surface on which you spun it and of any possibly relevant environmental considerations.
    - A description of your experimental protocol.
    - The results of your 100 spins. This means a list, in order, of what happened on each spin.
    - A summary of your results. This means (i) the total number of spins that resulted in either `Heads` or `Tails` (ideally, this number,

n, will equal 100) and (ii) the number of spins that resulted in `Heads` (y).

- The observed frequency of heads, y/n.

2. The Department of Mathematics at the College of William & Mary is housed in Jones Hall. To find the department, one passes through the building's main entrance, into its lobby, and immediately turns left. In Jones 131, the department's seminar room, is a long rectangular wood table. Let L denote the length of this table. The purpose of this experiment is to measure L using a standard (12-inch) ruler.[24]

   You will need a 12-inch ruler that is marked in increments of 1/16 inches. Groups of students may use the same ruler, but it is important that each student obtain his/her own measurement of L. Please do not attempt to obtain your measurement at a time when Jones 131 is being used for a seminar or faculty meeting!

   Your report of this experiment should include the following information:

   - A description of the ruler that you used. From what was it made? In what condition is it? Who owns it? What other students used the same ruler?
   - A description of your measuring protocol. How did you position the ruler initially? How did you reposition it? How did you ensure that you were measuring along a straight line?
   - An account of the experiment. When did you measure? How long did it take you? Please note any unusual circumstances that might bear on your results.
   - Your estimate (in inches, to the nearest 1/16 inch) of L.

3. Statisticians say that a procedure that tends to either underestimate or overestimate the quantity that it is being used to determine is *biased.*

   (a) In the preceding problem, suppose that you tried to measure the length of the table with a ruler that—unbeknownst to you—was really 11.9 inches long instead of the nominal 12 inches. Would you tend to underestimate or overestimate the true length of the table? Explain.

[24] Your instructor will substitute a suitable object for measurement.

(b) In the Lanarkshire milk experiment, would a tendency for well-fed children to wear heavier winter clothing than ill-nourished children cause weight gains due to milk supplements to be underestimated or overestimated? Explain.

4. To complete this exercise, either acquire the game of Pass the Pigs™ or play on-line at `http://www.fontface.com/games/pigs/`. Play involves rolling a pair of pig dice. Each die can land in one of six possible positions: side (no dot), side (dot), razorback, trotter, snouter, leaning jowler. Different combinations of these positions score different numbers of points.

   Play the game (or just roll an individual pig die). After each roll, record the position of each pig die. Play long enough to observe at least 120 outcomes. Tabulate the outcomes in the following format, where, for example, n_4 denotes the number of trotters that you observe.

Side (no dot)	Side (dot)	Razorback	Trotter	Snouter	Leaning Jowler
n_1	n_2	n_3	n_4	n_5	n_6

   Compare your results to the results reported in Wikipedia's Pass the Pigs™ entry. Do the two sets of results differ substantially?

Chapter 2

Mathematical Preliminaries

This chapter collects some fundamental mathematical concepts that we will use in our study of probability and statistics. Most of these concepts should seem familiar, although our presentation of them may be a bit more formal than you have previously encountered. This formalism will be quite useful as we study probability, but it will tend to recede into the background as we progress to the study of statistics.

2.1 Sets

We begin by supposing the existence of a designated *universe* of possible objects. In this book, we will often denote the universe by S. By a *set*, we mean a collection of objects with the property that each object in the universe either does or does not belong to the collection. We will tend to denote sets by uppercase Roman letters toward the beginning of the alphabet, e.g., A, B, C, etc. The set of objects that do not belong to a designated set A is called the *complement* of A. We will denote complements by A^c, B^c, C^c, etc. The complement of the universe is the *empty set*, denoted $S^c = \emptyset$.

An object that belongs to a designated set is called an *element* or *member* of that set. We will tend to denote elements by lower case Roman letters and write expressions such as $x \in A$, pronounced "x is an element of the set A." Sets with a small number of elements are often identified by simple enumeration, i.e., by writing down a list of elements. When we enumerate elements, we will enclose the list in braces and separate the elements by commas or semicolons. For example, the set of all feature films directed by Sergio Leone is

{ *A Fistful of Dollars*;
For a Few Dollars More;
The Good, the Bad, and the Ugly;
Once Upon a Time in the West;
Duck, You Sucker!;
Once Upon a Time in America }

In this book, we usually will be concerned with sets defined by certain mathematical properties. Some familiar sets to which we will refer repeatedly include:

- The set of *natural numbers*, $\boldsymbol{N} = \{1, 2, 3, \ldots\}$.
- The set of *integers*, $\boldsymbol{Z} = \{\ldots, -3, -2, -1, 0, 1, 2, 3, \ldots\}$.
- The set of *real numbers*, $\Re = (-\infty, \infty)$.

If A and B are sets and each element of A is also an element of B, then we say that A is a *subset* of B and write $A \subset B$. For example,

$$\boldsymbol{N} \subset \boldsymbol{Z} \subset \Re.$$

Quite often, a set A is defined to be those elements of another set B that satisfy a specified mathematical property. In such cases, we often specify A by writing a generic element of B to the left of a colon, the property to the right of the colon, and enclosing this syntax in braces. For example,

$$A = \{x \in \boldsymbol{Z} \,:\, x^2 < 5\} = \{-2, -1, 0, 1, 2\},$$

is pronounced "A is the set of integers x such that x^2 is less than 5."

Given sets A and B, there are several important sets that can be constructed from them. The *union* of A and B is the set

$$A \cup B = \{x \in S \,:\, x \in A \text{ or } x \in B\}$$

and the *intersection* of A and B is the set

$$A \cap B = \{x \in S \,:\, x \in A \text{ and } x \in B\}.$$

For example, if A is as above and

$$B = \{x \in \boldsymbol{Z} \,:\, |x - 2| \le 1\} = \{1, 2, 3\},$$

then $A \cup B = \{-2, -1, 0, 1, 2, 3\}$ and $A \cap B = \{1, 2\}$. Notice that unions and intersections are symmetric constructions, i.e., $A \cup B = B \cup A$ and $A \cap B = B \cap A$.

If $A \cap B = \emptyset$, i.e., if A and B have no elements in common, then A and B are *disjoint* or *mutually exclusive*. By convention, the empty set is a subset of every set, so

$$\emptyset \subset A \cap B \subset A \subset A \cup B \subset S$$

and

$$\emptyset \subset A \cap B \subset B \subset A \cup B \subset S.$$

These facts are illustrated by the *Venn diagram* in Figure 2.1, in which sets are qualitatively indicated by connected subsets of the plane. We will make frequent use of Venn diagrams as we develop basic facts about probabilities.

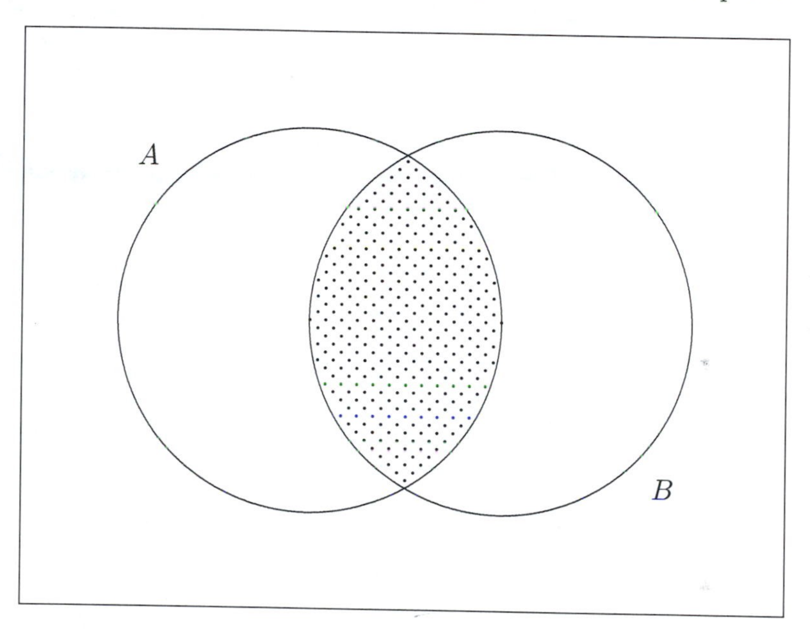

Figure 2.1: A Venn diagram. The shaded region represents the intersection of the nondisjoint sets A and B.

It is often useful to extend the concepts of union and intersection to more than two sets. Let $\{A_k\}$ denote an arbitrary collection of sets, where k is an index that identifies the set. Then $x \in S$ is an element of the union of $\{A_k\}$, denoted

$$\bigcup_k A_k,$$

if and only if there exists some k_0 such that $x \in A_{k_0}$. Also, $x \in S$ is an element of the intersection of $\{A_k\}$, denoted

$$\bigcap_k A_k,$$

if and only if $x \in A_k$ for every k. For example, if $A_k = \{0, 1, \ldots, k\}$ for $k = 1, 2, 3, \ldots$, then

$$\bigcup_k A_k = \{0, 1, 2, 3, \ldots\}$$

and

$$\bigcap_k A_k = \{0, 1\}.$$

It will be important to distinguish collections of sets with the following property:

Definition 2.1 *A collection of sets is* pairwise disjoint *if and only if each pair of sets in the collection has an empty intersection.*

Unions and intersections are related to each other by two distributive laws:

$$B \cap \left(\bigcup_k A_k \right) = \bigcup_k (B \cap A_k)$$

and

$$B \cup \left(\bigcap_k A_k \right) = \bigcap_k (B \cup A_k) \, .$$

Furthermore, unions and intersections are related to complements by DeMorgan's laws:

$$\left(\bigcup_k A_k \right)^c = \bigcap_k A_k^c$$

and

$$\left(\bigcap_k A_k \right)^c = \bigcup_k A_k^c.$$

The first law states that an object is not in any of the sets in the collection if and only if it is in the complement of each set; the second law states that an object is not in every set in the collection if it is in the complement of at least one set.

Finally, we consider another important set that can be constructed from A and B.

Definition 2.2 *The Cartesian product of two sets A and B, denoted $A \times B$, is the set of ordered pairs whose first component is an element of A and whose second component is an element of B, i.e.,*

$$A \times B = \{(a, b) \ : \ a \in A, \, b \in B\}.$$

For example, if $A = \{-2, -1, 0, 1, 2\}$ and $B = \{1, 2, 3\}$, then the set $A \times B$ contains the following elements:

$$\begin{array}{ccccc} (-2,1) & (-1,1) & (0,1) & (1,1) & (2,1) \\ (-2,2) & (-1,2) & (0,2) & (1,2) & (2,2) \\ (-2,3) & (-1,3) & (0,3) & (1,3) & (2,3) \end{array}$$

A familiar example of a Cartesian product is the Cartesian coordinatization of the plane,

$$\Re^2 = \Re \times \Re = \{(x, y) \ : \ x, y \in \Re\}.$$

Of course, this construction can also be extended to more than two sets, e.g.,

$$\Re^3 = \{(x, y, z) \ : \ x, y, z \in \Re\}.$$

2.2 Counting

This section is concerned with determining the number of elements in a specified set. One of the fundamental concepts that we will exploit in our brief study of counting is the notion of a *one-to-one correspondence* between two sets. We begin by illustrating this notion with an elementary example.

Example 2.1 Define two sets,

$$A_1 = \{\text{diamond}, \text{emerald}, \text{ruby}, \text{sapphire}\}$$

and

$$B = \{\text{blue}, \text{green}, \text{red}, \text{white}\}.$$

The elements of these sets can be paired in such a way that to each element of A_1 there is assigned a unique element of B and to each element of B there is assigned a unique element of A_1. Such a pairing can be accomplished in various ways; a natural assignment is the following:

$$\begin{array}{rcl} \text{diamond} & \leftrightarrow & \text{white} \\ \text{emerald} & \leftrightarrow & \text{green} \\ \text{ruby} & \leftrightarrow & \text{red} \\ \text{sapphire} & \leftrightarrow & \text{blue} \end{array}$$

This assignment exemplifies a one-to-one correspondence.

Now suppose that we augment A_1 by forming

$$A_2 = A_1 \cup \{\text{aquamarine}\}.$$

Although we can still assign a color to each gemstone, we *cannot* do so in such a way that each gemstone corresponds to a different color. There does not exist a one-to-one correspondence between A_2 and B.

From Example 2.1, we abstract

Definition 2.3 *Two sets can be placed in* one-to-one correspondence *if and only if their elements can be paired in such a way that each element of either set is associated with a unique element of the other set.*

The concept of one-to-one correspondence can then be exploited to obtain a formal definition of a familiar concept:

Definition 2.4 *A set A is* finite *if and only if there exists a natural number N such that the elements of A can be placed in one-to-one correspondence with the elements of $\{1, 2, \ldots, N\}$.*

If A is finite, then the natural number N that appears in Definition 2.4 is unique. It is, in fact, the number of elements in A. We will denote this quantity, sometimes called the *cardinality* of A, by $\#(A)$. In Example 2.1 above, $\#(A_1) = \#(B) = 4$ and $\#(A_2) = 5$.

The Multiplication Principle Most of our counting arguments will rely on a fundamental principle, which we illustrate with an example.

Example 2.2 *Suppose that each gemstone in Example 2.1 has been mounted on a ring. You desire to wear one of these rings on your left hand and another on your right hand. How many ways can this be done?*

First, suppose that you wear the diamond ring on your left hand. Then there are three rings available for your right hand: emerald, ruby, sapphire.

Next, suppose that you wear the emerald ring on your left hand. Again there are three rings available for your right hand: diamond, ruby, sapphire.

Suppose that you wear the ruby ring on your left hand. Once again there are three rings available for your right hand: diamond, emerald, sapphire.

Finally, suppose that you wear the sapphire ring on your left hand. Once more there are three rings available for your right hand: diamond, emerald, ruby.

We have counted a total of $3 + 3 + 3 + 3 = 12$ ways to choose a ring for each hand. Enumerating each possibility is rather tedious, but it reveals a useful shortcut. There are 4 ways to choose a ring for the left hand and, for each such choice, there are three ways to choose a ring for the right hand. Hence, there are $4 \cdot 3 = 12$ ways to choose a ring for each hand. This is an instance of a general principle:

> *Suppose that two decisions are to be made and that there are n_1 possible outcomes of the first decision. If, for each outcome of the first decision, there are n_2 possible outcomes of the second decision, then there are $n_1 n_2$ possible outcomes of the pair of decisions.*

Permutations and Combinations We now consider two more concepts that are often employed when counting the elements of finite sets. We motivate these concepts with an example.

Example 2.3 *A fast-food restaurant offers a single entree that comes with a choice of* 3 *side dishes from a total of* 15*. To address the perception that it serves only one dinner, the restaurant conceives an advertisement that identifies each choice of side dishes as a distinct dinner. Assuming that each entree must be accompanied by* 3 *distinct side dishes, e.g.,* {stuffing, mashed potatoes, green beans} *is permitted but* {stuffing, stuffing, mashed potatoes} *is not, how many distinct dinners are available?*[1]

Answer 2.3a The restaurant reasons that a customer, asked to choose 3 side dishes, must first choose 1 side dish from a total of 15. There are 15 ways of making this choice. Having made it, the customer must then choose a second side dish that is different from the first. For each choice of the first side dish, there are 14 ways of choosing the second; hence 15×14 ways of choosing the pair. Finally, the customer must choose a third side dish that is different from the first two. For each choice of the first two, there are 13 ways of choosing the third; hence $15 \times 14 \times 13$ ways of choosing

[1] This example is based on an actual incident involving the Boston Chicken (now Boston Market) restaurant chain and a high school math class in Denver, CO. Boston Market does not require patrons to order distinct side dishes.

the triple. Accordingly, the restaurant advertises that it offers a total of $15 \times 14 \times 13 = 2730$ possible dinners.

Answer 2.3b A high school math class considers the restaurant's claim and notes that the restaurant has counted side dishes of

{	stuffing,	mashed potatoes,	green beans	},
{	stuffing,	green beans,	mashed potatoes	},
{	mashed potatoes,	stuffing,	green beans	},
{	mashed potatoes,	green beans,	stuffing	},
{	green beans,	stuffing,	mashed potatoes	}, and
{	green beans,	mashed potatoes,	stuffing	}

as distinct dinners. Thus, the restaurant has counted dinners that differ only with respect to the order in which the side dishes were chosen as distinct. Reasoning that what matters is what is on one's plate, not the order in which the choices were made, the math class concludes that the restaurant has overcounted. As illustrated above, each triple of side dishes can be ordered in 6 ways: the first side dish can be any of 3, the second side dish can be any of the remaining 2, and the third side dish must be the remaining 1 ($3 \times 2 \times 1 = 6$). The math class writes a letter to the restaurant, arguing that the restaurant has overcounted by a factor of 6 and that the correct count is $2730 \div 6 = 455$. The restaurant cheerfully agrees and donates \$1000 to the high school's math club.

From Example 2.3 we abstract the following definitions:

Definition 2.5 *The number of* permutations *(ordered choices) of r objects from n objects is*

$$P(n,r) = n \times (n-1) \times \cdots \times (n-r+1).$$

Definition 2.6 *The number of* combinations *(unordered choices) of r objects from n objects is*

$$C(n,r) = P(n,r) \div P(r,r).$$

In Example 2.3, the restaurant claimed that it offered $P(15,3)$ dinners, while the math class argued that a more plausible count was $C(15,3)$. There, as always, the distinction was made on the basis of whether the order of the choices is or is not relevant.

Permutations and combinations are often expressed using factorial notation. Let

$$0! = 1$$

and let k be a natural number. Then the expression $k!$, pronounced "k factorial" is defined recursively by the formula

$$k! = k \times (k-1)!.$$

For example,

$$3! = 3 \times 2! = 3 \times 2 \times 1! = 3 \times 2 \times 1 \times 0! = 3 \times 2 \times 1 \times 1 = 3 \times 2 \times 1 = 6.$$

Because

$$\begin{aligned} n! &= n \times (n-1) \times \cdots \times (n-r+1) \times (n-r) \times \cdots \times 1 \\ &= P(n,r) \times (n-r)!, \end{aligned}$$

we can write

$$P(n,r) = \frac{n!}{(n-r)!}$$

and

$$C(n,r) = P(n,r) \div P(r,r) = \frac{n!}{(n-r)!} \div \frac{r!}{(r-r)!} = \frac{n!}{r!(n-r)!}.$$

Finally, we note (and will sometimes use) the popular notation

$$C(n,r) = \binom{n}{r},$$

pronounced "n choose r".

Example 2.4 *A coin is tossed* 10 *times. How many sequences of* 10 *tosses result in a total of exactly* 2 `Heads`*?*

Solution A sequence of `Heads` and `Tails` is completely specified by knowing which tosses resulted in `Heads`. To count how many sequences result in 2 `Heads`, we simply count how many ways there are to choose the pair of tosses on which `Heads` result. This is choosing 2 tosses from 10, or

$$\binom{10}{2} = \frac{10!}{2!(10-2)!} = \frac{10 \cdot 9}{2 \cdot 1} = 45.$$

Example 2.5 *Consider the hypothetical example described in Section 1.2. In a class of* 40 *students, how many ways can one choose* 20 *students to receive exam* `A`*? Assuming that the class comprises* 30 *freshmen and* 10 *nonfreshmen, how many ways can one choose* 15 *freshmen and* 5 *nonfreshmen to receive exam* `A`*?*

Solution There are

$$\binom{40}{20} = \frac{40!}{20!(40-20)!} = \frac{40 \cdot 39 \cdots 22 \cdot 21}{20 \cdot 19 \cdots 2 \cdot 1} = 137,846,528,820$$

ways to choose 20 students from 40. There are

$$\binom{30}{15} = \frac{30!}{15!(30-15)!} = \frac{30 \cdot 29 \cdots 17 \cdot 16}{15 \cdot 14 \cdots 2 \cdot 1} = 155,117,520$$

ways to choose 15 freshmen from 30 and

$$\binom{10}{5} = \frac{10!}{5!(10-5)!} = \frac{10 \cdot 9 \cdot 8 \cdot 7 \cdot 6}{5 \cdot 4 \cdot 3 \cdot 2 \cdot 1} = 252$$

ways to choose 5 nonfreshmen from 10; hence,

$$155,117,520 \cdot 252 = 39,089,615,040$$

ways to choose 15 freshmen and 5 nonfreshmen to receive exam `A`. Notice that, of all the ways to choose 20 students to receive exam `A`, about 28% result in exactly 15 freshmen and 5 nonfreshmen.

Countability Thus far, our study of counting has been concerned exclusively with finite sets. However, our subsequent study of probability will require us to consider sets that are not finite. Toward that end, we introduce the following definitions:

Definition 2.7 *A set is* infinite *if and only if it is not finite.*

Definition 2.8 *A set is* denumerable *if and only if its elements can be placed in one-to-one correspondence with the natural numbers.*

Definition 2.9 *A set is* countable *if and only if it is either finite or denumerable.*

Definition 2.10 *A set is* uncountable *if and only if it is not countable.*

Like Definition 2.4, Definition 2.8 depends on the notion of a one-to-one correspondence between sets. However, whereas this notion is completely straightforward when at least one of the sets is finite, it can be rather elusive when both sets are infinite. Accordingly, we provide some examples of denumerable sets. In each case, we superscript each element of the set in question with the corresponding natural number.

Example 2.6 Consider the set of even natural numbers, which excludes one of every two consecutive natural numbers It might seem that this set cannot be placed in one-to-one correspondence with the natural numbers in their entirety; however, infinite sets often possess counterintuitive properties. Here is a correspondence that demonstrates that this set is denumerable:

$$2^1, 4^2, 6^3, 8^4, 10^5, 12^6, 14^7, 16^8, 18^9, \ldots$$

Example 2.7 Consider the set of integers. It might seem that this set, which includes both a positive and a negative copy of each natural number, cannot be placed in one-to-one correspondence with the natural numbers; however, here is a correspondence that demonstrates that this set is denumerable:

$$\ldots, -4^9, -3^7, -2^5, -1^3, 0^1, 1^2, 2^4, 3^6, 4^8, \ldots$$

Example 2.8 Consider the Cartesian product of the set of natural numbers with itself. This set contains one copy of the entire set of natural numbers for each natural number—surely it cannot be placed in one-to-one correspondence with a single copy of the set of natural numbers! In fact, the following correspondence demonstrates that this set is also denumerable:

$$\begin{array}{cccccc}
(1,1)^1 & (1,2)^2 & (1,3)^6 & (1,4)^7 & (1,5)^{15} & \cdots \\
(2,1)^3 & (2,2)^5 & (2,3)^8 & (2,4)^{14} & (2,5)^{17} & \cdots \\
(3,1)^4 & (3,2)^9 & (3,3)^{13} & (3,4)^{18} & (3,5)^{26} & \cdots \\
(4,1)^{10} & (4,2)^{12} & (4,3)^{19} & (4,4)^{25} & (4,5)^{32} & \cdots \\
(5,1)^{11} & (5,2)^{20} & (5,3)^{24} & (5,4)^{33} & (5,5)^{41} & \cdots \\
\vdots & \vdots & \vdots & \vdots & \vdots & \ddots
\end{array}$$

In light of Examples 2.6–2.8, the reader may wonder what is required to construct a set that is not countable. We conclude this section by remarking that the following intervals are uncountable sets, where $a, b \in \Re$ and $a < b$.

$$\begin{aligned}
(a,b) &= \{x \in \Re : a < x < b\} \\
[a,b) &= \{x \in \Re : a \le x < b\} \\
(a,b] &= \{x \in \Re : a < x \le b\} \\
[a,b] &= \{x \in \Re : a \le x \le b\}
\end{aligned}$$

We will make frequent use of such sets, often referring to (a, b) as an *open* interval and $[a, b]$ as a *closed* interval.

2.3 Functions

A function is a rule that assigns a unique element of a set of possible labels, B, to each element of another set, A. A familiar example is the rule that assigns to each real number x the real number $y = x^2$, e.g., that assigns $y = 4$ to $x = 2$. Notice that each real number has a unique square ($y = 4$ is the only number that this rule assigns to $x = 2$), but that more than one number may have the same square ($y = 4$ is assigned to both $x = 2$ and $x = -2$).

The set A is the function's *domain* and the set B is the function's *range*. Notice that each element of A must be assigned some element of B, but that an element of B need not be assigned to any element of A. Thus, in the preceding example, every $x \in A = \Re$ has a squared value $y \in B = \Re$, but not every $y \in B$ is the square of some number $x \in A$. (For example, $y = -4$ is not the square of any real number.) The elements of B that are assigned to elements of A constitute the *range* of the function. In the preceding example, the range of the function is $[0, \infty)$.

We will use a variety of letters to denote various types of functions. Examples include $P, X, Y, f, g, F, G, \phi$. If ϕ is a function with domain A and possible labels B, then we write $\phi : A \to B$, often pronounced "ϕ maps A into B". If ϕ assigns $b \in B$ to $a \in A$, then we say that b is the value of ϕ at a and we write $b = \phi(a)$. In a natural abuse of notation, we write the range of ϕ as $\phi(A)$.

If $\phi : A \to B$, then for each $b \in B$ there is a subset (possibly empty) of A comprising those elements of A at which ϕ has value b. We denote this set by

$$\phi^{-1}(b) = \{a \in A : \phi(a) = b\}.$$

For example, if $\phi : \Re \to \Re$ is the function defined by $\phi(x) = x^2$, then

$$\phi^{-1}(4) = \{-2, 2\}.$$

More generally, if $B_0 \subset B$, then

$$\phi^{-1}(B_0) = \{a \in A : \phi(a) \in B_0\}.$$

Using the same example,

$$\phi^{-1}([4, 9]) = \left\{x \in \Re : x^2 \in [4, 9]\right\} = [-3, -2] \cup [2, 3].$$

The object ϕ^{-1} is called the *inverse* of ϕ and $\phi^{-1}(B_0)$ is called the inverse image of B_0.

2.4 Limits

In Section 2.2 we examined several examples of denumerable sets of real numbers. In each of these examples, we imposed an order on the set when we placed it in one-to-one correspondence with the natural numbers. Once an order has been specified, we can inquire how the set behaves as we progress through its values in the prescribed sequence. For example, the real numbers in the ordered denumerable set

$$\left\{1, \frac{1}{2}, \frac{1}{3}, \frac{1}{4}, \frac{1}{5}, \ldots\right\} \tag{2.1}$$

steadily decrease as one progresses through them. Furthermore, as in Zeno's famous paradoxes, the numbers seem to approach the value zero without ever actually attaining it. To describe such sets, it is helpful to introduce some specialized terminology and notation.

We begin with

Definition 2.11 *A sequence of real numbers is an ordered denumerable subset of* $\Re$.

Sequences are often denoted using a dummy variable that is specified or understood to index the natural numbers. For example, we might identify the sequence (2.1) by writing $\{1/n\}$ for $n = 1, 2, 3, \ldots$.

Next we consider the phenomenon that $1/n$ approaches 0 as n increases, although each $1/n > 0$. Let ϵ denote any strictly positive real number. What we have noticed is the fact that, no matter how small ϵ may be, eventually n becomes so large that $1/n < \epsilon$. We formalize this observation in

Definition 2.12 *Let* $\{y_n\}$ *denote a sequence of real numbers. We say that* $\{y_n\}$ *converges to a constant value* $c \in \Re$ *if and only if, for every* $\epsilon > 0$, *there exists a natural number* N *such that* $y_n \in (c-\epsilon, c+\epsilon)$ *for each* $n \geq N$.

If the sequence of real numbers $\{y_n\}$ converges to c, then we say that c is the *limit* of $\{y_n\}$ and we write either $y_n \to c$ as $n \to \infty$ or $\lim_{n\to\infty} y_n = c$. In particular,

$$\lim_{n\to\infty} \frac{1}{n} = 0.$$

2.5 Exercises

1. Suppose that $S = \boldsymbol{N}$ and consider the following three sets:

$$A = \{1, 2, 3, 4, 5\} \quad B = \{4, 5, 6, 7, 8\} \quad C = \{7, 8, 9, 10, 11\}$$

   (a) List the elements of $A \cup B \cup C$.

   (b) List the elements of $A \cap B \cap C$.

   (c) Is the collection $\{A, B, C\}$ pairwise disjoint? Why or why not?

2. Refer to Exercise 2.5.1 and list the elements of each of the following sets.

   (a) $A \cap (B \cup C)$

   (b) $A \cup (B \cap C)$

(c) $B \cap (A \cup C)$

(d) $B \cup (A \cap C)$

(e) $(C \cap A) \cup (C \cap B)$

(f) $(C \cup A) \cap (C \cup B)$

3. Let $S = \{1, 2, 3, 4, 5, 6, 7, 8, 9, 10\}$, let $A = \{1, 3, 5, 7, 9\}$, and let $B = \{1, 2, 3, 5, 7\}$. List the elements of each of the following sets.

   (a) A^c

   (b) B^c

   (c) $(A \cup B)^c$

   (d) $(A \cap B)^c$

4. A classic riddle inquires:

   As I was going to St. Ives,
   I met a man with seven wives.
   Each wife had seven sacks,
   Each sack had seven cats,
   Each cat had seven kits.
   Kits, cats, sacks, wives—
   How many were going to St. Ives?

   (a) How many creatures (human and feline) were in the entourage that the narrator encountered?

   (b) What is the answer to the riddle?

5. A well-known carol, "The Twelve Days of Christmas," describes a progression of gifts that the singer receives from her true love:

   On the first day of Christmas, my true love gave to me:
   A partridge in a pear tree.
   On the second day of Christmas, my true love gave to me:
   Two turtle doves, and a partridge in a pear tree.
   Et cetera.[2]

   How many birds did the singer receive from her true love?

6. The throw of an astragalus (see Section 1.4) has four possible outcomes, $\{1, 3, 4, 6\}$. When throwing four astragali,

[2]You should be able to find the complete lyrics by doing a web search.

(a) How many ways are there to obtain a dog, i.e., for each astragalus to produce a 1?

(b) How many ways are there to obtain a venus, i.e., for each astragalus to produce a different outcome?

Hint: Label each astragalus (e.g., antelope, bison, cow, deer) and keep track of the outcome of each distinct astragalus.

7. When throwing five astragali,

   (a) How many ways are there to obtain the throw of child-eating Cronos, i.e., to obtain three fours and two sixes?

   (b) How many ways are there to obtain the throw of Saviour Zeus, i.e., to obtain one one, two threes, and two fours?

8. The throw of one die has six possible outcomes, $\{1, 2, 3, 4, 5, 6\}$. A medieval poem, "The Chance of the Dyse," enumerates the fortunes that could be divined from casting three dice. Order does not matter, e.g., the fortune associated with 6-5-3 is also associated with 3-5-6. How many fortunes does the poem enumerate?

9. Suppose that five cards are dealt from a standard deck of playing cards.

   (a) How many hands are possible?

   (b) How many straight-flush hands are possible?

   (c) How many 4-of-a-kind hands are possible?

   (d) Why do you suppose that a straight flush beats 4-of-a-kind?

10. In the television reality game show *Survivor*, 16 contestants (the "castaways") compete for $1 million. The castaways are stranded in a remote location, e.g., an uninhabited island in the China Sea. Initially, the castaways are divided into two tribes. The tribes compete in a sequence of immunity challenges. After each challenge, the losing tribe must vote out one of its members and that person is eliminated from the game. Eventually, the tribes merge and the surviving castaways compete in a sequence of individual immunity challenges. The winner receives immunity and the merged tribe must then vote out one of its other members. After the merged tribe has been reduced to two members, a jury of the last 7 castaways to have been eliminated votes on who should be the Sole Survivor and win $1 million. (Technically, the

jury votes *for* the Sole Survivor, but this is equivalent to eliminating one of the final two castaways.)

(a) Suppose that we define an outcome of *Survivor* to be the name of the Sole Survivor. In any given game of *Survivor*, how many outcomes are possible?

(b) Suppose that we define an outcome of *Survivor* to be a list of the castaways' names, arranged in the order in which they were eliminated. In any given game of *Survivor*, how many outcomes are possible?

11. The final eight castaways in *Survivor 2: Australian Outback* included four men (Colby, Keith, Nick, and Rodger) and four women (Amber, Elisabeth, Jerri, and Tina). They participated in a reward challenge that required them to form four teams of two persons, one male and one female. (The teams raced over an obstacle course, recording the time of the slower team member.) The castaways elected to pair off by drawing lots.

    (a) How many ways were there for the castaways to form four teams?

    (b) Jerri was opposed to drawing lots—she wanted to team with Colby. How many ways are there for the castaways to form four male-female teams if one of the teams is Colby-Jerri?

    (c) If all pairings (male-male, male-female, female-female) are allowed, then how many ways are there for the castaways to form four teams?

12. In Major League Baseball's World Series, the winners of the National (`N`) and American (`A`) League pennants play a sequence of games. The first team to win four games wins the Series. Thus, the Series must last at least four games and can last no more than seven games. Let us define an *outcome* of the World Series by identifying which League's pennant winner won each game. For example, the outcome of the 1975 World Series, in which the Cincinnati Reds represented the National League and the Boston Red Sox represented the American League, was `ANNANAN`. How many World Series outcomes are possible?

13. Let A denote the set of all feature films directed by Sergio Leone. Consider a function, ϕ, with domain A and range $\boldsymbol{N}$ that assigns to each film in A the year in which it was released. The following table displays the value of this function for each element of A.

A Fistful of Dollars	1964
For a Few Dollars More	1965
The Good, the Bad, and the Ugly	1966
Once Upon a Time in the West	1968
Duck, You Sucker!	1972
Once Upon a Time in America	1984

(a) What is the range of ϕ?

(b) What is $\phi^{-1}(1968)$?

(c) What is $\phi^{-1}(1970)$?

(d) What is the inverse image of the subset of $\boldsymbol{N}$ known as The Sixties?

14. Consider the function $\phi : \Re \rightarrow \Re$ defined by $\phi(x) = 2^x$.

(a) What is $\phi(6)$?

(b) What is $\phi(-3)$?

(c) What is $\phi(\Re)$?

(d) What is $\phi^{-1}(16)$?

(e) What is $\phi^{-1}(1/4)$?

(f) What is $\phi^{-1}([2, 32])$?

15. We say that an integer y is a power of 10 if there exists an integer k for which $y = 10^k$. For example, $y = 1000$ is a power of 10 because $1000 = 10^3$. Let C denote the set of natural numbers that are powers of 10.

(a) Is C finite or denumerable? Why?

(b) For $n = 1, 2, 3, \ldots$, let

$$y_n = \left\{ \begin{array}{cl} 1/n & \text{if } n \notin C \\ 1 & \text{if } n \in C \end{array} \right\}.$$

Does the sequence $\{y_n\}$ converge to a limit? Why or why not?

16. For $n = 0, 1, 2, \ldots$, let

$$y_n = \sum_{k=0}^{n} 2^{-k} = 2^{-0} + 2^{-1} + \cdots + 2^{-n}.$$

(a) Compute y_0, y_1, y_2, y_3, and y_4.

(b) The sequence $\{y_0, y_1, y_2, \ldots\}$ is an example of a *sequence of partial sums.* Guess the value of its limit, usually written

$$\lim_{n\to\infty} y_n = \lim_{n\to\infty} \sum_{k=0}^{n} 2^{-k} = \sum_{k=0}^{\infty} 2^{-k}.$$

Chapter 3

Probability

The goal of statistical inference is to draw conclusions about a population from "representative information" about it. In future chapters, we will discover that a powerful way to obtain representative information about a population is through the planned introduction of chance. Thus, probability is the foundation of statistical inference—to study the latter, we must first study the former. Fortunately, the theory of probability is an especially beautiful branch of mathematics. Although our purpose in studying probability is to provide the reader with some tools that will be needed when we study statistics, we also hope to impart some of the beauty of those tools.

3.1 Interpretations of Probability

Probabilistic statements can be interpreted in different ways. For example, how would you interpret the following statement?

> There is a 40 percent chance of rain today.

Your interpretation is apt to vary depending on the context in which the statement is made. If the statement was made as part of a forecast by the National Weather Service, then something like the following interpretation might be appropriate:

> In the recent history of this locality, of all days on which present atmospheric conditions have been experienced, rain has occurred on approximately 40 percent of them.

This is an example of the *frequentist* interpretation of probability. With this interpretation, a probability is a long-run average proportion of occurrence.

Suppose, however, that you had just peered out a window, wondering if you should carry an umbrella to school, and asked your roommate if she thought that it was going to rain. Unless your roommate is studying metereology, it is not plausible that she possesses the knowledge required to make a frequentist statement! If her response was a casual "I'd say that there's a 40 percent chance," then something like the following interpretation might be appropriate:

> I believe that it might very well rain, but that it's a little less likely to rain than not.

This is an example of the *subjectivist* interpretation of probability. With this interpretation, a probability expresses the strength of one's belief.

The philosopher I. Hacking has observed that dual notions of probability, one aleatory (frequentist) and one epistemological (subjectivist) have co-existed throughout history, and that "philosophers seem singularly unable to put [them] asunder..."[1] We shall not attempt so perilous an undertaking. But however we decide to interpret probabilities, we will need a formal mathematical description of probability to which we can appeal for insight and guidance. The remainder of this chapter provides an introduction to the most commonly adopted approach to *axiomatic probability*. The chapters that follow tend to emphasize a frequentist interpretation of probability, but the mathematical formalism can also be used with a subjectivist interpretation.

3.2 Axioms of Probability

The mathematical model that has dominated the study of probability was formalized by the Russian mathematician A. N. Kolmogorov in a monograph published in 1933. The central concept in this model is a *probability space*, assumed to have three components:

S A *sample space*, a universe of "possible" outcomes for the experiment in question.

$\mathcal{C}$ A designated collection of "observable" subsets (called *events*) of the sample space.

P A *probability measure*, a function that assigns real numbers (called *probabilities*) to events.

[1] I. Hacking, *The Emergence of Probability*, Cambridge University Press, 1975, Chapter 2: Duality.

We describe each of these components in turn.

The Sample Space The sample space is a set. Depending on the nature of the experiment in question, it may or may not be easy to decide upon an appropriate sample space.

Example 3.1 *A coin is tossed once.*

A plausible sample space for this experiment will comprise two outcomes, `Heads` and `Tails`. Denoting these outcomes by `H` and `T`, we have

$$S = \{\mathtt{H}, \mathtt{T}\}.$$

Remark We have discounted the possibility that the coin will come to rest on edge. This is the first example of a theme that will recur throughout this text, that mathematical models are rarely—if ever—completely faithful representations of nature. As described by Mark Kac,

> Models are, for the most part, caricatures of reality, but if they are good, then, like good caricatures, they portray, though perhaps in distorted manner, some of the features of the real world. The main role of models is not so much to explain and predict—though ultimately these are the main functions of science—as to polarize thinking and to pose sharp questions.[2]

In Example 3.1, and in most of the other elementary examples that we will use to illustrate the fundamental concepts of axiomatic probability, the fidelity of our mathematical descriptions to the physical phenomena described should be apparent. Practical applications of inferential statistics, however, often require imposing mathematical assumptions that may be suspect. Data analysts must constantly make judgments about the plausibility of their assumptions, not so much with a view to whether or not the assumptions are completely correct (they almost never are), but with a view to whether or not the assumptions are sufficient for the analysis to be meaningful.

[2]M. Kac (1969), Some mathematical models in science, *Science*, 166:695–699.

Example 3.2 *A coin is tossed twice.*

A plausible sample space for this experiment will comprise four outcomes, two outcomes per toss. Here,

$$S = \left\{ \begin{array}{cc} \texttt{HH} & \texttt{TH} \\ \texttt{HT} & \texttt{TT} \end{array} \right\}.$$

Example 3.3 *An individual's height is measured.*

In this example, it is less clear what outcomes are possible. All human heights fall within certain bounds, but precisely what bounds should be specified? And what of the fact that heights are not measured exactly?

Only rarely would one address these issues when choosing a sample space. For this experiment, most statisticians would choose as the sample space the set of all real numbers, then worry about which real numbers were actually observed. Thus, the phrase "possible outcomes" refers to conceptual rather than practical possibility. The sample space is usually chosen to be mathematically convenient and all-encompassing.

The Collection of Events Events are subsets of the sample space, but how do we decide which subsets of S should be designated as events? If the outcome $s \in S$ was observed and $E \subset S$ is an event, then we say that E *occurred* if and only if $s \in E$. A subset of S is *observable* if it is always possible for the experimenter to determine whether or not it occurred. Our intent is that the collection of events should be the collection of observable subsets. This intent is often tempered by our desire for mathematical convenience and by our need for the collection to possess certain mathematical properties. In practice, the issue of observability is rarely considered and certain conventional choices are automatically adopted. For example, when S is a finite set, one usually designates *all* subsets of S to be events.

Whether or not we decide to grapple with the issue of observability, the collection of events *must* satisfy the following properties:

1. The sample space is an event.

2. If E is an event, then E^c is an event.

3. The union of any countable collection of events is an event.

A collection of subsets with these properties is sometimes called a *sigma-field*.

Taken together, the first two properties imply that both S and $\emptyset$ must be events. If S and $\emptyset$ are the only events, then the third property holds; hence, the collection $\{S, \emptyset\}$ is a sigma-field. It is not, however, a very useful collection of events, as it describes a situation in which the experimental outcomes cannot be distinguished!

Example 3.1 (continued) To distinguish `Heads` from `Tails`, we must assume that each of these individual outcomes is an event. Thus, the only plausible collection of events for this experiment is the collection of all subsets of S, i.e.,

$$\mathcal{C} = \{S, \{\mathtt{H}\}, \{\mathtt{T}\}, \emptyset\}.$$

Example 3.2 (continued) If we designate all subsets of S as events, then we obtain the following collection:

$$\mathcal{C} = \left\{ \begin{array}{l} S, \\ \{\mathtt{HH},\mathtt{HT},\mathtt{TH}\},\ \{\mathtt{HH},\mathtt{HT},\mathtt{TT}\}, \\ \{\mathtt{HH},\mathtt{TH},\mathtt{TT}\},\ \{\mathtt{HT},\mathtt{TH},\mathtt{TT}\}, \\ \{\mathtt{HH},\mathtt{HT}\},\ \{\mathtt{HH},\mathtt{TH}\},\ \{\mathtt{HH},\mathtt{TT}\}, \\ \{\mathtt{HT},\mathtt{TH}\},\ \{\mathtt{HT},\mathtt{TT}\},\ \{\mathtt{TH},\mathtt{TT}\}, \\ \{\mathtt{HH}\},\ \{\mathtt{HT}\},\ \{\mathtt{TH}\},\ \{\mathtt{TT}\}, \\ \emptyset \end{array} \right\}.$$

This is perhaps the most plausible collection of events for this experiment, but others are also possible. For example, suppose that we were unable to distinguish the order of the tosses, so that we could not distinguish between the outcomes `HT` and `TH`. Then the collection of events should not include any subsets that contain one of these outcomes but not the other, e.g., $\{\mathtt{HH},\mathtt{TH},\mathtt{TT}\}$. Thus, the following collection of events might be deemed appropriate:

$$\mathcal{C} = \left\{ \begin{array}{l} S, \\ \{\mathtt{HH},\mathtt{HT},\mathtt{TH}\},\ \{\mathtt{HT},\mathtt{TH},\mathtt{TT}\}, \\ \{\mathtt{HH},\mathtt{TT}\}, \{\mathtt{HT},\mathtt{TH}\}, \\ \{\mathtt{HH}\},\ \{\mathtt{TT}\}, \\ \emptyset \end{array} \right\}.$$

The interested reader should verify that this collection is indeed a sigma-field.

The Probability Measure Once the collection of events has been designated, each event $E \in \mathcal{C}$ can be assigned a probability $P(E)$. This must be done according to specific rules; in particular, the probability measure P *must* satisfy the following properties:

1. If E is an event, then $0 \leq P(E) \leq 1$.

2. $P(S) = 1$.

3. If $\{E_1, E_2, E_3, \ldots\}$ is a countable collection of pairwise disjoint events, then
$$P\left(\bigcup_{i=1}^{\infty} E_i\right) = \sum_{i=1}^{\infty} P(E_i).$$

We discuss each of these properties in turn.

The first property states that probabilities are nonnegative and finite. Thus, neither the statement that "the probability that it will rain today is $-.5$" nor the statement that "the probability that it will rain today is infinity" are meaningful. These restrictions have certain mathematical consequences. The further restriction that probabilities are no greater than unity is actually a consequence of the second and third properties.

The second property states that the probability that an outcome occurs, that *something* happens, is unity. This is a convention that simplifies formulae and facilitates interpretation.

The third property, called *countable additivity*, is the most interesting. Consider Example 3.2, supposing that $\{\texttt{HT}\}$ and $\{\texttt{TH}\}$ are events and that we want to compute the probability that exactly one `Head` is observed, i.e., the probability of
$$\{\texttt{HT}\} \cup \{\texttt{TH}\} = \{\texttt{HT}, \texttt{TH}\}.$$

Because $\{\texttt{HT}\}$ and $\{\texttt{TH}\}$ are events, their union is an event and therefore has a probability. Because they are mutually exclusive, we would like that probability to be
$$P\left(\{\texttt{HT}, \texttt{TH}\}\right) = P\left(\{\texttt{HT}\}\right) + P\left(\{\texttt{TH}\}\right).$$

We ensure this by requiring that the probability of the union of any two disjoint events is the sum of their respective probabilities.

Having assumed that
$$A \cap B = \emptyset \Rightarrow P(A \cup B) = P(A) + P(B), \tag{3.1}$$

it is easy to compute the probability of any finite union of pairwise disjoint events. For example, if A, B, C, and D are pairwise disjoint events, then
$$P(A \cup B \cup C \cup D) \;=\; P(A \cup (B \cup C \cup D))$$

$$\begin{aligned} &= P(A) + P(B \cup C \cup D) \\ &= P(A) + P(B \cup (C \cup D)) \\ &= P(A) + P(B) + P(C \cup D) \\ &= P(A) + P(B) + P(C) + P(D). \end{aligned}$$

Thus, from (3.1) can be deduced the following implication:

If $E_1, \ldots, E_n$ are pairwise disjoint events, then

$$P\left(\bigcup_{i=1}^{n} E_i\right) = \sum_{i=1}^{n} P(E_i).$$

This implication is known as *finite additivity.* Notice that the union of $E_1, \ldots, E_n$ must be an event (and hence have a probability) because each E_i is an event.

An extension of finite additivity, countable additivity is the following implication:

If $E_1, E_2, E_3, \ldots$ are pairwise disjoint events, then

$$P\left(\bigcup_{i=1}^{\infty} E_i\right) = \sum_{i=1}^{\infty} P(E_i).$$

The reason for insisting upon this extension has less to do with applications than with theory. Although some axiomatic theories of probability assume only finite additivity, it is generally felt that the stronger assumption of countable additivity results in a richer theory. Again, notice that the union of $E_1, E_2, \ldots$ must be an event (and hence have a probability) because each E_i is an event.

Finally, we emphasize that *probabilities are assigned to events.* It may or may not be that the individual experimental outcomes are events. If they are, then they will have probabilities. In some such cases (see Chapter 4), the probability of any event can be deduced from the probabilities of the individual outcomes; in other such cases (see Chapter 5), this is not possible.

All of the facts about probability that we will use in studying statistical inference are consequences of the assumptions of the Kolmogorov probability model. It is not the purpose of this book to present derivations of these facts; however, three elementary (and useful) propositions suggest how one might proceed along such lines. In each case, a Venn diagram helps to illustrate the proof.

Theorem 3.1 *If E is an event then*

$$P(E^c) = 1 - P(E).$$

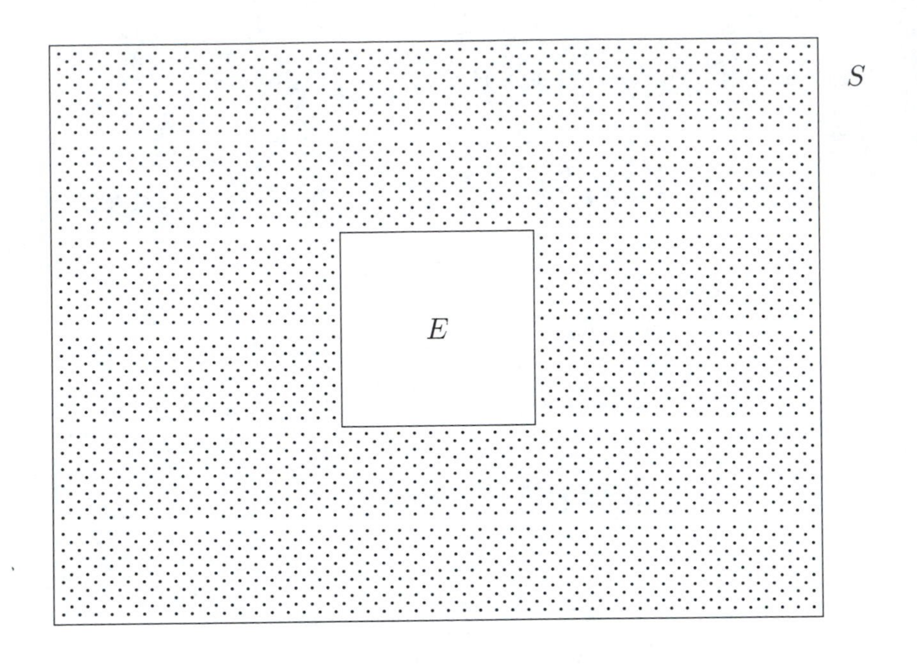

Figure 3.1: A Venn diagram for the probability of E^c.

Proof Refer to Figure 3.1. E^c is an event because E is an event. By definition, E and E^c are disjoint events whose union is S. Hence,

$$1 = P(S) = P(E \cup E^c) = P(E) + P(E^c)$$

and the theorem follows upon subtracting $P(E)$ from both sides. □

Theorem 3.2 *If A and B are events and $A \subset B$, then*

$$P(A) \leq P(B).$$

Proof Refer to Figure 3.2. A^c and B^c are events because A and B are events. Hence, the shaded region $B \cap A^c = B \cup A^c$, is an event and

$$B = A \cup (B \cap A^c).$$

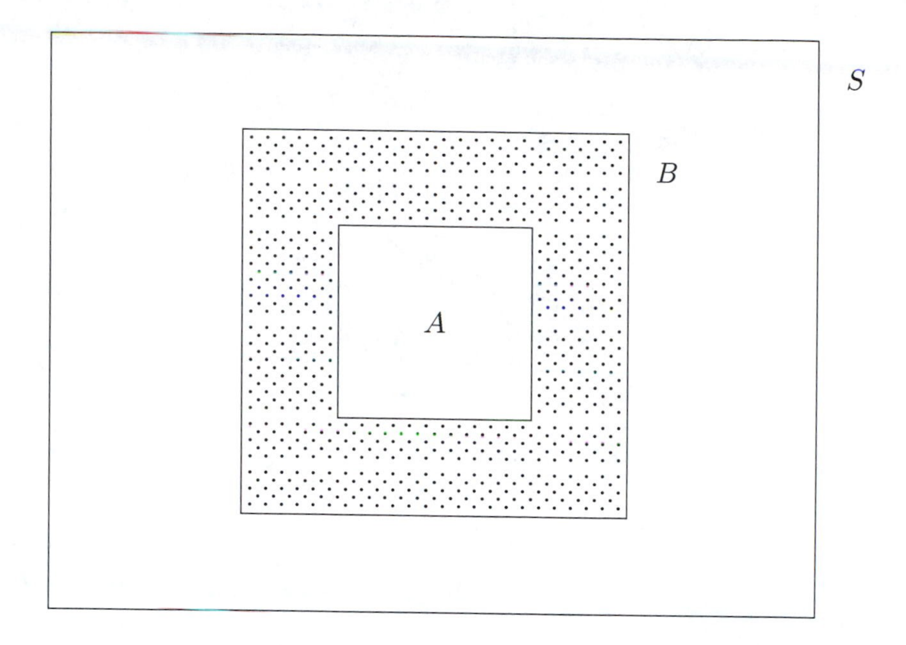

Figure 3.2: A Venn diagram for the probability of $A \subset B$.

Because A and $B \cap A^c$ are disjoint events,

$$P(B) = P(A) + P(B \cap A^c) \geq P(A),$$

as claimed. □

Theorem 3.3 *If A and B are events, then*

$$P(A \cup B) = P(A) + P(B) - P(A \cap B).$$

Proof Refer to Figure 3.3. Both $A \cup B$ and $A \cap B = (A^c \cup B^c)^c$ are events because A and B are events. Similarly, $A \cap B^c$ and $B \cap A^c$ are also events.

Notice that $A \cap B^c$, $B \cap A^c$, and $A \cap B$ are pairwise disjoint events. Hence,

$$\begin{aligned}
&P(A) + P(B) - P(A \cap B) \\
&\quad = P((A \cap B^c) \cup (A \cap B)) + P((B \cap A^c) \cup (A \cap B)) - P(A \cap B) \\
&\quad = P(A \cap B^c) + P(A \cap B) + P(B \cap A^c) + P(A \cap B) - P(A \cap B)
\end{aligned}$$

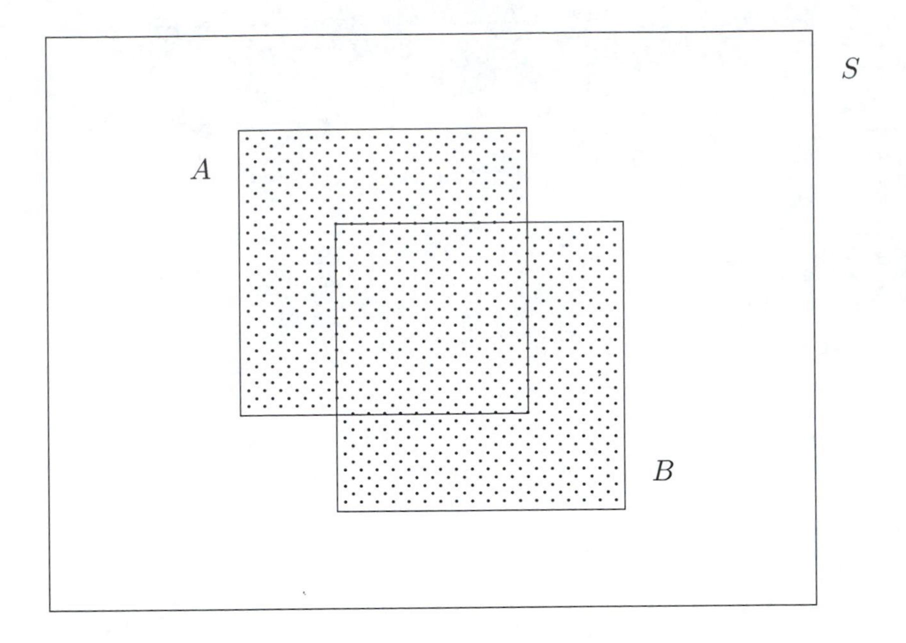

Figure 3.3: A Venn diagram for the probability of $A \cup B$.

$$
\begin{aligned}
&= P(A \cap B^c) + P(A \cap B) + P(B \cap A^c) \\
&= P((A \cap B^c) \cup (A \cap B) \cup (B \cap A^c)) \\
&= P(A \cup B),
\end{aligned}
$$

as claimed. □

Theorem 3.3 provides a general formula for computing the probability of the union of two sets. Notice that, if A and B are in fact disjoint, then

$$P(A \cap B) = P(\emptyset) = P(S^c) = 1 - P(S) = 1 - 1 = 0$$

and we recover our original formula for that case.

3.3 Finite Sample Spaces

Let

$$S = \{s_1, \ldots, s_N\}$$

denote a sample space that contains N outcomes and suppose that every subset of S is an event. For notational convenience, let

$$p_i = P(\{s_i\})$$

denote the probability of outcome i, for $i = 1, \ldots, N$. Then, for any event A, we can write

$$P(A) = P\left(\bigcup_{s_i \in A} \{s_i\}\right) = \sum_{s_i \in A} P(\{s_i\}) = \sum_{s_i \in A} p_i. \tag{3.2}$$

Thus, if the sample space is finite, then the probabilities of the individual outcomes determine the probability of any event. The same reasoning applies if the sample space is denumerable.

In this section, we focus on an important special case of finite probability spaces, the case of "equally likely" outcomes. By a fair coin, we mean a coin that when tossed is equally likely to produce `Heads` or `Tails`, i.e., the probability of each of the two possible outcomes is 1/2. By a fair die, we mean a die that when tossed is equally likely to produce any of six possible outcomes, i.e., the probability of each outcome is 1/6. In general, we say that the outcomes of a finite sample space are equally likely if

$$p_i = \frac{1}{N} \tag{3.3}$$

for $i = 1, \ldots, N$.

In the case of equally likely outcomes, we substitute (3.3) into (3.2) and obtain

$$P(A) = \sum_{s_i \in A} \frac{1}{N} = \frac{\sum_{s_i \in A} 1}{N} = \frac{\#(A)}{\#(S)}. \tag{3.4}$$

This equation reveals that, when the outcomes in a finite sample space are equally likely, calculating probabilities is just a matter of counting. The *counting* may be quite difficult, but the *probabilty* is trivial. We illustrate this point with some examples.

Example 3.4 *A fair coin is tossed twice. What is the probability of observing exactly one* `Head`*?*

The sample space for this experiment was described in Example 3.2. Because the coin is fair, each of the four outcomes in S is equally likely. Let A denote the event that exactly one `Head` is observed. Then $A = \{\texttt{HT}, \texttt{TH}\}$ and

$$P(A) = \frac{\#(A)}{\#(S)} = \frac{2}{4} = \frac{1}{2} = 0.5.$$

Example 3.5 *A fair die is tossed once. What is the probability that the number of dots on the top face of the die is divisible by 3?*

The sample space for this experiment is $S = \{1,2,3,4,5,6\}$. Because the die is fair, each of the six outcomes in S is equally likely. Let $A = \{3,6\}$ denote the event that a number divisible by 3 is observed. Then

$$P(A) = \frac{\#(A)}{\#(S)} = \frac{2}{6} = \frac{1}{3}.$$

Example 3.6 *A deck of 40 cards, labelled 1,2,3,...,40, is shuffled and cards are dealt as specified in each of the following scenarios.*

(a) *One hand of four cards is dealt to Arlen. What is the probability that Arlen's hand contains four even numbers?*

Let S denote the possible hands that might be dealt. Because the order in which the cards are dealt is not important,

$$\#(S) = \binom{40}{4}.$$

Let A denote the event that the hand contains four even numbers. There are 20 even cards, so the number of ways of dealing 4 even cards is

$$\#(A) = \binom{20}{4}.$$

Substituting these expressions into (3.4), we obtain

$$P(A) = \frac{\#(A)}{\#(S)} = \frac{\binom{20}{4}}{\binom{40}{4}} = \frac{51}{962} \doteq 0.0530.^3$$

(b) *One hand of four cards is dealt to Arlen. What is the probability that this hand is a straight, i.e., that it contains four consecutive numbers?*

Let S denote the possible hands that might be dealt. Again,

$$\#(S) = \binom{40}{4}.$$

Let A denote the event that the hand is a straight. The possible straights are:

[3] We use the symbol $\doteq$ to indicate numerical approximation, as in $\pi \doteq 3.14159$.

$$\begin{array}{c} 1\text{-}2\text{-}3\text{-}4 \\ 2\text{-}3\text{-}4\text{-}5 \\ 3\text{-}4\text{-}5\text{-}6 \\ \vdots \\ 37\text{-}38\text{-}39\text{-}40 \end{array}$$

By simple enumeration (just count the number of ways of choosing the smallest number in the straight), there are 37 such hands. Hence,

$$P(A) = \frac{\#(A)}{\#(S)} = \frac{37}{\binom{40}{4}} = \frac{1}{2470} \doteq 0.0004.$$

(c) *One hand of four cards is dealt to Arlen and a second hand of four cards is dealt to Mike. What is the probability that Arlen's hand is a straight and Mike's hand contains four even numbers?*

Let S denote the possible pairs of hands that might be dealt. Dealing the first hand requires choosing 4 cards from 40. After this hand has been dealt, the second hand requires choosing an additional 4 cards from the remaining 36. Hence,

$$\#(S) = \binom{40}{4} \cdot \binom{36}{4}.$$

Let A denote the event that Arlen's hand is a straight and Mike's hand contains four even numbers. There are 37 ways for Arlen's hand to be a straight. Each straight contains 2 even numbers, leaving 18 even numbers available for Mike's hand. Thus, for each way of dealing a straight to Arlen, there are $\binom{18}{4}$ ways of dealing 4 even numbers to Mike. Hence,

$$P(A) = \frac{\#(A)}{\#(S)} = \frac{37 \cdot \binom{18}{4}}{\binom{40}{4} \cdot \binom{36}{4}} \doteq 2.1032 \times 10^{-5}.$$

Example 3.7 *Five fair dice are tossed simultaneously.* Let S denote the possible outcomes of this experiment. Each die has 6 possible outcomes, so

$$\#(S) = 6 \cdot 6 \cdot 6 \cdot 6 \cdot 6 = 6^5.$$

(a) *What is the probability that the top faces of the dice all show the same number of dots?*

Let A denote the specified event; then A comprises the following outcomes:

1-1-1-1-1
2-2-2-2-2
3-3-3-3-3
4-4-4-4-4
5-5-5-5-5
6-6-6-6-6

By simple enumeration, $\#(A) = 6$. (Another way to obtain $\#(A)$ is to observe that the first die might result in any of six numbers, after which only one number is possible for each of the four remaining dice. Hence, $\#(A) = 6 \cdot 1 \cdot 1 \cdot 1 \cdot 1 = 6$.) It follows that

$$P(A) = \frac{\#(A)}{\#(S)} = \frac{6}{6^5} = \frac{1}{1296} \doteq 0.0008.$$

(b) *What is the probability that the top faces of the dice show exactly four different numbers?*

Let A denote the specified event. If there are exactly 4 different numbers, then exactly 1 number must appear twice. There are 6 ways to choose the number that appears twice and $\binom{5}{2}$ ways to choose the two dice on which this number appears. There are $5 \cdot 4 \cdot 3$ ways to choose the 3 different numbers on the remaining dice. Hence,

$$P(A) = \frac{\#(A)}{\#(S)} = \frac{6 \cdot \binom{5}{2} \cdot 5 \cdot 4 \cdot 3}{6^5} = \frac{25}{54} \doteq 0.4630.$$

(c) *What is the probability that the top faces of the dice show exactly three 6s or exactly two 5s?*

Let A denote the event that exactly three 6s are observed and let B denote the event that exactly two 5s are observed. We must calculate

$$P(A \cup B) = P(A) + P(B) - P(A \cap B) = \frac{\#(A) + \#(B) - \#(A \cap B)}{\#(S)}.$$

There are $\binom{5}{3}$ ways of choosing the three dice on which a 6 appears and $5 \cdot 5$ ways of choosing a different number for each of the two remaining dice. Hence,

$$\#(A) = \binom{5}{3} \cdot 5^2.$$

There are $\binom{5}{2}$ ways of choosing the two dice on which a 5 appears and $5 \cdot 5 \cdot 5$ ways of choosing a different number for each of the three remaining dice. Hence,

$$\#(B) = \binom{5}{2} \cdot 5^3.$$

There are $\binom{5}{3}$ ways of choosing the three dice on which a 6 appears and only 1 way in which a 5 can then appear on the two remaining dice. Hence,

$$\#(A \cap B) = \binom{5}{3} \cdot 1.$$

Thus,

$$P(A \cup B) = \frac{\binom{5}{3} \cdot 5^2 + \binom{5}{2} \cdot 5^3 - \binom{5}{3}}{6^5} = \frac{1490}{6^5} \doteq 0.1916.$$

Example 3.8 (The Birthday Problem) *In a class of k students, what is the probability that at least two students share a common birthday?*

As is inevitably the case with constructing mathematical models of actual phenomena, some simplifying assumptions are required to make this problem tractable. We begin by assuming that there are 365 possible birthdays, i.e., we ignore February 29. Then the sample space, S, of possible birthdays for k students comprises 365^k outcomes.

Next we assume that each of the 365^k outcomes is equally likely. This is not literally correct, as slightly more babies are born in some seasons than in others. Furthermore, if the class contains twins, then only certain pairs of birthdays are possible outcomes for those two students! In most situations, however, the assumption of equally likely outcomes is reasonably plausible.

Let A denote the event that at least two students in the class share a birthday. We might attempt to calculate

$$P(A) = \frac{\#(A)}{\#(S)},$$

but a moment's reflection should convince the reader that counting the number of outcomes in A is an extremely difficult undertaking. Instead, we invoke Theorem 3.1 and calculate

$$P(A) = 1 - P(A^c) = 1 - \frac{\#(A^c)}{\#(S)}.$$

This is considerably easier, because we count the number of outcomes in which each student has a different birthday by observing that 365 possible birthdays are available for the oldest student, after which 364 possible birthdays remain for the next oldest student, after which 363 possible birthdays remain for the next, etc. The formula is

$$\#\left(A^c\right) = 365 \cdot 364 \cdots (366 - k)$$

and so

$$P(A) = 1 - \frac{365 \cdot 364 \cdots (366 - k)}{365 \cdot 365 \cdots 365}.$$

The reader who computes $P(A)$ for several choices of k may be astonished to discover that a class of just $k = 23$ students is required to obtain $P(A) > 0.5$!

3.4 Conditional Probability

Consider a sample space with 10 equally likely outcomes, together with the events indicated in the Venn diagram that appears in Figure 3.4. Applying the methods of Section 3.3, we find that the (unconditional) probability of A is

$$P(A) = \frac{\#(A)}{\#(S)} = \frac{3}{10} = 0.3.$$

Suppose, however, that we know that we can restrict attention to the experimental outcomes that lie in B. Then the *conditional probability* of the event A given the occurrence of the event B is

$$P(A|B) = \frac{\#(A \cap B)}{\#(S \cap B)} = \frac{1}{5} = 0.2,$$

where $P(A|B)$ is pronounced "the probability of A given B." Notice that (for this example) the conditional probability, $P(A|B)$, differs from the unconditional probability, $P(A)$.

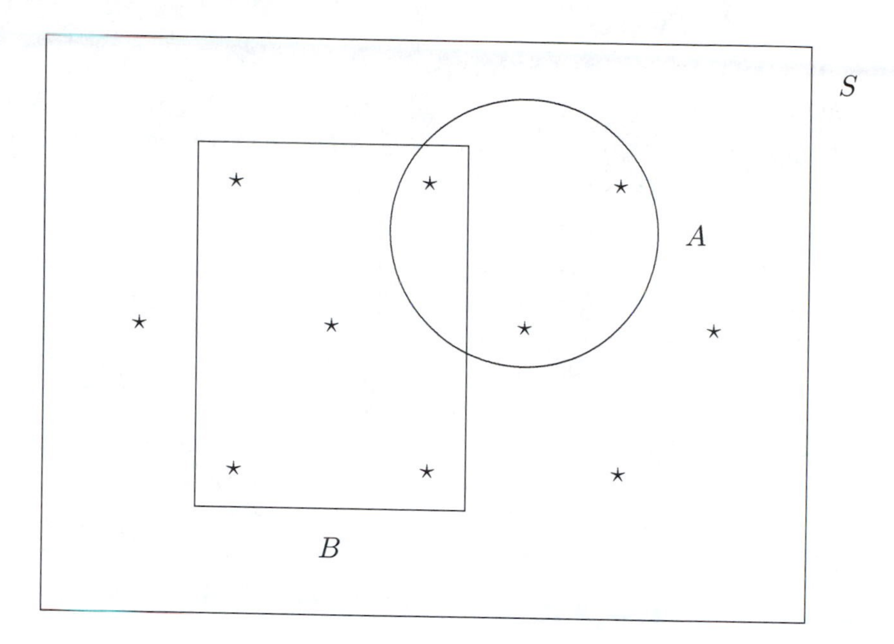

Figure 3.4: A Venn diagram that illustrates conditional probability. Each $\star$ represents an individual outcome.

To develop a definition of conditional probability that is not specific to finite sample spaces with equally likely outcomes, we now write

$$P(A|B) = \frac{\#(A \cap B)}{\#(S \cap B)} = \frac{\#(A \cap B)/\#(S)}{\#(B)/\#(S)} = \frac{P(A \cap B)}{P(B)}.$$

We take this as a definition:

Definition 3.1 *If A and B are events, and $P(B) > 0$, then*

$$P(A|B) = \frac{P(A \cap B)}{P(B)}. \tag{3.5}$$

The following consequence of Definition 3.1 is extremely useful. Upon multiplication of equation (3.5) by $P(B)$, we obtain

$$P(A \cap B) = P(B)P(A|B)$$

when $P(B) > 0$. Furthermore, upon interchanging the roles of A and B, we obtain

$$P(A \cap B) = P(B \cap A) = P(A)P(B|A)$$

when $P(A) > 0$. We will refer to these equations as the *multiplication rule* for conditional probability.

Used in conjunction with *tree diagrams*, the multiplication rule provides a powerful tool for analyzing situations that involve conditional probabilities.

Example 3.9 *Consider three fair coins, identical except that one coin (*`HH`*) is* `Heads` *on both sides, one coin (*`HT`*) is* `Heads` *on one side and* `Tails` *on the other, and one coin (*`TT`*) is* `Tails` *on both sides. A coin is selected at random and tossed. The face-up side of the coin is* `Heads`*. What is the probability that the face-down side of the coin is* `Heads`*?*

This problem was once considered by Marilyn vos Savant in her syndicated column, *Ask Marilyn*. As have many of the probability problems that she has considered, it generated a good deal of controversy. Many readers reasoned as follows:

1. The observation that the face-up side of the tossed coin is `Heads` means that the selected coin was not `TT`. Hence the selected coin was either `HH` or `HT`.

2. If `HH` was selected, then the face-down side is `Heads`; if `HT` was selected, then the face-down side is `Tails`.

3. Hence, there is a 1 in 2, or 50 percent, chance that the face-down side is `Heads`.

At first glance, this reasoning seems perfectly plausible and readers who advanced it were dismayed that Marilyn insisted that 0.5 is not the correct probability. How did these readers err?

A tree diagram of this experiment is depicted in Figure 3.5. The branches represent possible outcomes and the numbers associated with the branches are the respective probabilities of those outcomes. The initial triple of branches represents the initial selection of a coin—we have interpreted "at random" to mean that each coin is equally likely to be selected. The second level of branches represents the toss of the coin by identifying its resulting up-side. For `HH` and `TT`, only one outcome is possible; for `HT`, there are two equally likely outcomes. Finally, the third level of branches represents the down-side of the tossed coin. In each case, this outcome is determined by the up-side.

The multiplication rule for conditional probability makes it easy to calculate the probabilities of the various paths through the tree. The probability

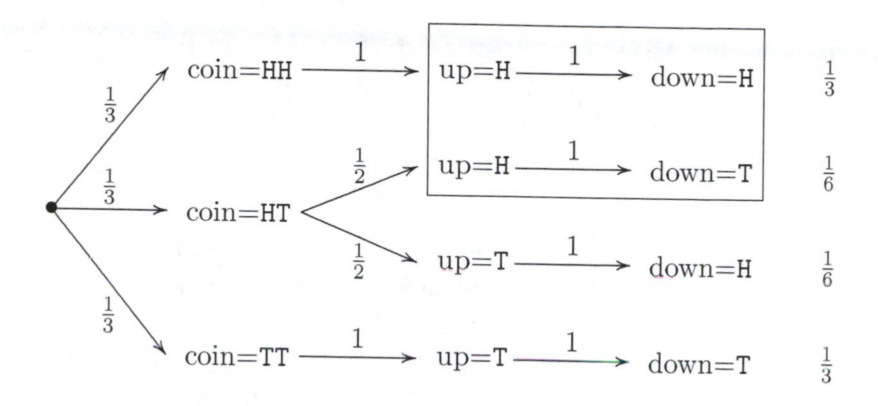

Figure 3.5: A tree diagram for Example 3.9.

that HT is selected and the up-side is Heads and the down-side is Tails is

$$\begin{aligned} P(\mathtt{HT} \cap \text{up=}\mathtt{H} \cap \text{down=}\mathtt{T}) &= P(\mathtt{HT} \cap \text{up=}\mathtt{H}) \cdot P(\text{down=}\mathtt{T}|\mathtt{HT} \cap \text{up=}\mathtt{H}) \\ &= P(\mathtt{HT}) \cdot P(\text{up=}\mathtt{H}|\mathtt{HT}) \cdot 1 \\ &= (1/3) \cdot (1/2) \cdot 1 \\ &= 1/6 \end{aligned}$$

and the probability that HH is selected and the up-side is Heads and the down-side is Heads is

$$\begin{aligned} P(\mathtt{HH} \cap \text{up=}\mathtt{H} \cap \text{down=}\mathtt{H}) &= P(\mathtt{HH} \cap \text{up=}\mathtt{H}) \cdot P(\text{down=}\mathtt{H}|\mathtt{HH} \cap \text{up=}\mathtt{H}) \\ &= P(\mathtt{HH}) \cdot P(\text{up=}\mathtt{H}|\mathtt{HH}) \cdot 1 \\ &= (1/3) \cdot 1 \cdot 1 \\ &= 1/3. \end{aligned}$$

Once these probabilities have been computed, it is easy to answer the original question:

$$P(\text{down=}\mathtt{H}|\text{up=}\mathtt{H}) = \frac{P(\text{down=}\mathtt{H} \cap \text{up=}\mathtt{H})}{P(\text{up=}\mathtt{H})} = \frac{1/3}{(1/3)+(1/6)} = \frac{2}{3},$$

which was Marilyn's answer.

From the tree diagram, we can discern the fallacy in our first line of reasoning. Having narrowed the possible coins to HH and HT, we claimed that HH and HT were equally likely candidates to have produced the observed

Head. In fact, HH was twice as likely as HT. Once this fact is noted it seems completely intuitive (HH has twice as many Heads as HT), but it is easily overlooked. This is an excellent example of how the use of tree diagrams may prevent subtle errors in reasoning.

Example 3.10 (Bayes Theorem) An important application of conditional probability can be illustrated by considering a population of patients at risk for contracting the HIV virus. The population can be partitioned into two sets: those who have contracted the virus and developed antibodies to it, and those who have not contracted the virus and lack antibodies to it. We denote the first set by D and the second set by D^c.

A test designed to detect the presence of HIV antibodies in human blood also partitions the population into two sets: those who test positive for HIV antibodies and those who test negative for HIV antibodies. We denote the first set by $+$ and the second set by $-$.

Together, the partitions induced by the true disease state and by the observed test outcome partition the population into four sets, as in the following Venn diagram:

$D \cap +$	$D \cap -$
$D^c \cap +$	$D^c \cap -$

(3.6)

In two of these cases, $D \cap +$ and $D^c \cap -$, the test provides the correct diagnosis; in the other two cases, $D^c \cap +$ and $D \cap -$, the test results in a diagnostic error. We call $D^c \cap +$ a *false positive* and $D \cap -$ a *false negative*.

In such situations, several quantities are likely to be known, at least approximately. The medical establishment is likely to have some notion of $P(D)$, the probability that a patient selected at random from the population is infected with HIV. This is the proportion of the population that is infected—it is called the *prevalence* of the disease. For the calculations that follow, we will assume that $P(D) = 0.001$.

Because diagnostic procedures undergo extensive evaluation before they are approved for general use, the medical establishment is likely to have a fairly precise notion of the probabilities of false positive and false negative test results. These probabilities are conditional: a false positive is a positive test result within the set of patients who are not infected and a false negative is a negative test result within the set of patients who are infected. Thus, the probability of a false positive is $P(+|D^c)$ and the probability of a false negative is $P(-|D)$. For the calculations that follow, we will assume that

$P(+|D^c) = 0.015$ and $P(-|D) = 0.003$.[4]

Now suppose that a randomly selected patient has a positive test result. The patient has an extreme interest in properly assessing the probability that a diagnosis of HIV is correct. This probability can be expressed as $P(D|+)$, the conditional probability that a patient has HIV given a positive test result. This quantity is sometimes called the *predictive value* of the test.

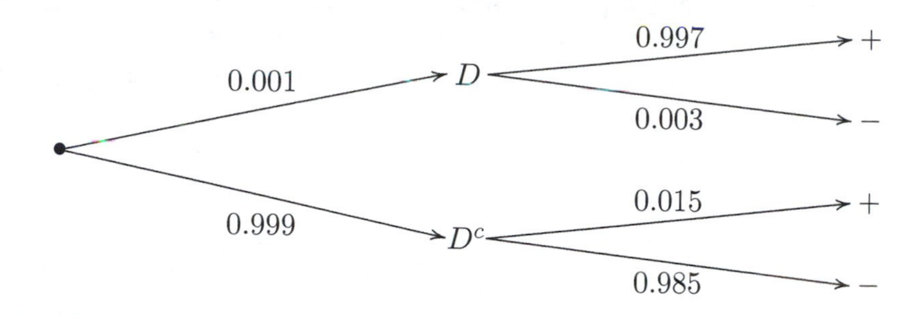

Figure 3.6: A tree diagram for Example 3.10.

To motivate our calculation of $P(D|+)$, it is again helpful to construct a tree diagram, as in Figure 3.6. This diagram was constructed so that the branches depicted in the tree have known probabilities, i.e., we first branch on the basis of disease state because $P(D)$ and $P(D^c)$ are known, then on the basis of test result because $P(+|D)$, $P(-|D)$, $P(+|D^c)$, and $P(-|D^c)$ are known. Notice that each of the four paths in the tree corresponds to exactly one of the four sets in (3.6). Furthermore, we can calculate the probability of each set by multiplying the probabilities that occur along its corresponding path:

$$P(D \cap +) = P(D) \cdot P(+|D) = 0.001 \cdot 0.997,$$

$$P(D \cap -) = P(D) \cdot P(-|D) = 0.001 \cdot 0.003,$$

$$P(D^c \cap +) = P(D^c) \cdot P(+|D^c) = 0.999 \cdot 0.015,$$

$$P(D^c \cap -) = P(D^c) \cdot P(-|D^c) = 0.999 \cdot 0.985.$$

The predictive value of the test is now obtained by computing

$$P(D|+) = \frac{P(D \cap +)}{P(+)} = \frac{P(D \cap +)}{P(D \cap +) + P(D^c \cap +)}$$

[4]See F. M. Sloan et al. (1991), HIV testing: state of the art, *Journal of the American Medical Association*, 266:2861–2866.

$$= \frac{0.001 \cdot 0.997}{0.001 \cdot 0.997 + 0.999 \cdot 0.015} \doteq 0.0624.$$

This probability may seem quite small, but consider that a positive test result can be obtained in two ways. If the person has the HIV virus, then a positive result is obtained with high probability, but very few people actually have the virus. If the person does not have the HIV virus, then a positive result is obtained with low probability, but so many people do not have the virus that the combined number of false positives is quite large relative to the number of true positives. This is a common phenomenon when screening for diseases.

The preceding calculations can be generalized and formalized in a formula known as Bayes Theorem; however, because such calculations will not play an important role in this book, we prefer to emphasize the use of tree diagrams to derive the appropriate calculations on a case-by-case basis.

Independence We now introduce a concept that is of fundamental importance in probability and statistics. The intuitive notion that we wish to formalize is the following:

> Two events are independent if the occurrence of either is unaffected by the occurrence of the other.

This notion can be expressed mathematically using the concept of conditional probability. Let A and B denote events and assume for the moment that the probability of each is strictly positive. If A and B are to be regarded as independent, then the occurrence of A is not affected by the occurrence of B. This can be expressed by writing

$$P(A|B) = P(A). \tag{3.7}$$

Similarly, the occurrence of B is not affected by the occurrence of A. This can be expressed by writing

$$P(B|A) = P(B). \tag{3.8}$$

Substituting the definition of conditional probability into (3.7) and multiplying by $P(B)$ leads to the equation

$$P(A \cap B) = P(A) \cdot P(B).$$

Substituting the definition of conditional probability into (3.8) and multiplying by $P(A)$ leads to the same equation. We take this equation, called the multiplication rule for independence, as a definition:

Definition 3.2 *Two events A and B are independent if and only if*

$$P(A \cap B) = P(A) \cdot P(B).$$

We proceed to explore some consequences of this definition.

Example 3.11 Notice that we did not require $P(A) > 0$ or $P(B) > 0$ in Definition 3.2. Suppose that $P(A) = 0$ or $P(B) = 0$, so that $P(A) \cdot P(B) = 0$. Because $A \cap B \subset A$, $P(A \cap B) \leq P(A)$; similarly, $P(A \cap B) \leq P(B)$. It follows that

$$0 \leq P(A \cap B) \leq \min(P(A), P(B)) = 0$$

and therefore that

$$P(A \cap B) = 0 = P(A) \cdot P(B).$$

Thus, if either of two events has probability zero, then the events are necessarily independent.

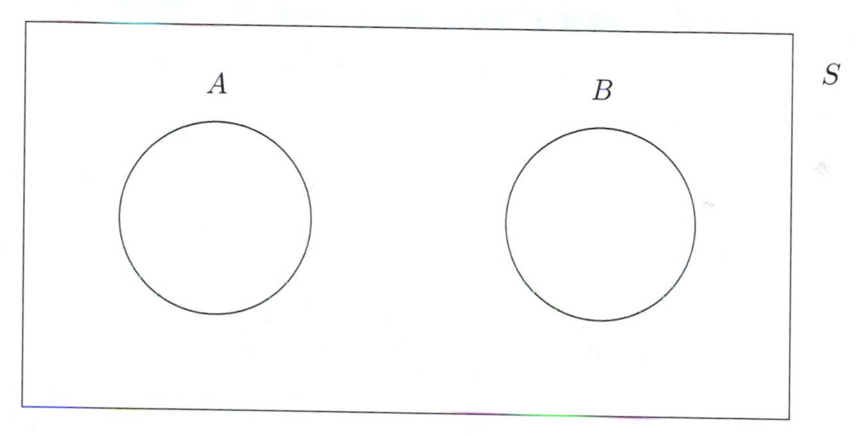

Figure 3.7: A Venn diagram for Example 3.12.

Example 3.12 Consider the disjoint events depicted in Figure 3.7 and suppose that $P(A) > 0$ and $P(B) > 0$. Are A and B independent? Many students instinctively answer that they are, but independence is very different from mutual exclusivity. In fact, if A occurs then B does not (and vice versa), so Figure 3.7 is actually a fairly extreme example of *dependent* events. This can also be deduced from Definition 3.2: $P(A) \cdot P(B) > 0$, but

$$P(A \cap B) = P(\emptyset) = 0$$

so A and B are not independent.

Example 3.13 *For each of the following, explain why the events A and B are or are not independent.*

(a) $P(A) = 0.4$, $P(B) = 0.5$, $P([A \cup B]^c) = 0.3$.

It follows that

$$P(A \cup B) = 1 - P([A \cup B]^c) = 1 - 0.3 = 0.7$$

and, because $P(A \cup B) = P(A) + P(B) - P(A \cap B)$, that

$$P(A \cap B) = P(A) + P(B) - P(A \cup B) = 0.4 + 0.5 - 0.7 = 0.2.$$

Then, since

$$P(A) \cdot P(B) = 0.5 \cdot 0.4 = 0.2 = P(A \cap B),$$

it follows that A and B are independent events.

(b) $P(A \cap B^c) = 0.3$, $P(A^c \cap B) = 0.2$, $P(A^c \cap B^c) = 0.1$.

Refer to the Venn diagram in Figure 3.8 to see that

$$P(A) \cdot P(B) = 0.7 \cdot 0.6 = 0.42 \neq 0.40 = P(A \cap B)$$

and hence that A and B are dependent events.

Thus far we have verified that two events are independent by verifying that the multiplication rule for independence holds. In applications, however, we usually reason somewhat differently. Using our intuitive notion of independence, we appeal to common sense, our knowledge of science, etc., to decide if independence is a property that we wish to incorporate into our mathematical model of the experiment in question. If it is, then we *assume* that two events are independent and the multiplication rule for independence becomes available to us for use as a computational formula.

Example 3.14 *Consider an experiment in which a typical penny is first tossed, then spun. Let A denote the event that the toss results in* `Heads` *and let B denote the event that the spin results in* `Heads`. *What is the probability of observing two* `Heads`*?*

We assume that, for a typical penny, $P(A) = 0.5$ and $P(B) = 0.3$ (see Section 1.1.1). Common sense tells us that the occurrence of either event is unaffected by the occurrence of the other. (Time is not reversible, so

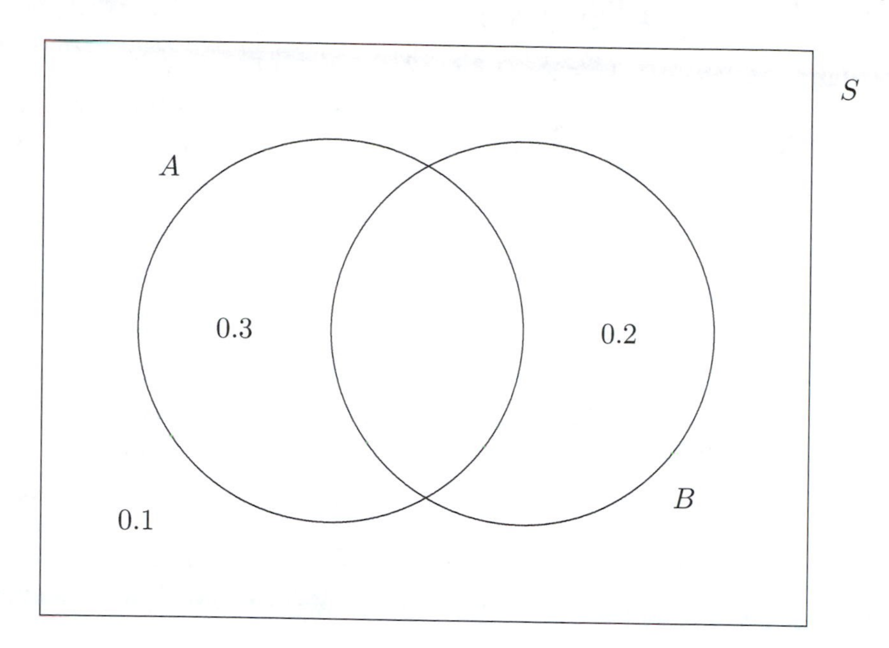

Figure 3.8: A Venn diagram for Example 3.13.

obviously the occurrence of A is not affected by the occurrence of B. One might argue that tossing the penny so that A occurs results in wear that is slightly different than the wear that results if A^c occurs, thereby slightly affecting the subsequent probability that B occurs. However, this argument strikes most students as completely preposterous. Even if it has a modicum of validity, the effect is undoubtedly so slight that we can safely neglect it in constructing our mathematical model of the experiment.) Therefore, we *assume* that A and B are independent and calculate that

$$P(A \cap B) = P(A) \cdot P(B) = 0.5 \cdot 0.3 = 0.15.$$

Example 3.15 *For each of the following, explain why the events A and B are or are not independent.*

(a) *Consider the population of William & Mary undergraduate students, from which one student is selected at random. Let A denote the event that the student is female and let B denote the event that the student is concentrating in elementary education.*

I'm told that $P(A)$ is roughly 60 percent, while it appears to me that $P(A|B)$ exceeds 90 percent. Whatever the exact probabilities, it is evident that the probability that a random elementary education concentrator is female is considerably greater than the probability that a random student is female. Hence, A and B are dependent events.

(b) *Consider the population of registered voters, from which one voter is selected at random. Let A denote the event that the voter belongs to a country club and let B denote the event that the voter is a Republican.*

It is generally conceded that one finds a greater proportion of Republicans among the wealthy than in the general population. Since one tends to find a greater proportion of wealthy persons at country clubs than in the general population, it follows that the probability that a random country club member is a Republican is greater than the probability that a randomly selected voter is a Republican. Hence, A and B are dependent events.[5]

Before progressing further, we ask what it should mean for A, B, and C to be three *mutually independent* events. Certainly each pair should comprise two independent events, but we would also like to write

$$P(A \cap B \cap C) = P(A) \cdot P(B) \cdot P(C).$$

It turns out that this equation cannot be deduced from the pairwise independence of A, B, and C, so we have to include it in our definition of mutual independence. Similar equations must be included when defining the mutual independence of more than three events. Here is a general definition:

Definition 3.3 *Let $\{A_\alpha\}$ be an arbitrary collection of events. These events are mutually independent if and only if, for every finite choice of events $A_{\alpha_1}, \ldots, A_{\alpha_k}$,*

$$P(A_{\alpha_1} \cap \cdots \cap A_{\alpha_k}) = P(A_{\alpha_1}) \cdots P(A_{\alpha_k}).$$

[5]This phenomenon may seem obvious, but it was overlooked by the respected *Literary Digest* poll. Their embarrassingly awful prediction of the 1936 presidential election resulted in the previously popular magazine going out of business. George Gallup's relatively accurate prediction of the outcome (and his uncannily accurate prediction of what the *Literary Digest* poll would predict) revolutionized polling practices.

Example 3.16 In the preliminary hearing for the criminal trial of O. J. Simpson, the prosecution presented conventional blood-typing evidence that blood found at the murder scene possessed three characteristics also possessed by Simpson's blood. The prosecution also presented estimates of the prevalence of each characteristic in the general population, i.e., of the probabilities that a person selected at random from the general population would possess these characteristics. Then, to obtain the estimated probability that a randomly selected person would possess all three characteristics, the prosecution multiplied the three individual probabilities, resulting in an estimate of 0.005.

In response to this evidence, defense counsel Gerald Uehlman objected that the prosecution had not established that the three events in question were independent and therefore had not justified their use of the multiplication rule. The prosecution responded that it was standard practice to multiply such probabilities and Judge Kennedy-Powell admitted the 0.005 estimate on that basis. No attempt was made to assess whether or not the standard practice was proper; it was inferred from the fact that the practice was standard that it must be proper. In this example, science and law diverge. From a scientific perspective, Gerald Uehlman was absolutely correct in maintaining that an assumption of independence must be justified.

3.5 Random Variables

Informally, a *random variable* is a rule for assigning real numbers to experimental outcomes. By convention, random variables are usually denoted by upper case Roman letters near the end of the alphabet, e.g., X, Y, Z.

Example 3.17 *A coin is tossed once and* `Heads` *(*`H`*) or* `Tails` *(*`T`*) is observed.*

The sample space for this experiment is $S = \{\mathtt{H}, \mathtt{T}\}$. For reasons that will become apparent, it is often convenient to assign the real number 1 to `Heads` and the real number 0 to `Tails`. This assignment, which we denote by the random variable X, can be depicted as follows:

$$\boxed{\begin{array}{c}\mathtt{H}\\ \mathtt{T}\end{array}} \xrightarrow{X} \boxed{\begin{array}{c}1\\ 0\end{array}}$$

In functional notation, $X : S \to \Re$ and the rule of assignment is defined by

$$\begin{aligned} X(\mathtt{H}) &= 1, \\ X(\mathtt{T}) &= 0. \end{aligned}$$

Example 3.18 *A coin is tossed twice and the number of* Heads *is counted.*

The sample space for this experiment is $S = \{\texttt{HH}, \texttt{HT}, \texttt{TH}, \texttt{TT}\}$. We want to assign the real number 2 to the outcome HH, the real number 1 to the outcomes HT and TH, and the real number 0 to the outcome TT. Several representations of this assignment are possible:

(a) Direct assignment, which we denote by the random variable Y, can be depicted as follows:

$$\boxed{\begin{array}{cc} \texttt{HH} & \texttt{HT} \\ \texttt{TH} & \texttt{TT} \end{array}} \xrightarrow{Y} \boxed{\begin{array}{cc} 2 & 1 \\ 1 & 0 \end{array}}$$

In functional notation, $Y : S \rightarrow \Re$ and the rule of assignment is defined by

$$\begin{aligned} Y(\texttt{HH}) &= 2, \\ Y(\texttt{HT}) = Y(\texttt{TH}) &= 1, \\ Y(\texttt{TT}) &= 0. \end{aligned}$$

(b) Instead of directly assigning the counts, we might take the intermediate step of assigning an ordered pair of numbers to each outcome. As in Example 3.17, we assign 1 to each occurrence of Heads and 0 to each occurrence of Tails. We denote this assignment by $X : S \rightarrow \Re^2$. In this context, $X = (X_1, X_2)$ is called a *random vector*. Each component of the random vector X is a random variable.

Next, we define a function $g : \Re^2 \rightarrow \Re$ by

$$g(x_1, x_2) = x_1 + x_2.$$

The composition $g(X)$ is equivalent to the random variable Y, as revealed by the following depiction:

$$\boxed{\begin{array}{cc} \texttt{HH} & \texttt{HT} \\ \texttt{TH} & \texttt{TT} \end{array}} \xrightarrow{X} \boxed{\begin{array}{cc} (1,1) & (1,0) \\ (0,1) & (0,0) \end{array}} \xrightarrow{g} \boxed{\begin{array}{cc} 2 & 1 \\ 1 & 0 \end{array}}$$

(c) The preceding representation suggests defining two random variables, X_1 and X_2, as in the following depiction:

$$\boxed{\begin{array}{cc} 1 & 1 \\ 0 & 0 \end{array}} \xleftarrow{X_1} \boxed{\begin{array}{cc} \texttt{HH} & \texttt{HT} \\ \texttt{TH} & \texttt{TT} \end{array}} \xrightarrow{X_2} \boxed{\begin{array}{cc} 1 & 0 \\ 1 & 0 \end{array}}$$

As in the preceding representation, the random variable X_1 counts the number of Heads observed on the first toss and the random variable

X_2 counts the number of `Heads` observed on the second toss. The sum of these random variables, $X_1 + X_2$, is evidently equivalent to the random variable Y.

The primary reason that we construct a random variable, X, is to replace the probability space that is naturally suggested by the experiment in question with a familiar probability space in which the possible outcomes are real numbers. Thus, we replace the original sample space, S, with the familiar number line, $\Re$. To complete the transference, we must decide which subsets of $\Re$ will be designated as events and we must specify how the probabilities of these events are to be calculated.

It is an interesting fact that it is impossible to construct a probability space in which the set of outcomes is $\Re$ and every subset of $\Re$ is an event. For this reason, we define the collection of events to be the smallest collection of subsets that satisfies the assumptions of the Kolmogorov probability model and that contains every interval of the form $(-\infty, y]$. This collection is called the *Borel sets* and it is a very large collection of subsets of $\Re$. In particular, it contains every interval of real numbers and every set that can be constructed by applying a countable number of set operations (union, intersection, complementation) to intervals. Most students will never see a set that is not a Borel set!

Finally, we must define a probability measure that assigns probabilities to Borel sets. Of course, we want to do so in a way that preserves the probability structure of the experiment in question. The only way to do so is to define the probability of each Borel set B to be the probability of the set of outcomes to which X assigns a value in B. This set of outcomes is denoted by

$$X^{-1}(B) = \{s \in S : X(s) \in B\}$$

and is depicted in Figure 3.9.

How do we know that the set of outcomes to which X assigns a value in B is an event and therefore has a probability? We don't, so we guarantee that it is by including this requirement in our formal definition of a random variable.

Definition 3.4 *A function $X : S \to \Re$ is a random variable if and only if*

$$P\left(\{s \in S : X(s) \leq y\}\right)$$

exists for all choices of $y \in \Re$.

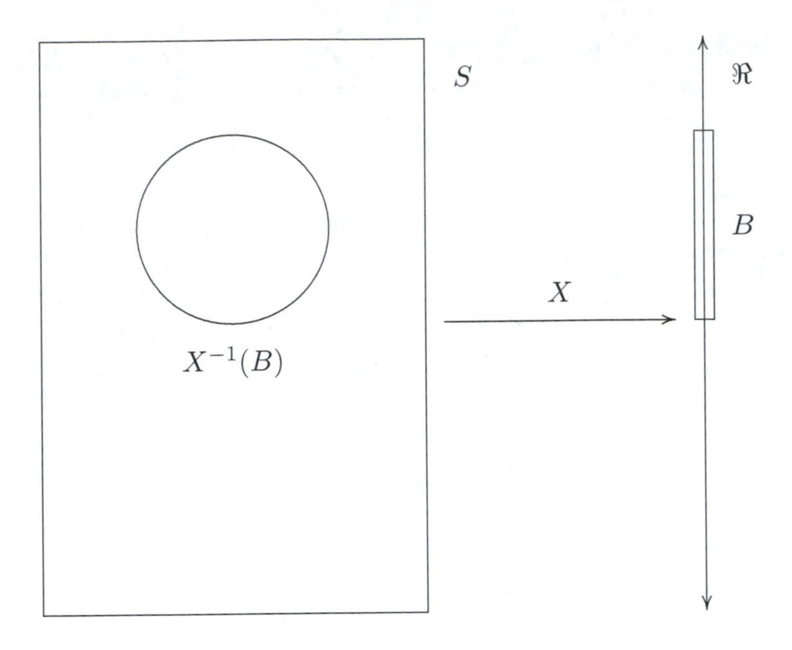

Figure 3.9: The inverse image of a Borel set.

We will denote the probability measure induced by the random variable X by P_X. The following equation defines various representations of P_X:

$$\begin{aligned} P_X\left((-\infty, y]\right) &= P\left(X^{-1}\left((-\infty, y]\right)\right) \\ &= P\left(\{s \in S : X(s) \in (-\infty, y]\}\right) \\ &= P\left(-\infty < X \leq y\right) \\ &= P\left(X \leq y\right) \end{aligned}$$

A probability measure on the Borel sets is called a *probability distribution* and P_X is called the distribution of the random variable X. A hallmark feature of probability theory is that we study the distributions of random variables rather than arbitrary probability measures. One important reason for this emphasis is that many different experiments may result in identical distributions. For example, the random variable in Example 3.17 might have the same distribution as a random variable that assigns 1 to male newborns and 0 to female newborns.

Cumulative Distribution Functions Our construction of the probability measure induced by a random variable suggests that the following function will be useful in describing the properties of random variables.

Definition 3.5 *The cumulative distribution function (cdf) of a random variable X is the function $F : \Re \to \Re$ defined by*

$$F(y) = P(X \leq y).$$

Example 3.17 (continued) We consider two probability structures that might obtain in the case of a typical penny.

(a) *A typical penny is tossed.*

For this experiment, $P(\mathtt{H}) = P(\mathtt{T}) = 0.5$, and the following values of the cdf are easily determined:

- If $y < 0$, e.g., $y = -9.1185$ or $y = -0.3018$, then

$$F(y) = P(X \leq y) = P(\emptyset) = 0.$$

- $F(0) = P(X \leq 0) = P(\{\mathtt{T}\}) = 0.5.$
- If $y \in (0, 1)$, e.g., $y = 0.6241$ or $y = 0.9365$, then

$$F(y) = P(X \leq y) = P(\{\mathtt{T}\}) = 0.5.$$

- $F(1) = P(X \leq 1) = P(\{\mathtt{T}, \mathtt{H}\}) = 1.$
- If $y > 1$, e.g., $y = 1.5248$ or $y = 7.7397$, then

$$F(y) = P(X \leq y) = P(\{\mathtt{T}, \mathtt{H}\}) = 1.$$

The entire cdf is plotted in Figure 3.10.

(b) *A typical penny is spun.*

For this experiment, we assume that $P(\mathtt{H}) = 0.3$ and $P(\mathtt{T}) = 0.7$ (see Section 1.1.1). Then the following values of the cdf are easily determined:

- If $y < 0$, e.g., $y = -1.6633$ or $y = -0.5485$, then

$$F(y) = P(X \leq y) = P(\emptyset) = 0.$$

- $F(0) = P(X \leq 0) = P(\{\mathtt{T}\}) = 0.7.$
- If $y \in (0, 1)$, e.g., $y = 0.0685$ or $y = 0.4569$, then

$$F(y) = P(X \leq y) = P(\{\mathtt{T}\}) = 0.7.$$

- $F(1) = P(X \leq 1) = P(\{\mathtt{T}, \mathtt{H}\}) = 1.$
- If $y > 1$, e.g., $y = 1.4789$ or $y = 2.6117$, then

$$F(y) = P(X \leq y) = P(\{\mathtt{T}, \mathtt{H}\}) = 1.$$

The entire cdf is plotted in Figure 3.11.

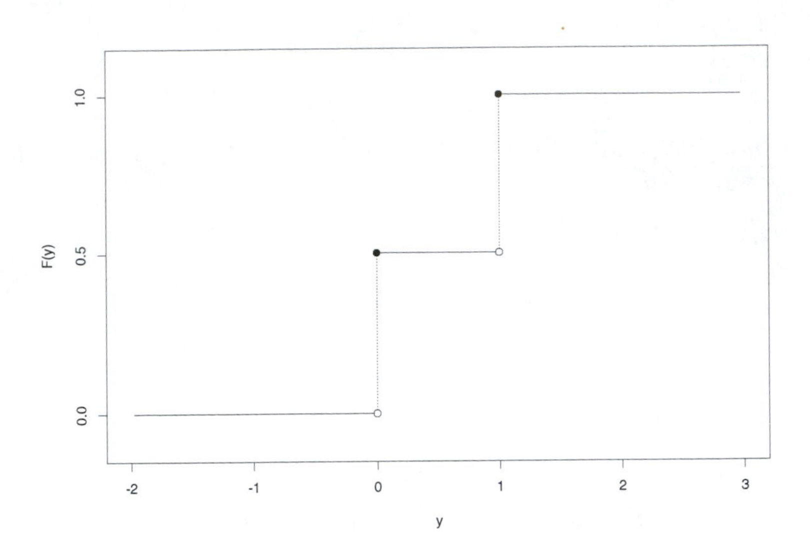

Figure 3.10: The cumulative distribution function for tossing a penny with $P(\texttt{Heads}) = 0.5$.

Example 3.18 (continued) Suppose that the coin is fair, so that each of the four possible outcomes in S is equally likely, i.e., has probability 0.25. Then the following values of the cdf are easily determined:

- If $y < 0$, e.g., $y = -4.2132$ or $y = -0.5615$, then

$$F(y) = P(X \leq y) = P(\emptyset) = 0.$$

- $F(0) = P(X \leq 0) = P(\{\texttt{TT}\}) = 0.25.$
- If $y \in (0, 1)$, e.g., $y = 0.3074$ or $y = 0.6924$, then

$$F(y) = P(X \leq y) = P(\{\texttt{TT}\}) = 0.25.$$

- $F(1) = P(X \leq 1) = P(\{\texttt{TT}, \texttt{HT}, \texttt{TH}\}) = 0.75.$
- If $y \in (1, 2)$, e.g., $y = 1.4629$ or $y = 1.5159$, then

$$F(y) = P(X \leq y) = P(\{\texttt{TT}, \texttt{HT}, \texttt{TH}\}) = 0.75.$$

- $F(2) = P(X \leq 2) = P(\{\texttt{TT}, \texttt{HT}, \texttt{TH}, \texttt{HH}\}) = 1.$

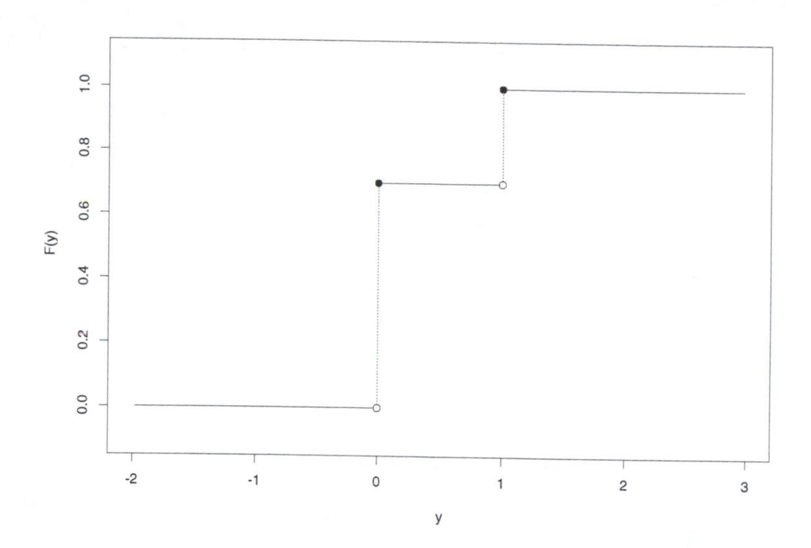

Figure 3.11: The cumulative distribution function for spinning a penny with $P(\texttt{Heads}) = 0.3$.

- If $y > 2$, e.g., $y = 2.1252$ or $y = 3.7790$, then

$$F(y) = P(X \le y) = P(\{\texttt{TT}, \texttt{HT}, \texttt{TH}, \texttt{HH}\}) = 1.$$

The entire cdf is plotted in Figure 3.12.

Let us make some observations about the cdfs that we have plotted. First, each cdf assumes its values in the unit interval, $[0, 1]$. This is a general property of cdfs: each $F(y) = P(X \le y)$, and probabilities necessarily assume values in $[0, 1]$.

Second, each cdf is nondecreasing; i.e., if $y_2 > y_1$, then $F(y_2) \ge F(y_1)$. This is also a general property of cdfs, for suppose that we observe an outcome s such that $X(s) \le y_1$. Because $y_1 < y_2$, it follows that $X(s) \le y_2$. Thus, $\{X \le y_1\} \subset \{X \le y_2\}$ and therefore

$$F(y_1) = P(X \le y_1) \le P(X \le y_2) = F(y_2).$$

Finally, each cdf equals 1 for sufficiently large y and 0 for sufficiently small y. This is *not* a general property of cdfs—it occurs in our examples because $X(S)$ is a bounded set; i.e., there exist finite real numbers a and b

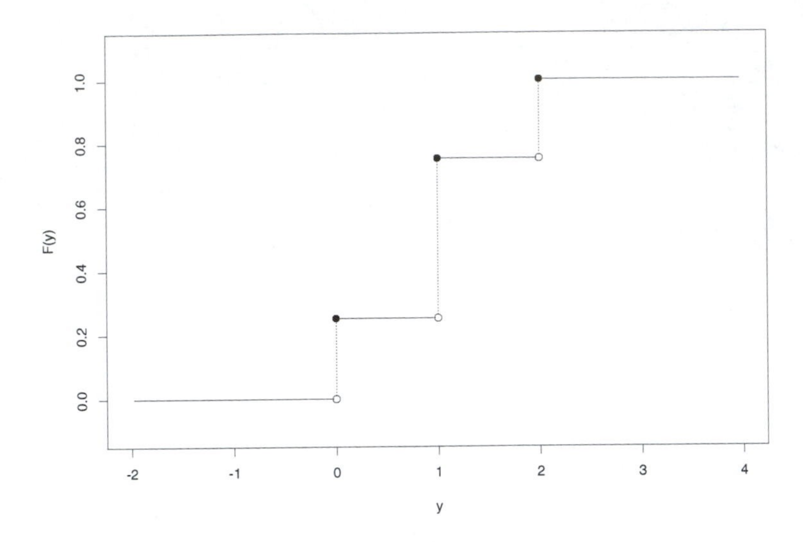

Figure 3.12: The cumulative distribution function for tossing two pennies with $P(\texttt{Heads}) = 0.5$ and counting the number of `Heads`.

such that every $x \in X(S)$ satisfies $a \leq x \leq b$. However, all cdfs do satisfy the following properties:

$$\lim_{y \to \infty} F(y) = 1 \quad \text{and} \quad \lim_{y \to -\infty} F(y) = 0.$$

Independence We say that two random variables, X_1 and X_2, are independent if each event defined by X_1 is independent of each event defined by X_2. More precisely,

Definition 3.6 *Let $X_1 : S \to \Re$ and $X_2 : S \to \Re$ be random variables. X_1 and X_2 are independent if and only if, for each $y_1 \in \Re$ and each $y_2 \in \Re$,*

$$P(X_1 \leq y_1, X_2 \leq y_2) = P(X_1 \leq y_1) \cdot P(X_2 \leq y_2).$$

This definition can be extended to mutually independent collections of random variables in precisely the same way that we extended Definition 3.2 to Definition 3.3.

Intuitively, two random variables are independent if the distribution of each does not depend on the value of the other. As we discussed in Section

3.4, in most applications we will appeal to common sense, our knowledge of science, etc., to decide if independence is a property that we wish to incorporate into our mathematical model of the experiment in question. If it is, then we will *assume* that the appropriate random variables are independent. This assumption will allow us to apply many powerful theorems from probability and statistics that are true only of independent random variables.

3.6 Case Study: Padrolling in Milton Murayama's *All I asking for is my body*

The American dice game Craps evolved from the English dice game Hazard:

> According to tradition, blacks living around New Orleans tried their hand at Hazard.... In the course of time they modifed the rules and playing procedures so greatly that they ended up inventing the game of Craps (in the U.S. idiom known as Crapshooting or Shooting Craps and here identified as Private Craps to distinguish it from Open Craps and the more formalized variants offered in gambling casinos). ...The popularity of the private game of Craps with the U.S. military personnel during World Wars I and II helped to spread that game to many parts of the world.[6]

Craps is played with two fair dice, each marked in a specific way. According to Hoyle,

> Each face of [each] die is marked with one to six dots, opposite faces representing... numbers adding to seven; if the vertical face toward you is 5, and the horizontal face on top of the die is 6, [then] the 3 should be on the vertical face to your right."[7]

The *shooter* rolls the pair of dice, resulting in one of $6 \times 6 = 36$ possible outcomes. Of interest is the combined number of dots on the horizontal faces atop the two dice, a number that we denote by the random variable X. The possible values of X are displayed in Figure 3.13.

Let x denote the value of X produced by the first roll. The game ends immediately if $x \in \{2, 3, 7, 11, 12\}$. If $x \in \{7, 11\}$, then x is a *natural* and

---

[6]Dice and dice games. *The New Encyclopædia Britannica in 30 Volumes*, Macropædia, Volume 5, 1974, pp. 702–706.

[7]R. L. Frey (1970). *According to Hoyle*. Fawcett Publications, Greenwich, CT, 1970, p. 266.

	1	2	3	4	5	6
1	2	3	4	5	6	7
2	3	4	5	6	7	8
3	4	5	6	7	8	9
4	5	6	7	8	9	10
5	6	7	8	9	10	11
6	7	8	9	10	11	12

Figure 3.13: The possible outcomes of rolling two standard dice.

the shooter wins; if $x \in \{2, 3, 12\}$, then x is *craps* and the shooter loses; otherwise, x becomes the shooter's *point*. If the first roll is not decisive, then the shooter continues to roll until he either (a) again rolls x (*makes his point*), in which case he wins, or (b) rolls 7 (*craps out*), in which case he loses.

A game of craps is fair when each of the 36 outcomes in Figure 3.13 is equally likely. Fairness is usually ensured by tossing the dice from a cup, or, more crudely, by tossing them against a wall. In a fair game of craps, we have the following probabilities:

$$\begin{aligned} P(X = 7) &= 6/36 \\ P(X = 6) = P(X = 8) &= 5/36 \\ P(X = 5) = P(X = 9) &= 4/36 \\ P(X = 4) = P(X = 10) &= 3/36 \\ P(X = 3) = P(X = 11) &= 2/36 \\ P(X = 2) = P(X = 12) &= 1/36 \end{aligned}$$

Let us begin by calculating the probability that the shooter wins a fair game of craps.

There are several ways for the shooter to win. We will calculate the probability of each, then sum these probabilities.

- Roll a natural.
$$P(X \in \{7, 11\}) = \frac{6+2}{36} = \frac{2}{3^2}.$$

- Roll $x = 6$ or $x = 8$, then make point.

  First,
$$P(X \in \{6, 8\}) = \frac{5+5}{36} = \frac{5}{18}.$$

Then, the shooter must roll x before rolling 7. Other outcomes are ignored. There are 5 ways to roll x versus 6 ways to roll 7, so the conditional probability of making point is 5/11. Hence, the probability of the shooter winning in this way is

$$\frac{5}{18} \cdot \frac{5}{11} = \frac{25}{2 \cdot 3^2 \cdot 11}.$$

- Roll $x = 5$ or $x = 9$, then make point.

  First,

  $$P(X \in \{5, 9\}) = \frac{4+4}{36} = \frac{2}{9}.$$

  Then, the shooter must roll x before rolling 7. Other outcomes are ignored. There are 4 ways to roll x versus 6 ways to roll 7, so the conditional probability of making point is 4/10. Hence, the probability of the shooter winning in this way is

  $$\frac{2}{9} \cdot \frac{4}{10} = \frac{4}{3^2 \cdot 5}.$$

- Roll $x = 4$ or $x = 10$, then make point.

  First,

  $$P(X \in \{4, 10\}) = \frac{3+3}{36} = \frac{1}{6}.$$

  Then, the shooter must roll x before rolling 7. Other outcomes are ignored. There are 3 ways to roll x versus 6 ways to roll 7, so the conditional probability of making point is 3/9. Hence, the probability of the shooter winning in this way is

  $$\frac{1}{6} \cdot \frac{3}{9} = \frac{1}{2 \cdot 3^2}.$$

The probability that the shooter wins is

$$\frac{2}{3^2} + \frac{25}{2 \cdot 3^2 \cdot 11} + \frac{4}{3^2 \cdot 5} + \frac{1}{2 \cdot 3^2} = \frac{244}{495} \doteq 0.4929.$$

Thus, the shooter is slightly more likely to lose than to win a fair game of craps.

Milton Murayama's 1959 novel, *All I asking for is my body*, is a brilliant evocation of *nisei* (second-generation Japanese American) life on Hawaiian sugar plantations in the 1930s.[8] One of its central concerns is the concept of Japanese honor and its implicatons for the young protagonist/narrator, Kiyoshi, and his siblings. Years earlier, Kiyoshi's parents had sacrificed their future to pay Kiyoshi's grandfather's debts; now they owe the impossible sum of $6000 and they expect their children to sacrifice for them. Toward the novel's end, Japan attacks Pearl Harbor and Kiyoshi subsequently volunteers for an all-*nisei* regiment that will fight in Europe. In the final chapter, he contrives to win $6000 by playing Craps.

Kiyoshi had watched a former classmate, Hiroshi Sakai, play Craps at the Citizens' Quarters in Kahana.

> It was weird the way he kept winning. Whenever he rolled, the dice rolled in unison like the wheels of a cart, and even when one die rolled ahead of the other, neither flipped on its side. The Kahana players finally refused to fade [bet against] him, and he stopped coming.

We subsequently learn that Hiroshi's technique is called *padrolling*.

In the Army,

> Everybody had money and every third guy was a crapshooter. The sight of all that money drove me mad. There was $25,000 at least floating around in the crap games.... Most of the games were played on blankets on barrack floors, the dice rolled by hand. There were a few guys who rolled the dice the way Hiroshi did at the Citizens' Quarters in Kahana. The dice didn't bounce but rolled out in unison like the wheels of a cart. There had to be an advantage to that.

Kiyoshi buys a pair of dice and examines them carefully. He realizes that, by rolling the dice "like the wheels of a cart," he can keep the sides of the dice that form the axis of the wheels from appearing. Then, by combining certain numbers to form the axis, he can improve his chance of winning.

Kiyoshi teaches himself to padroll and develops the following system for choosing the axis:

1. For the initial roll, use the 1-6 axis for each die.

   Padrolling this axis has the effect of eliminating the first and sixth rows and columns in Figure 3.13, resulting in the following set of possible

[8] M. Murayama (1988). *All I asking for is my body.* University of Hawaii Press. I am indebted to M. Lynn Weiss for bringing this novel to my attention.

outcomes:

	2	3	4	5
2	4	5	6	7
3	5	6	7	8
4	6	7	8	9
5	7	8	9	10

Notice that this choice eliminates the possibility of crapping out! Furthermore, assuming that the 16 remaining outcomes are equally likely, it also improves the chance of rolling a natural from 4/18 to 4/16.

2. If $x \in \{6, 8\}$, then use the 1-6 axis on one die and the 2-5 axis on the other.

   Padrolling this axis results in the following set of possible outcomes:

	1	3	4	6
2	3	5	6	8
3	4	6	7	9
4	5	7	8	10
5	6	8	9	11

   With this choice, there are 3 ways to roll x versus 2 ways to roll 7. Again assuming that the 16 remaining outcomes are equally likely, this choice improves the conditional probability of making point from 5/11 to 3/5.

3. If $x \in \{4, 5, 9, 10\}$, then use the 1-6 axis on one die and the 3-4 axis on the other.

   Padrolling this axis results in the following set of possible outcomes:

	1	2	5	6
2	3	4	7	8
3	4	5	8	9
4	5	6	9	10
5	6	7	10	11

   With this choice, there are 2 ways to roll x versus 2 ways to roll 7. Again, assume that the 16 remaining outcomes are equally likely. If $x \in \{5, 9\}$, then this choice improves the conditional probability of making point from 4/10 to 2/4. If $x \in \{4, 10\}$, then this choice improves the conditional probability of making point from 3/9 to 2/4.

If a shooter padrolls successfully, then the probability that he will win using Kiyoshi's system is

$$\frac{4}{16} + \frac{6}{16} \cdot \frac{3}{5} + \frac{4}{16} \cdot \frac{2}{4} + \frac{2}{16} \cdot \frac{2}{4} = \frac{53}{80} = 0.6625,$$

a substantial improvement on his chance of winning a fair game. "And," Kiyoshi rationalizes, "it wasn't really cheating. The others had the option of stopping any of your rolls, or they could play with a cup, or have the roller bang the dice against the wall, or use a canvas or the bare floor instead of a blanket." So, Kiyoshi padrolls. I leave to you the pleasure of discovering whether or not he succeeds in winning the $6000 his family needs.

3.7 Exercises

1. Consider three events that might occur when Arlen digs a new mine in the Cleveland National Forest in San Diego County, California:

$$\begin{aligned} \text{A} &= \{ \text{ quartz specimens are found } \} \\ \text{B} &= \{ \text{ tourmaline specimens are found } \} \\ \text{C} &= \{ \text{ aquamarine specimens are found } \} \end{aligned}$$

   Assume the following probabilities:

$$\begin{array}{ccc} P(A) = 0.80 & P(B) = 0.36 & P(C) = 0.28 \\ P(A \cap B) = 0.29 & P(A \cap C) = 0.24 & P(B \cap C) = 0.16 \\ & P(A \cap B \cap C) = 0.13 & \end{array}$$

   (a) Draw a suitable Venn diagram for this situation.
   (b) Calculate the probability that both quartz and tourmaline will be found, but not aquamarine.
   (c) Calculate the probability that quartz will be found, but not tourmaline or aquamarine.
   (d) Calculate the probability that none of these types of specimens will be found.
   (e) Calculate the probability of $A^c \cap (B \cup C)$.

2. Consider two urns, one containing four tickets labelled $\{1, 3, 4, 6\}$; the other containing ten tickets, labelled $\{1, 3, 3, 3, 3, 4, 4, 4, 4, 6\}$.

(a) What is the probability of drawing a 3 from the first urn?

(b) What is the probability of drawing a 3 from the second urn?

(c) Which urn is a better model for throwing an astragalus? Why?

3. Suppose that five cards are dealt from a standard deck of playing cards.

   (a) What is the probability of drawing a straight flush?

   (b) What is the probability of drawing 4 of a kind?

   Hint: Use the results of Exercise 2.5.9.

4. Thirteen finalists competed in the ninth cycle of the reality television show *America's Next Top Model.* After the finalists were revealed, but before any were eliminated, Case correctly predicted that the final five models would be (in alphabetical order) Bianca, Chantal, Heather, Jenah, and Saleisha.

   Suppose that I write the name of each finalist on a slip of paper, place the slips in an urn, and randomly draw five slips (without replacement).

   (a) What is the probability that I will draw the names of the five final models?

   (b) What is the probability that one of the names that I draw will be Saleisha?

5. Suppose that four fair dice are thrown simultaneously.

   (a) How many outcomes are possible?

   (b) What is the probability that each top face shows a different number?

   (c) What is the probability that the top faces show four numbers that sum to five?

   (d) What is the probability that at least one of the top faces shows an odd number?

   (e) What is the probability that three of the top faces show the same odd number and the other top face shows an even number?

6. A *dreidl* is a four-sided top that contains a Hebrew letter on each side: nun, gimmel, heh, shin. These letters are an acronym for the Hebrew phrase *nes gadol hayah sham* (a great miracle happened there), which refers to the miracle of the temple light that burned for eight days with only one day's supply of oil—the miracle celebrated at Chanukah. Here we suppose that a fair dreidl (one that is equally likely to fall on each of its four sides) is to be spun ten times. Compute the probability of each of the following events:

   (a) Five gimmels and five hehs;

   (b) No nuns or shins;

   (c) Two letters are absent and two letters are present;

   (d) At least two letters are absent.

7. Suppose that $P(A) = 0.7$, $P(B) = 0.6$, and $P(A^c \cap B) = 0.2$.

   (a) Draw a Venn diagram that describes this experiment.

   (b) Is it possible for A and B to be disjoint events? Why or why not?

   (c) What is the probability of $A \cup B^c$?

   (d) Is it possible for A and B to be independent events? Why or why not?

   (e) What is the conditional probability of A given B?

8. Pre-eclampsia is a disorder of pregnancy in which the interaction between mother and placenta is disrupted, causing a hypertensive response. Various studies have investigated the accuracy of using blood pressure to predict pre-eclampsia.[9]

   Consider a test that predicts pre-eclampsia when mean arterial pressure is at least 90 mm Hg. Assume that the *sensitivity* of this test is $P(+|D) = 0.62$ and that the *specificity* of this test is $P(-|D^c) = 0.82$.

   (a) What is the conditional probability of a false positive test result?

   (b) What is the conditional probability of a false negative test result?

   (c) Suppose that a pregnant woman is selected at random from a population in which the prevalence of pre-eclampsia is $P(D) = 0.05$. Construct a tree diagram that describes this experiment.

[9] See J. S. Cnossen et al., Accuracy of mean arterial pressure and blood pressure measurements in predicting pre-eclampsia: systematic review and meta-analysis, *BMJ*, 2008 May 17; 336(7653):1117–1120, from which the numbers used in this exercise where taken.

(d) What is the probability that the above test will diagnose the selected woman as having pre-eclampsia?

(e) Suppose that the above test does diagnose the selected woman as having pre-eclampsia. What then is the probability that this woman actually does have pre-eclampsia?

9. Comment on the following passage in whatever way seems appropriate to you.

> And in a study slated to appear in COGNITION, Cosmides and Tooby confront a cognitive bias known as the "base-rate fallacy." As an illustration, they cite a 1978 study in which 60 staff and students at Harvard Medical School attempted to solve this problem: "If a test to detect a disease whose prevalence is 1/1,000 has a false positive rate of 5%, what is the chance that a person found to have a positive result actually has the disease, assuming you know nothing about the person's symptoms or signs?"
>
> Nearly half the sample estimated this probability as 95 percent; only 11 gave the correct response of 2 percent. Most participants neglected the base rate of the disease (it strikes 1 in 1,000 people) and formed a judgment solely from the characteristics of the test.[10]

10. Mike owns a box that contains 6 pairs of 14-carat gold, cubic zirconia earrings. The earrings are of three sizes: 3mm, 4mm, and 5mm. There are 2 pairs of each size.

    Each time that Mike needs an inexpensive gift for a female friend, he randomly selects a pair of earrings from the box. If the selected pair is 4mm, then he buys an identical pair to replace it. If the selected pair is 3mm, then he does not replace it. If the selected pair is 5mm, then he tosses a fair coin. If he observes `Heads`, then he buys two identical pairs of earrings to replace the selected pair; if he observes `Tails`, then he does not replace the selected pair.

    (a) What is the probability that the second pair selected will be 4mm?

    (b) If the second pair was not 4mm, then what is the probability that the first pair was 5mm?

[10] B. Bower, Roots of reason, *Science News*, 145:72–75, January 29, 1994.

11. It is a curious fact that approximately 85% of all U.S. residents who are struck by lightning are men. Consider the population of U.S. residents, from which a person is randomly selected. Let A denote the event that the person is male and let B denote the event that the person will be struck by lightning.

    (a) Estimate $P(A|B)$ and $P(A^c|B)$.

    (b) Compare $P(A|B)$ and $P(A)$. Are A and B independent events?

    (c) Suggest reasons why $P(A|B)$ is so much larger than $P(A^c|B)$. It is tempting to joke that men don't know enough to come in out of the rain! Why might there be some truth to this possibility; i.e., why might men be more reluctant to take precautions than women? Can you suggest other explanations?

12. For each of the following pairs of events, explain why A and B are dependent or independent.

    (a) Consider the population of U.S. citizens, from which a person is randomly selected. Let A denote the event that the person is a member of a chess club and let B denote the event that the person is a woman.

    (b) Consider the population of male U.S. citizens who are 30 years of age. A man is selected at random from this population. Let A denote the event that he will be bald before reaching 40 years of age and let B denote the event that his father went bald before reaching 40 years of age.

    (c) Consider the population of students who attend high school in the U.S. A student is selected at random from this population. Let A denote the event that the student speaks Spanish and let B denote the event that the student lives in Texas.

    (d) Consider the population of months in the 20th century. A month is selected at random from this population. Let A denote the event that a hurricane crossed the North Carolina coastline during this month and let B denote the event that it snowed in Denver, Colorado, during this month.

    (e) Consider the population of Hollywood feature films produced during the 20th century. A movie is selected at random from this population. Let A denote the event that the movie was filmed in color and let B denote the event that the movie is a western.

(f) Consider the population of U.S. college freshmen, from which a student is randomly selected. Let A denote the event that the student attends the College of William & Mary, and let B denote the event that the student graduated from high school in Virginia.

(g) Consider the population of all persons (living or dead) who have earned a Ph.D. from an American university, from which one is randomly selected. Let A denote the event that the person's Ph.D. was earned before 1950 and let B denote the event that the person is female.

(h) Consider the population of persons who resided in New Orleans before Hurricane Katrina. A person is selected at random from this population. Let A denote the event that the person left New Orleans before Katrina arrived, and let B denote the event that the person belonged to a household whose 2004 income was below the federal poverty line.

(i) Consider the population of all couples who married in the United States in 1995. A couple is selected at random from this population. Let A denote the event that the couple cohabited (lived together) before marrying, and let B denote the event that the couple had divorced by 2005.

13. Two graduate students are renting a house. Before leaving town for winter break, each writes a check for her share of the rent. Emilie writes her check on December 16. By chance, it happens that the number of her check ends with the digits 16. Anne writes her check on December 18. By chance, it happens that the number of her check ends with the digits 18. What is the probability of such a coincidence, i.e., that both students would use checks with numbers that end in the same two digits as the date?

14. Suppose that X is a random variable with cdf

$$F(y) = \left\{ \begin{array}{cc} 0 & y \leq 0 \\ y/3 & y \in [0,1) \\ 2/3 & y \in [1,2] \\ y/3 & y \in [2,3] \\ 1 & y \geq 3 \end{array} \right\}.$$

Graph F and compute the following probabilities:

(a) $P(X > 0.5)$

(b) $P(2 < X \leq 3)$

(c) $P(0.5 < X \leq 2.5)$

(d) $P(X = 1)$

15. In Section 3.6, we calculated the probability that the shooter will win a fair game of craps. In so doing, we glossed a subtle point.

    Suppose that the shooter's first roll results in $x = 8$. Now the shooter must roll until he rolls another 8, in which case he makes his point and wins, or until he rolls a 7, in which case he craps out and loses. We argued that "there are 5 ways to roll 8 versus 6 ways to roll 7, so the conditional probability of making point is 5/11." This argument appears to ignore the possibility that the shooter might roll indefinitely, never rolling 8 or 7. The following calculations eliminate that possibility.

    For $i = 1, 2, 3, \ldots$, let X_i denote the result of roll i in a fair game of craps. Assume that we have observed $X_1 = x = 8$.

    (a) Calculate the probability that $X_2 \notin \{7, 8\}$.

    (b) Calculate the probability that $X_2 \notin \{7, 8\}$ *and* that $X_3 \notin \{7, 8\}$.

    (c) Calculate the probability that $X_2 \notin \{7, 8\}$ *and* that $X_3 \notin \{7, 8\}$ *and* that $X_4 \notin \{7, 8\}$.

    (d) What is the probability that the shooter will never roll another 7 or 8?

16. In the final chapter of *All I asking for is my body*, Kiyoshi places an initial, double-or-nothing bet of \$200. If he wins, he will have \$400. If he then wins a second double-or-nothing bet of \$400, he will have \$800. And so on. If he wins five consecutive times, he will have \$6400, enough to pay his family's debt.

    (a) Calculate the probability that the shooter will win five consecutive games of Craps if each of the games is fair.

    (b) Calculate the probability that the shooter will win five consecutive games of Craps if the shooter is allowed to use Kiyoshi's padrolling system.

    (c) Kiyoshi recalls that "Hiroshi never lost." Does this seem plausible?

Chapter 4

Discrete Random Variables

Our introduction of random variables in Section 3.5 was completely general; i.e., the principles that we discussed apply to *all* random variables. In this chapter we will study an important special class of random variables, the *discrete* random variables. One of the advantages of restricting attention to discrete random variables is that the mathematics required to define various fundamental concepts for this class is fairly minimal.

4.1 Basic Concepts

We begin with a formal definition.

Definition 4.1 *A random variable X is* discrete *if and only if $X(S)$, the set of possible values of X, is countable.*

Our primary interest will be in random variables for which $X(S)$ is finite; however, there are many important random variables for which $X(S)$ is denumerable. The methods described in this chapter apply to both possibilities.

In contrast to the cumulative distribution function (cdf) defined in Section 3.5, we now introduce the probability mass function (pmf).

Definition 4.2 *Let X be a discrete random variable. The* probability mass function *(pmf) of X is the function $f : \Re \to \Re$ defined by*

$$f(x) = P(X = x).$$

If f is the pmf of X, then f necessarily possesses several properties worth noting:

1. $f(x) \geq 0$ for every $x \in \Re$.
2. If $x \notin X(S)$, then $f(x) = 0$.
3. By the definition of $X(S)$,

$$\begin{aligned}\sum_{x \in X(S)} f(x) &= \sum_{x \in X(S)} P(X = x) = P\left(\bigcup_{x \in X(S)} \{x\}\right) \\ &= P\left(X \in X(S)\right) = 1.\end{aligned}$$

There is an important relation between the pmf and the cdf. For each $y \in \Re$, let

$$L(y) = \{x \in X(S) \; : \; x \leq y\}$$

denote the values of X that are less than or equal to y. Then

$$\begin{aligned}F(y) &= P(X \leq y) = P\left(X \in L(y)\right) \\ &= \sum_{x \in L(y)} P(X = x) = \sum_{x \in L(y)} f(x). \end{aligned} \tag{4.1}$$

Thus, the value of the cdf at y can be obtained by summing the values of the pmf at all values $x \leq y$.

More generally, we can compute the probability that X assumes its value in *any* set $B \subset \Re$ by summing the values of the pmf over all values of X that lie in B. Here is the formula:

$$P(X \in B) = \sum_{x \in X(S) \cap B} P(X = x) = \sum_{x \in X(S) \cap B} f(x). \tag{4.2}$$

We now turn to some elementary examples of discrete random variables and their pmfs.

4.2 Examples

Example 4.1 *A fair coin is tossed and the outcome is* `Heads` *or* `Tails`*. Define a random variable X by $X(\texttt{Heads}) = 1$ and $X(\texttt{Tails}) = 0$.*

The pmf of X is the function f defined by

$$\begin{aligned}f(0) &= P(X = 0) = 0.5, \\ f(1) &= P(X = 1) = 0.5,\end{aligned}$$

and $f(x) = 0$ for all $x \notin X(S) = \{0, 1\}$.

Example 4.2 *A typical penny is spun and the outcome is* `Heads` *or* `Tails`*. Define a random variable X by $X(\texttt{Heads}) = 1$ and $X(\texttt{Tails}) = 0$.*

Assuming that $P(\texttt{Heads}) = 0.3$ (see Section 1.1.1), the pmf of X is the function f defined by

$$\begin{aligned} f(0) &= P(X = 0) = 0.7, \\ f(1) &= P(X = 1) = 0.3, \end{aligned}$$

and $f(x) = 0$ for all $x \notin X(S) = \{0, 1\}$.

Example 4.3 *A fair die is tossed and the number of dots on the upper face is observed. The sample space is $S = \{1, 2, 3, 4, 5, 6\}$. Define a random variable X by $X(s) = 1$ if s is a prime number and $X(s) = 0$ if s is not a prime number.*

The pmf of X is the function f defined by

$$\begin{aligned} f(0) &= P(X = 0) = P(\{4, 6\}) = 1/3, \\ f(1) &= P(X = 1) = P(\{1, 2, 3, 5\}) = 2/3, \end{aligned}$$

and $f(x) = 0$ for all $x \notin X(S) = \{0, 1\}$.

Examples 4.1–4.3 have a common structure that we proceed to generalize.

Definition 4.3 *A random variable X is a Bernoulli trial if $X(S) = \{0, 1\}$.*

Traditionally, we call $X = 1$ a "success" and $X = 0$ a "failure".

The family of probability distributions of Bernoulli trials is parametrized (indexed) by a real number $p \in [0, 1]$, usually by setting $p = P(X = 1)$. We communicate that X is a Bernoulli trial with success probability p by writing $X \sim \text{Bernoulli}(p)$. The pmf of such a random variable is the function f defined by

$$\begin{aligned} f(0) &= P(X = 0) = 1 - p, \\ f(1) &= P(X = 1) = p, \end{aligned}$$

and $f(x) = 0$ for all $x \notin X(S) = \{0, 1\}$.

Several important families of random variables can be derived from Bernoulli trials. Consider, for example, the familiar experiment of tossing a fair coin twice and counting the number of `Heads`. In Section 4.4, we will generalize this experiment and count the number of successes in n Bernoulli trials. This will lead to the family of *binomial* probability distributions

Bernoulli trials are also a fundamental ingredient of the St. Petersburg Paradox, described in Example 4.12. In that experiment, a fair coin is tossed until `Heads` is observed and the number of `Tails` is counted. More generally, consider an experiment in which a sequence of independent Bernoulli trials, each with success probability p, is performed until the first success is observed. Let $X_1, X_2, X_3, \ldots$ denote the individual Bernoulli trials and let Y denote the number of failures that precede the first success. Then the possible values of Y are $Y(S) = \{0, 1, 2, \ldots\}$ and the pmf of Y is

$$\begin{aligned} f(j) = P(Y = j) &= P(X_1 = 0, \ldots, X_j = 0, X_{j+1} = 1) \\ &= P(X_1 = 0) \cdots P(X_j = 0) \cdot P(X_{j+1} = 1) \\ &= (1-p)^j p \end{aligned}$$

if $j \in Y(S)$ and $f(j) = 0$ if $j \notin Y(S)$. This family of probability distributions is also parametrized by a real number $p \in [0, 1]$. It is called the *geometric* family and a random variable with a geometric distribution is said to be a geometric random variable, written $Y \sim \text{Geometric}(p)$.

If $Y \sim \text{Geometric}(p)$ and $k \in Y(S)$, then

$$F(k) = P(Y \le k) = 1 - P(Y > k) = 1 - P(Y \ge k+1).$$

Because the event $\{Y \ge k+1\}$ occurs if and only if $X_1 = \cdots X_{k+1} = 0$, we conclude that

$$F(k) = 1 - (1-p)^{k+1}.$$

Example 4.4 *Gary is a college student who is determined to have a date for an approaching formal. He believes that each woman he asks is twice as likely to decline his invitation as to accept it, but he resolves to extend invitations until one is accepted. However, each of his first ten invitations is declined. Assuming that Gary's assumptions about his own desirability are correct, what is the probability that he would encounter such a run of bad luck?*

Gary evidently believes that he can model his invitations as a sequence of independent Bernoulli trials, each with success probability $p = 1/3$. If so, then the number of unsuccessful invitations that he extends is a random variable $Y \sim \text{Geometric}(1/3)$ and

$$P(Y \ge 10) = 1 - P(Y \le 9) = 1 - F(9) = 1 - \left[1 - \left(\frac{2}{3}\right)^{10}\right] \doteq 0.0173.$$

Either Gary is very unlucky or his assumptions are flawed. Perhaps his probability model is correct, but $p < 1/3$. Perhaps, as seems likely, the probability of success depends on whom he asks. Or perhaps the trials were not really independent.[1] If Gary's invitations cannot be modelled as independent and identically distributed Bernoulli trials, then the geometric distribution cannot be used.

Another important family of random variables is often derived by considering an *urn model*. Imagine an urn that contains m red balls and n black balls. The experiment of present interest involves selecting k balls from the urn in such a way that each of the $\binom{m+n}{k}$ possible outcomes that might be obtained are equally likely. Let X denote the number of red balls selected in this manner. If we observe $X = x$, then x red balls were selected from a total of m red balls and $k - x$ black balls were selected from a total of n black balls. Evidently, $x \in X(S)$ if and only if x is an integer such that $x \leq \min(m, k)$ and $k - x \leq \min(n, k)$. Furthermore, if $x \in X(S)$, then the pmf of X is

$$f(x) = P(X = x) = \frac{\#\{X = x\}}{\#(S)} = \frac{\binom{m}{x}\binom{n}{k-x}}{\binom{m+n}{k}}. \tag{4.3}$$

This family of probability distributions is parametrized by a triple of integers, (m, n, k), for which $m, n \geq 0$, $m + n \geq 1$, and $0 \leq k \leq m + n$. It is called the *hypergeometric* family, and a random variable with a hypergeometric distribution is said to be a hypergeometric random variable, written $Y \sim \text{Hypergeometric}(m, n, k)$.

The trick to using the hypergeometric distribution in applications is to recognize a correspondence between the actual experiment and an idealized urn model, as in the following example.

Example 4.5 *Consider the hypothetical example described in Section 1.2, in which* 30 *freshmen and* 10 *nonfreshmen are randomly assigned exam* `A` *or* `B`*. What is the probability that exactly* 15 *freshmen (and therefore exactly* 5 *nonfreshmen) receive exam* `A`*?*

In Example 2.5 we calculated that the probability in question is

$$\frac{\binom{30}{15}\binom{10}{5}}{\binom{40}{20}} = \frac{39,089,615,040}{137,846,528,820} \doteq 0.28. \tag{4.4}$$

Let us re-examine this calculation. Suppose that we write each student's name on a slip of paper, mix the slips in a jar, then draw 20 slips without

[1] In the alleged incident on which this example is based, the women all lived in the same residential college. It seems doubtful that each woman was completely unaware of the invitation that preceded hers.

replacement. These 20 students receive exam A; the remaining 20 students receive exam B. Now drawing slips of paper from a jar is exactly like drawing balls from an urn. There are $m = 30$ slips with freshman names (red balls) and $n = 10$ slips with nonfreshman names (black balls), of which we are drawing $k = 20$ without replacement. Using the hypergeometric pmf defined by (4.3), the probability of drawing exactly $x = 15$ freshman names is

$$\frac{\binom{m}{x}\binom{n}{k-x}}{\binom{m+n}{k}} = \frac{\binom{30}{15}\binom{10}{5}}{\binom{40}{20}},$$

the left-hand side of (4.4).

It is evident that calculations with the hypergeometric distribution can become rather tedious. Accordingly, this is a convenient moment to introduce computer software for the purpose of evaluating certain pmfs and cdfs. The statistical programming language R includes functions that evaluate pmfs and cdfs for a variety of distributions, including the geometric and hypergeometric.[2] For the geometric, these functions are `dgeom` and `pgeom`; for the hypergeometric, these functions are `dhyper` and `phyper`. We can calculate the probability in Example 4.4 as follows:

```
> 1-pgeom(q=9,prob=1/3)
[1] 0.01734153
```

Similarly, we can calculate the probability in Example 4.5 as follows:

```
> dhyper(15,m=30,n=10,k=20)
[1] 0.2835734
```

4.3 Expectation

Sometime in the early 1650s, the eminent theologian and amateur mathematician Blaise Pascal found himself in the company of the Chevalier de Méré.[3] De Méré posed to Pascal a famous problem: how to divide the pot of an interrupted dice game. Pascal communicated the problem to Pierre de Fermat in 1654, beginning a celebrated correspondence that established a foundation for the mathematics of probability.

---

[2]See Appendix R for information about obtaining, installing, and using R.

[3]This account of the origins of modern probability can be found in Chapter 6 of David Bergamini's *Mathematics*, Life Science Library, Time Inc., New York, 1963.

Pascal and Fermat began by agreeing that the pot should be divided according to each player's chances of winning it. For example, suppose that each of two players has selected a number from the set $S = \{1, 2, 3, 4, 5, 6\}$. For each roll of a fair die that produces one of their respective numbers, the corresponding player receives a token. The first player to accumulate five tokens wins a pot of \$100. Suppose that the game is interrupted with Player A having accumulated four tokens and Player B having accumulated only one. The probability that Player B would have won the pot had the game been completed is the probability that B's number would have appeared four more times before A's number appeared one more time. Because we can ignore rolls that produce neither number, this is equivalent to the probability that a fair coin will have a run of four consecutive `Heads`, i.e., $0.5 \cdot 0.5 \cdot 0.5 \cdot 0.5 = 0.0625$. Hence, according to Pascal and Fermat, Player B is entitled to $0.0625 \cdot \$100 = \$6.25$ from the pot and Player A is entitled to the remaining \$93.75.

The crucial concept in Pascal's and Fermat's analysis is the notion that each prospect should be weighted by the chance of realizing that prospect. This notion motivates

Definition 4.4 *The expected value of a discrete random variable X, which we will denote $E(X)$ or simply EX, is the probability-weighted average of the possible values of X, i.e.,*

$$EX = \sum_{x \in X(S)} xP(X = x) = \sum_{x \in X(S)} xf(x).$$

Remark The expected value of X, EX, is often called the *population mean* and denoted μ.

Example 4.6 If $X \sim \text{Bernoulli}(p)$, then

$$\mu = EX = \sum_{x \in \{0,1\}} xP(X = x) = 0 \cdot P(X = 0) + 1 \cdot P(X = 1) = P(X = 1) = p.$$

Notice that, in general, the expected value of X is *not* the average of its possible values. In this example, the possible values are $X(S) = \{0, 1\}$ and the average of these values is (always) 0.5. In contrast, the expected value depends on the probabilities of the values.

Fair Value The expected payoff of a game of chance is sometimes called the *fair value* of the game. For example, suppose that you own a slot machine that pays a jackpot of \$1000 with probability $p = 0.0005$ and \$0 with probability $1 - p = 0.9995$. How much should you charge a customer to play this machine? Letting X denote the payoff (in dollars), the expected payoff per play is

$$EX = 1000 \cdot 0.0005 + 0 \cdot 0.9995 = 0.5;$$

hence, if you want to make a profit, then you should charge more than \$0.50 per play. Suppose, however, that a rival owner of an identical slot machine attempted to compete for the same customers. According to standard microeconomic theory, competition would cause each of you to try to undercut the other, eventually resulting in an equilibrium price of exactly \$0.50 per play, the fair value of the game.

We proceed to illustrate both the mathematics and the psychology of fair value by considering several lotteries. A *lottery* is a choice between receiving a certain payoff and playing a game of chance. In each of the following examples, we emphasize that the value accorded the game of chance by a rational person may be very different from the game's expected value. In this sense, the phrase "fair value" is often a misnomer.

Example 4.7a *You are offered the choice between receiving a certain \$5 and playing the following game: a fair coin is tossed and you receive \$10 or \$0 according to whether* `Heads` *or* `Tails` *is observed.*

The expected payoff from the game (in dollars) is

$$EX = 10 \cdot 0.5 + 0 \cdot 0.5 = 5,$$

so your options are equivalent with respect to expected earnings. One might therefore suppose that a rational person would be indifferent to which option he or she selects. Indeed, in my experience, some students prefer to take the certain \$5 and some students prefer to gamble on perhaps winning \$10. For this example, the phrase "fair value" seems apt.

Example 4.7b *You are offered the choice between receiving a certain \$5000 and playing the following game: a fair coin is tossed and you receive \$10,000 or \$0 according to whether* `Heads` *or* `Tails` *is observed.*

The mathematical structure of this lottery is identical to that of the preceding lottery, except that the stakes are higher. Again, the options are equivalent with respect to expected earnings; again, one might suppose that a rational person would be indifferent to which option he or she selects. However, many students who opt to gamble on perhaps winning \$10 in Example 4.7a opt to take the certain \$5000 in Example 4.7b.

Example 4.7c *You are offered the choice between receiving a certain \$1 million and playing the following game: a fair coin is tossed and you receive \$2 million or \$0 according to whether* `Heads` *or* `Tails` *is observed.*

The mathematical structure of this lottery is identical to that of the preceding two lotteries, except that the stakes are now *much* higher. Again, the options are equivalent with respect to expected earnings; however, almost every student to whom I have presented this lottery has expressed a strong preference for taking the certain \$1 million.

Example 4.8 *You are offered the choice between receiving a certain \$1 million and playing the following game: a fair coin is tossed and you receive \$5 million or \$0 according to whether* `Heads` *or* `Tails` *is observed.*

The expected payoff from this game (in millions of dollars) is

$$EX = 5 \cdot 0.5 + 0 \cdot 0.5 = 2.5,$$

so playing the game is the more attractive option with respect to expected earnings. Nevertheless, most students opt to take the certain \$1 million. This should *not* be construed as an irrational decision. For example, the addition of \$1 million to my own modest estate would secure my eventual retirement. The addition of an extra \$4 million would be very pleasant indeed, allowing me to increase my current standard of living. However, I do not value the additional \$4 million nearly as much as I value the initial \$1 million. As Aesop observed, "A little thing in hand is worth more than a great thing in prospect." For this example, the phrase "fair value" introduces normative connotations that are not appropriate.

Example 4.9 Consider the following passage from an article about investing:

> ...it's human nature to overweight low probabilities that offer high returns. In one study, subjects were given a choice between a 1-in-1000 chance to win \$5000 or a sure thing to win \$5; or a 1-in-1000 chance of losing \$5000 versus a sure loss of \$5. In the first case, the expected value (mathematically speaking) is making \$5. In the second case, it's losing \$5. Yet in the first situation, which mimics a lottery,

> more than 70% of people asked chose to go for the $5000. In the second situation, more than 80% would take the $5 hit.[4]

The author evidently considered the reported preferences paradoxical, but are they really surprising? Plus or minus $5 will not appreciably alter the financial situations of most subjects, but plus or minus $5000 will. It is perfectly rational to risk a negligible amount on the chance of winning $5000 while declining to risk a negligible amount on the chance of losing $5000. The following examples further explicate this point.

Example 4.10 The same article advises, "To limit completely irrational risks, such as lottery tickets, try speculating only with money you would otherwise use for simple pleasures, such as your morning coffee."

Consider a hypothetical state lottery, in which 6 numbers are drawn (without replacement) from the set $\{1, 2, \ldots, 39, 40\}$. For $2, you can purchase a ticket that specifies 6 such numbers. If the numbers on your ticket match the numbers selected by the state, then you win $1 million; otherwise, you win nothing. (For the sake of simplicity, we ignore the possibility that you might have to split the jackpot with other winners and the possibility that you might win a lesser prize.) Is buying a lottery ticket "completely irrational"?

The probability of winning the lottery in question is

$$p = \frac{1}{\binom{40}{6}} = \frac{1}{3,838,380} \doteq 2.6053 \times 10^{-7},$$

so your expected prize (in dollars) is approximately

$$10^6 \cdot 2.6053 \times 10^{-7} \doteq 0.26,$$

which is considerably less than the cost of a ticket. Evidently, it is completely irrational to buy tickets for this lottery *as an investment strategy*. Suppose, however, that I buy one ticket per week and reason as follows: I will almost certainly lose $2 per week, but that loss will have virtually no impact on my standard of living; however, if by some miracle I win, then gaining $1 million will revolutionize my standard of living. This can hardly be construed as irrational behavior, although Frick's advice to speculate only with funds earmarked for entertainment is well-taken.

---

[4] R. Frick, The 7 Deadly Sins of Investing, *Kiplinger's Personal Finance Magazine*, March 1998, p. 138.

In most state lotteries, the fair value of the game is less than the cost of a lottery ticket. This is only natural—lotteries exist because they generate revenue for the state that runs them! (By the same reasoning, gambling must favor the house because casinos make money for their owners.) However, on very rare occasions a jackpot is so large that the typical situation is reversed. Several years ago, an Australian syndicate noticed that the fair value of a Florida state lottery exceeded the price of a ticket and purchased a large number of tickets as an (ultimately successful) investment strategy. And Voltaire once purchased every ticket in a raffle upon noting that the prize was worth more than the total cost of the tickets being sold!

Example 4.11 If the first case described in Example 4.9 mimics a lottery, then the second case mimics insurance. Mindful that insurance companies (like casinos) make money, Ambrose Bierce offered the following definition:

> INSURANCE, *n.* An ingenious modern game of chance in which the player is permitted to enjoy the comfortable conviction that he is beating the man who keeps the table.[5]

However, while it is certainly true that the fair value of an insurance policy is less than the premiums required to purchase it, it does not follow that buying insurance is irrational. I can easily afford to pay \$200 per year for homeowners insurance, but I would be ruined if all of my possessions were destroyed by fire and I received no compensation for them. My decision that a certain but affordable loss is preferable to an unlikely but catastrophic loss is an example of *risk-averse* behavior.

Before presenting our concluding example of fair value, we derive a useful formula. Suppose that $X : S \to \Re$ is a discrete random variable and $\phi : \Re \to \Re$ is a function. Let $Y = \phi(X)$. Then $Y : \Re \to \Re$ is a random variable and

$$\begin{aligned} E\phi(X) &= EY = \sum_{y \in Y(S)} yP(Y = y) \\ &= \sum_{y \in Y(S)} yP\left(\phi(X) = y\right) \\ &= \sum_{y \in Y(S)} yP\left(X \in \phi^{-1}(y)\right) \end{aligned}$$

[5] Ambrose Bierce, *The Devil's Dictionary*, 1881–1906. In *The Collected Writings of Ambrose Bierce*, Citadel Press, Secaucus, NJ, 1946.

$$
\begin{aligned}
&= \sum_{y \in Y(S)} y \left(\sum_{x \in \phi^{-1}(y)} P(X = x) \right) \\
&= \sum_{y \in Y(S)} \sum_{x \in \phi^{-1}(y)} y P(X = x) \\
&= \sum_{y \in Y(S)} \sum_{x \in \phi^{-1}(y)} \phi(x) P(X = x) \\
&= \sum_{x \in X(S)} \phi(x) P(X = x) \\
&= \sum_{x \in X(S)} \phi(x) f(x). \qquad (4.5)
\end{aligned}
$$

Example 4.12 (St. Petersburg Paradox) *Consider a game in which the jackpot starts at \$1 and doubles each time that* `Tails` *is observed when a fair coin is tossed. The game terminates when* `Heads` *is observed for the first time. How much would you pay for the privilege of playing this game? How much would you charge if you were responsible for making the payoff?*

This is a curious game. With high probability, the payoff will be rather small; however, there is a small chance of a very large payoff. In response to the first question, most students discount the latter possibility and respond that they would pay only a small amount, rarely more than \$4. In response to the second question, most students recognize the possibility of a large payoff and demand payment of a considerably greater amount. Let us consider if the notion of fair value provides guidance in reconciling these perspectives.

Let X denote the number of `Tails` that are observed before the game terminates. Then $X(S) = \{0, 1, 2, \ldots\}$ and the geometric random variable X has pmf

$$f(x) = P(x \text{ consecutive } \texttt{Tails}) = 0.5^x.$$

The payoff from this game (in dollars) is $Y = 2^X$; hence, the expected payoff is

$$E2^X = \sum_{x=0}^{\infty} 2^x \cdot 0.5^x = \sum_{x=0}^{\infty} 1 = \infty.$$

This is quite startling! The "fair value" of this game provides very little insight into the value that a rational person would place on playing it. This remarkable example is known as the St. Petersburg Paradox.

Properties of Expectation We now state (and sometimes prove) some useful consequences of Definition 4.4 and Equation 4.5.

Theorem 4.1 *Let X denote a discrete random variable and suppose that $P(X = c) = 1$. Then $EX = c$.*

Theorem 4.1 states that, if a random variable always assumes the same value c, then the probability-weighted average of the values that it assumes is c. This fact should be obvious.

Theorem 4.2 *Let X denote a discrete random variable and suppose that $c \in \Re$ is constant. Then*

$$E\left[c\phi(X)\right] = \sum_{x \in X(S)} c\phi(x)f(x) = c \sum_{x \in X(S)} \phi(x)f(x) = cE\left[\phi(X)\right].$$

Theorem 4.2 states that we can interchange the order of multiplying by a constant and computing the expected value. Notice that this property of expectation follows directly from the analogous property for summation.

Theorem 4.3 *Let X denote a discrete random variable. Then*

$$\begin{aligned} E\left[\phi_1(X) + \phi_2(X)\right] &= \sum_{x \in X(S)} [\phi_1(x) + \phi_2(x)]f(x) \\ &= \sum_{x \in X(S)} [\phi_1(x)f(x) + \phi_2(x)f(x)] \\ &= \sum_{x \in X(S)} \phi_1(x)f(x) + \sum_{x \in X(S)} \phi_2(x)f(x) \\ &= E\left[\phi_1(X)\right] + E\left[\phi_2(X)\right]. \end{aligned}$$

Theorem 4.3 states that we can interchange the order of adding functions of a random variable and computing the expected value. Again, this property of expectation follows directly from the analogous property for summation.

Theorem 4.4 *Let X_1 and X_2 denote discrete random variables. Then*

$$E\left[X_1 + X_2\right] = EX_1 + EX_2.$$

Theorem 4.4 states that the expected value of a sum equals the sum of the expected values.

Variance Now suppose that X is a discrete random variable, let $\mu = EX$ denote its expected value, or population mean, and define a function $\phi : \Re \to \Re$ by

$$\phi(x) = (x - \mu)^2.$$

For any $x \in \Re$, $\phi(x)$ is the squared deviation of x from the expected value of X. If X always assumes the value μ, then $\phi(X)$ always assumes the value 0; if X tends to assume values near μ, then $\phi(X)$ will tend to assume small values; if X often assumes values far from μ, then $\phi(X)$ will often assume large values. Thus, $E\phi(X)$, the expected squared deviation of X from its expected value, is a measure of the variability of the population $X(S)$. We summarize this observation in

Definition 4.5 *The variance of a discrete random variable X, which we will denote Var(X) or simply Var$\,X$, is the probability-weighted average of the squared deviations of X from $EX = \mu$, i.e.,*

$$\mathit{Var}\, X = E(X - \mu)^2 = \sum_{x \in X(S)} (x - \mu)^2 f(x).$$

Remark The variance of X, Var X, is often called the *population variance* and denoted σ^2.

Denoting the population variance by σ^2 may strike the reader as awkward notation, but there is an excellent reason for it. Because the variance measures squared deviations from the population mean, it is measured in different units than either the random variable itself or its expected value. For example, if X measures length in meters, then so does EX, but Var X is measured in meters squared. To recover a measure of population variability in the original units of measurement, we take the square root of the variance and obtain σ.

Definition 4.6 *The standard deviation of a random variable is the square root of its variance.*

Remark The standard deviation of X, often denoted σ, is often called the *population standard deviation.*

Example 4.1 (continued) If $X \sim \text{Bernoulli}(p)$, then

$$\begin{aligned}
\sigma^2 = \text{Var}\, X &= E(X-\mu)^2 \\
&= (0-\mu)^2 \cdot P(X=0) + (1-\mu)^2 \cdot P(X=1) \\
&= (0-p)^2(1-p) + (1-p)^2 p \\
&= p(1-p)(p+1-p) \\
&= p(1-p).
\end{aligned}$$

Before turning to a more complicated example, we establish a useful fact.

Theorem 4.5 *If X is a discrete random variable, then*

$$\begin{aligned}
Var\, X &= E(X-\mu)^2 \\
&= E(X^2 - 2\mu X + \mu^2) \\
&= EX^2 + E(-2\mu X) + E\mu^2 \\
&= EX^2 - 2\mu EX + \mu^2 \\
&= EX^2 - 2\mu^2 + \mu^2 \\
&= EX^2 - (EX)^2.
\end{aligned}$$

A straightforward way to calculate the variance of a discrete random variable that assumes a fairly small number of values is to exploit Theorem 4.5 and organize one's calculations in the form of a table.

Example 4.13 *Suppose that X is a random variable whose possible values are $X(S) = \{2, 3, 5, 10\}$. Suppose that the probability of each of these values is given by the formula $f(x) = P(X = x) = x/20$.*

(a) Calculate the expected value of X.

(b) Calculate the variance of X.

(c) Calculate the standard deviation of X.

Solution

x	$f(x)$	$xf(x)$	x^2	$x^2 f(x)$
2	0.10	0.20	4	0.40
3	0.15	0.45	9	1.35
5	0.25	1.25	25	6.25
10	0.50	5.00	100	50.00
		6.90		58.00

(a) $\mu = EX = 0.2 + 0.45 + 1.25 + 5 = 6.9$.

(b) $\sigma^2 = \text{Var}\, X = EX^2 - (EX)^2 = (0.4 + 1.35 + 6.25 + 50) - 6.9^2 = 58 - 47.61 = 10.39$.

(c) $\sigma = \sqrt{10.39} \doteq 3.2234$.

Now suppose that $X : S \to \Re$ is a discrete random variable and $\phi : \Re \to \Re$ is a function. Let $Y = \phi(X)$. Then Y is a discrete random variable and

$$\text{Var}\, \phi(X) = \text{Var}\, Y = E\left[Y - EY\right]^2 = E\left[\phi(X) - E\phi(X)\right]^2 . \tag{4.6}$$

We conclude this section by stating (and sometimes proving) some useful consequences of Definition 4.5 and Equation 4.6.

Theorem 4.6 *Let X denote a discrete random variable and suppose that $c \in \Re$ is constant. Then*

$$Var(X + c) = Var\, X.$$

Although possibly startling at first glance, this result is actually quite intuitive. The variance depends on the squared deviations of the values of X from the expected value of X. If we add a constant to each value of X, then we shift both the individual values of X and the expected value of X by the same amount, preserving the squared deviations. The *variability* of a population is not affected by shifting each of the values in the population by the same amount.

Theorem 4.7 *Let X denote a discrete random variable and suppose that $c \in \Re$ is constant. Then*

$$\begin{aligned} Var(cX) &= E\left[cX - E(cX)\right]^2 \\ &= E\left[cX - cEX\right]^2 \\ &= E\left[c(X - EX)\right]^2 \\ &= E\left[c^2(X - EX)^2\right] \\ &= c^2 E(X - EX)^2 \\ &= c^2\, Var\, X. \end{aligned}$$

To understand this result, recall that the variance is measured in the original units of measurement squared. If we take the square root of each expression in Theorem 4.7, then we see that one can interchange multiplying a random variable by a nonnegative constant with computing its *standard deviation*.

Theorem 4.8 *If the discrete random variables X_1 and X_2 are independent, then*

$$Var(X_1 + X_2) = Var X_1 + Var X_2.$$

Theorem 4.8 is analogous to Theorem 4.4. However, in order to ensure that the variance of a sum equals the sum of the variances, the random variables must be independent.

4.4 Binomial Distributions

Suppose that a fair coin is tossed twice and the number of **Heads** is counted. Let Y denote the total number of **Heads**. Because the sample space has four equally likely outcomes, viz.,

$$S = \{\mathtt{HH}, \mathtt{HT}, \mathtt{TH}, \mathtt{TT}\},$$

the pmf of Y is easily determined:

$$\begin{aligned} f(0) &= P(Y=0) = P(\{\mathtt{HH}\}) = 0.25, \\ f(1) &= P(Y=1) = P(\{\mathtt{HT}, \mathtt{TH}\}) = 0.5, \\ f(2) &= P(Y=2) = P(\{\mathtt{TT}\}) = 0.25, \end{aligned}$$

and $f(y) = 0$ if $y \notin Y(S) = \{0, 1, 2\}$.

Referring to representation (c) of Example 3.18, the above experiment has the following characteristics:

- Let X_1 denote the number of **Heads** observed on the first toss and let X_2 denote the number of **Heads** observed on the second toss. Then the random variable of interest is $Y = X_1 + X_2$.

- The random variables X_1 and X_2 are independent.

- The random variables X_1 and X_2 have the same distribution, viz.

$$X_1, X_2 \sim \text{Bernoulli}(0.5).$$

We proceed to generalize this example in two ways:

1. We allow any finite number of trials.

2. We allow any success probability $p \in [0, 1]$.

Definition 4.7 *Let $X_1, \ldots, X_n$ be mutually independent Bernoulli trials, each with success probability p. Then*

$$Y = \sum_{i=1}^{n} X_i$$

is a binomial random variable, denoted

$$Y \sim \text{Binomial}(n; p).$$

Applying Theorem 4.4, we see that the expected value of a binomial random variable is the product of the number of trials and the probability of success:

$$EY = E\left(\sum_{i=1}^{n} X_i\right) = \sum_{i=1}^{n} EX_i = \sum_{i=1}^{n} p = np.$$

Furthermore, because the trials are independent, we can apply Theorem 4.8 to calculate the variance:

$$\text{Var}\, Y = \text{Var}\left(\sum_{i=1}^{n} X_i\right) = \left(\sum_{i=1}^{n} \text{Var}\, X_i\right) = \left(\sum_{i=1}^{n} p(1-p)\right) = np(1-p).$$

Because Y counts the total number of successes in n Bernoulli trials, it should be apparent that $Y(S) = \{0, 1, \ldots, n\}$. Let f denote the pmf of Y. For fixed n, p, and $j \in Y(S)$, we wish to determine

$$f(j) = P(Y = j).$$

To illustrate the reasoning required to make this determination, suppose that there are $n = 6$ trials, each with success probability $p = 0.3$, and that we wish to determine the probability of observing exactly $j = 2$ successes. Some examples of experimental outcomes for which $Y = 2$ include the following:

110000 000011 010010

Because the trials are mutually independent, we see that

$$\begin{aligned} P(110000) &= 0.3 \cdot 0.3 \cdot 0.7 \cdot 0.7 \cdot 0.7 \cdot 0.7 = 0.3^2 \cdot 0.7^4, \\ P(000011) &= 0.7 \cdot 0.7 \cdot 0.7 \cdot 0.7 \cdot 0.3 \cdot 0.3 = 0.3^2 \cdot 0.7^4, \\ P(010010) &= 0.7 \cdot 0.3 \cdot 0.7 \cdot 0.7 \cdot 0.3 \cdot 0.7 = 0.3^2 \cdot 0.7^4. \end{aligned}$$

It should be apparent that the probability of each outcome for which $Y = 2$ is the product of $j = 2$ factors of $p = 0.3$ and $n - j = 4$ factors of $1 - p = 0.7$.

Furthermore, the number of such outcomes is the number of ways of choosing $j = 2$ successes from a total of $n = 6$ trials. Thus,

$$f(2) = P(Y = 2) = \binom{6}{2} 0.3^2 0.7^4$$

for the specific example in question and the general formula for the binomial pmf is

$$f(j) = P(Y = j) = \binom{n}{j} p^j (1-p)^{n-j}.$$

It follows, of course, that the general formula for the binomial cdf is

$$\begin{aligned} F(k) = P(Y \leq k) &= \sum_{j=0}^{k} P(Y = j) = \sum_{j=0}^{k} f(j) \\ &= \sum_{j=0}^{k} \binom{n}{j} p^j (1-p)^{n-j}. \end{aligned} \tag{4.7}$$

Except for very small numbers of trials, direct calculation of (4.7) is rather tedious. Fortunately, we can use the R function `pbinom` to compute values of binomial cdfs.

As the following examples should make clear, the trick to evaluating binomial probabilities is to write them in expressions that involve only probabilities of the form $P(Y \leq k)$.

Example 4.15 *In* 10 *trials with success probability* 0.5, *what is the probability that no more than* 4 *successes will be observed?*

Here, $n = 10$, $p = 0.5$, and we want to calculate

$$P(Y \leq 4) = F(4).$$

We do so in R as follows:

```
> pbinom(4,size=10,prob=.5)
[1] 0.3769531
```

Example 4.16 *In* 12 *trials with success probability* 0.3, *what is the probability that more than* 6 *successes will be observed?*

Here, $n = 12$, $p = 0.3$, and we want to calculate

$$P(Y > 6) = 1 - P(Y \leq 6) = 1 - F(6).$$

We do so in R as follows:

```
> 1-pbinom(6,12,.3)
[1] 0.03860084
```

Example 4.17 *In 15 trials with success probability 0.6, what is the probability that at least 5 but no more than 10 successes will be observed?*

Here, $n = 15$, $p = 0.6$, and we want to calculate

$$P(5 \leq Y \leq 10) = P(Y \leq 10) - P(Y \leq 4) = F(10) - F(4).$$

We do so in R as follows:

```
> pbinom(10,15,.6)-pbinom(4,15,.6)
[1] 0.7733746
```

Example 4.18 *In 20 trials with success probability 0.9, what is the probability that exactly 16 successes will be observed?*

Here, $n = 20$, $p = 0.9$, and we want to calculate

$$P(Y = 16) = P(Y \leq 16) - P(Y \leq 15) = F(16) - F(15).$$

We do so in R as follows:

```
> pbinom(16,20,.9)-pbinom(15,20,.9)
[1] 0.08977883
```

Example 4.19 *In 81 trials with success probability 0.64, what is the probability that the proportion of observed successes will be between 60 and 70 percent?*

Here, $n = 81$, $p = 0.64$, and we want to calculate

$$\begin{aligned} P(0.6 < Y/81 < 0.7) &= P(0.6 \cdot 81 < Y < 0.7 \cdot 81) \\ &= P(48.6 < Y < 56.7) \\ &= P(49 \leq Y \leq 56) \\ &= P(Y \leq 56) - P(Y \leq 48) \\ &= F(56) - F(48). \end{aligned}$$

We do so in R as follows:

```
> pbinom(56,81,.64)-pbinom(48,81,.64)
[1] 0.6416193
```

Many practical situations can be modelled using a binomial distribution. Doing so typically requires one to perform the following steps.

1. Identify what constitutes a Bernoulli trial and what constitutes a success. Verify or assume that the trials are mutually independent with a common probability of success.

2. Identify the number of trials (n) and the common probability of success (p).

3. Identify the event whose probability is to be calculated.

4. Calculate the probability of the event in question, e.g., by using the `pbinom` function in `R`.

Example 4.20 *RD Airlines flies planes that seat* 58 *passengers. Years of experience have revealed that* 20 *percent of the persons who purchase tickets fail to claim their seat. (Such persons are called "no-shows".) Because of this phenomenon, RD routinely overbooks its flights, i.e., RD typically sells more than* 58 *tickets per flight. If more than* 58 *passengers show, then the "extra" passengers are "bumped" to another flight. Suppose that RD sells* 64 *tickets for a certain flight from Washington to New York. How might RD estimate the probability that at least one passenger will have to be bumped?*

1. Each person who purchased a ticket must decide whether or not to claim his or her seat. This decision represents a Bernoulli trial, for which we will declare a decision to claim the seat a success. Strictly speaking, the Bernoulli trials in question are neither mutually independent nor identically distributed. Some individuals, e.g., families, travel together and make a common decision as to whether or not to claim their seats. Furthermore, some travellers are more likely to change their plans than others. Nevertheless, absent more detailed information, we should be able to compute an approximate answer by assuming that the total number of persons who claim their seats has a binomial distribution.

2. The problem specifies that $n = 64$ persons have purchased tickets. Appealing to past experience, we assume that the probability that each person will show is $p = 1 - 0.2 = 0.8$.

3. At least one passenger will have to be bumped if more than 58 passengers show, so the desired probability is

$$P(Y > 58) = 1 - P(Y \le 58) = 1 - F(58).$$

4. The necessary calculation can be performed in R as follows:

```
> 1-pbinom(58,64,.8)
[1] 0.006730152
```

4.5 Exercises

1. Suppose that $X(S) = \{1, 3, 4, 6\}$ with $P(X = 1) = P(X = 6) = 0.1$ and $P(X = 3) = P(X = 4) = 0.4$. Determine each of the following:
   (a) The probability mass function of X.
   (b) The cumulative distribution function of X.
   (c) The expected value of X.
   (d) The variance of X.
   (e) The standard deviation of X.

2. Suppose that a weighted die is tossed. Let X denote the number of dots that appear on the upper face of the die, and suppose that $P(X = x) = (7 - x)/20$ for $x = 1, 2, 3, 4, 5$ and $P(X = 6) = 0$. Determine each of the following:
   (a) The probability mass function of X.
   (b) The cumulative distribution function of X.
   (c) The expected value of X.
   (d) The variance of X.
   (e) The standard deviation of X.

3. Consider an urn that contains 10 tickets, labelled

$$\{1, 1, 1, 1, 2, 5, 5, 10, 10, 10\}.$$

   From this urn, I propose to draw a ticket. Let X denote the value of the ticket I draw. Determine each of the following:
   (a) The probability mass function of X.
   (b) The cumulative distribution function of X.
   (c) The expected value of X.
   (d) The variance of X.

(e) The standard deviation of X.

4. Suppose that a jury of 12 persons is to be selected from a pool of 25 persons who were called for jury duty. The pool comprises 12 retired persons, 6 employed persons, 5 unemployed persons, and 2 students. Assuming that each person is equally likely to be selected, answer the following:

   (a) What is the probability that both students will be selected?

   (b) What is the probability that the jury will contain exactly twice as many retired persons as employed persons?

5. When casting four astragali, a throw that results in four different uppermost sides is called a *venus*. (See Section 1.4.) Suppose that four astragali, $\{A, B, C, D\}$ each have the following probabilities of producing the four possible uppermost faces: $P(1) = P(6) = 0.1$, $P(3) = P(4) = 0.4$.

   (a) Suppose that we write $A = 1$ to indicate the event that A produces side 1, etc. Compute $P(A = 1, B = 3, C = 4, D = 6)$.

   (b) Compute $P(A = 1, B = 6, C = 3, D = 4)$.

   (c) What is the probability that one throw of these four astragali will produce a venus?

   Hint: See Exercise 2.5.6.

   (d) For $k = 2$, $k = 3$, and $k = 100$, what is the probability that k throws of these four astragali will produce a run of k venuses?

6. Suppose that each of five astragali have the probabilities specified in the previous exercise. When throwing these five astagali,

   (a) What is the probability of obtaining the throw of child-eating Cronos, i.e., of obtaining three fours and two sixes?

   (b) What is the probability of obtaining the throw of Saviour Zeus, i.e., of obtaining one one, two threes, and two fours?

   Hint: See Exercise 2.5.7.

7. Koko (a cat) is trying to catch a mouse who lives under Susan's house. The mouse has two exits, one outside and one inside, and randomly selects the outside exit 60% of the time. Each midnight, the mouse emerges for a constitutional. If Koko waits outside and the mouse

chooses the outside exit, then Koko has a 20% chance of catching the mouse. If Koko waits inside, then there is a 30% chance that he will fall asleep. However, if he stays awake and the mouse chooses the inside exit, then Koko has a 40% chance of catching the mouse.

(a) Is Koko more likely to catch the mouse if he waits inside or outside? Why?

Hint: Draw a tree diagram for each strategy.

(b) If Koko decides to wait outside each midnight, then what is the probability that he will catch the mouse within a week (no more than 7 nights)?

8. The game called Pass the Pigs™ appears to have been inspired by a dice game called Pig. In one version of Pig, a player rolls a single fair die. Each turn, the player rolls until either (a) s/he elects to stop, or (b) the die produces 1 dot. Points are accumulated turn by turn. If the player elects to stop rolling before a 1 appears, then the point total for that turn is the sum of all the dots produced on that turn; if a 1 appears, then the point total for that turn is 0.

   Suppose that Wilbur is about to take a turn in a game of Pig. He decides that he will roll the die a second time if he is permitted to do so, but that he will stop after two rolls regardless of the outcome. How many points should Wilbur expect to accumulate on his turn?

   Hint: Draw a tree diagram.

9. Three urns each contain ten gems:

   - Urn 1 contains 6 rubies and 4 emeralds.
   - Urn 2r contains 8 rubies and 2 emeralds.
   - Urn 2e contains 4 rubies and 6 emeralds.

   The following procedure is used to select two gems. First, one gem is drawn at random from urn 1. If this first gem is a ruby, then a second gem is drawn at random from urn 2r; however, if the first gem is an emerald, then the second gem is drawn at random from urn 2e.

   (a) Construct a tree diagram that describes this procedure.

   (b) What is the probability that a ruby is obtained on the second draw?

(c) Suppose that the second gem is a ruby. What then is the probability that the first gem was also a ruby?

(d) Suppose that this procedure is independently replicated three times. What is the probability that a ruby is obtained on the second draw exactly once?

(e) Suppose that this procedure is independently replicated three times and that a ruby is obtained on the second draw each time. What then is the probability that the first gem was a ruby each time?

10. Arlen is planning a dinner party at which he will be able to accommodate seven guests. From past experience, he knows that each person invited to the party will accept his invitation with probability 0.5. He also knows that each person who accepts will actually attend with probability 0.8. Suppose that Arlen invites twelve people. Assuming that they behave independently of one another, what is the probability that he will end up with more guests than he can accommodate?

11. Hotels that host conferences routinely overbook their rooms because some people who plan to attend conferences fail to arrive. A common assumption is that 10 percent of the hotel rooms reserved by conference attendees will not be claimed. In contrast, only 4 percent of the persons who reserve hotel rooms for the annual Joint Statistical Meetings (JSM) fail to claim them.

    Suppose that a certain hotel has 100 rooms. Incorrectly believing that statisticians behave like normal people, the hotel accepts 110 room reservations for JSM. What is the probability that the hotel will have to turn away statisticians who have reserved rooms?

12. A small liberal arts college receives applications for admission from 1000 high school seniors. The college has dormitory space for a freshman class of 95 students and will have to arrange for off-campus housing for any additional freshmen. In previous years, an average of 64 percent of the students that the college has accepted have elected to attend another school. Clearly the college should accept more than 95 students, but its administration does not want to take too big a chance that it will have to accommodate more than 95 students. After some delibration, the administrators decide to accept 225 students. Answer the following questions as well as you can with the information provided.

(a) How many freshmen do you expect that the college will have to accommodate?

(b) What is the probability that the college will have to arrange for some freshmen to live off-campus?

13. In NCAA tennis matches, line calls are made by the players. If an umpire is observing the match, then a player can challenge an opponent's call. The umpire will either affirm or overrule the challenged call. In one of their recent team matches, the William & Mary women's tennis team challenged 38 calls by its opponents. The umpires overruled 12 of the challenged calls. This struck Nina and Delphine as significant, as it is their impression that approximately 20 percent of all challenged calls in NCAA tennis matches are overruled. Let us assume that their impression is correct.

    (a) What is the probability that chance variation would result in at least 12 of 38 challenged calls being overruled?

    (b) Suppose that the William & Mary women's tennis team plays 25 team matches next year and challenges exactly 38 calls in each match. (In fact, the number of challenged calls varies from match to match.) What is the probability that they will play at least one team match in which at least 12 challenged calls are overruled?

14. The Association for Research and Enlightenment (ARE) in Virginia Beach, VA, offers daily demonstrations of a standard technique for testing extrasensory perception (ESP). A "sender" is seated before a box on which one of five symbols (plus, square, star, circle, wave) can be illuminated. A random mechanism selects symbols in such a way that each symbol is equally likely to be illuminated. When a symbol is illuminated, the sender concentrates on it and a "receiver" attempts to identify which symbol has been selected. The receiver indicates a symbol on the receiver's box, which sends a signal to the sender's box that cues it to select and illuminate another symbol. This process of illuminating, sending, and receiving a symbol is repeated 25 times. Each selection of a symbol to be illuminated is independent of the others. The receiver's score (for a set of 25 trials) is the number of symbols that s/he correctly identifies. For the purpose of this exercise, please suppose that ESP does not exist.

    (a) How many symbols should we expect the receiver to identify correctly?

(b) The ARE considers a score of more than 7 matches to be indicative of ESP. What is the probability that the receiver will provide such an indication?

(c) The ARE provides all audience members with scoring sheets and invites them to act as receivers. Suppose that, as on August 31, 2002, there are 21 people in attendance: 1 volunteer sender, 1 volunteer receiver, and 19 additional receivers in the audience. What is the probability that at least one of the 20 receivers will attain a score indicative of ESP?

15. Mike teaches two sections of Applied Statistics each year for thirty years, for a total of 1500 students. Each of his students spins a penny 89 times and counts the number of `Heads`. Assuming that each of these 1500 pennies has $P(\texttt{Heads}) = 0.3$ for a single spin, what is the probability that Mike will encounter at least one student who observes no more than two `Heads`?

Chapter 5

Continuous Random Variables

5.1 A Motivating Example

Some of the concepts that were introduced in Chapter 4 pose technical difficulties when the random variable is not discrete. We illustrate some of these difficulties by considering a random variable X whose set of possible values is the unit interval, i.e., $X(S) = [0, 1]$. Specifically, we ask the following question:

> *What probability distribution formalizes the notion of "equally likely" outcomes in the unit interval* $[0, 1]$*?*

When studying finite sample spaces in Section 3.3, we formalized the notion of "equally likely" by assigning the same probability to each individual outcome in the sample space. Thus, if $S = \{s_1, \ldots, s_N\}$, then $P(\{s_i\}) = 1/N$. This construction sufficed to define probabilities of events: if $E \subset S$, then

$$E = \{s_{i_1}, \ldots, s_{i_k}\};$$

and consequently

$$P(E) = P\left(\bigcup_{j=1}^{k} \left\{s_{i_j}\right\}\right) = \sum_{j=1}^{k} P\left(\left\{s_{i_j}\right\}\right) = \sum_{j=1}^{k} \frac{1}{N} = \frac{k}{N}.$$

Unfortunately, the present example does not work out quite so neatly.

How should we assign $P(X = 0.5)$? Of course, we must have $0 \leq P(X = 0.5) \leq 1$. If we try $P(X = 0.5) = \epsilon$ for any real number $\epsilon > 0$, then a difficulty arises. Because we are assuming that every value in the unit interval is equally likely, it must be that $P(X = x) = \epsilon$ for *every* $x \in [0, 1]$. Consider the event

$$E = \left\{ \frac{1}{2}, \frac{1}{3}, \frac{1}{4}, \ldots \right\}. \tag{5.1}$$

Then we must have

$$P(E) = P\left(\bigcup_{j=2}^{\infty} \left\{ \frac{1}{j} \right\} \right) = \sum_{j=2}^{\infty} P\left(\left\{ \frac{1}{j} \right\} \right) = \sum_{j=2}^{\infty} \epsilon = \infty, \tag{5.2}$$

which we cannot allow. Hence, we *must* assign a probability of zero to the outcome $x = 0.5$ and, because all outcomes are equally likely, $P(X = x) = 0$ for every $x \in [0, 1]$.

Because every $x \in [0, 1]$ is a possible outcome, our conclusion that $P(X = x) = 0$ is initially somewhat startling. However, it is a mistake to identify impossibility with zero probability. In Section 3.2, we established that the impossible event (empty set) has probability zero, but we did *not* say that it is the only such event. To avoid confusion, we now emphasize:

> *If an event is impossible, then it necessarily has probability zero; however, having probability zero does not necessarily mean that an event is impossible.*

If $P(X = x) = \epsilon = 0$, then the calculation in (5.2) reveals that the event defined by (5.1) has probability zero. Furthermore, there is nothing special about this particular event—the probability of *any* countable event must be zero! Hence, to obtain positive probabilities, e.g., $P(X \in [0, 1]) = 1$, we must consider events whose cardinality is more than countable.

Consider the events $[0, 0.5]$ and $[0.5, 1]$. Let us construe the phrase "equally likely outcomes" to imply that these events have the same probability, i.e.,

$$P(X \in [0, 0.5]) = P(X \in [0.5, 1]).$$

Then, because $[0, 0.5] \cup [0.5, 1] = [0, 1]$ and $P(X = 0.5) = 0$, we must have

$$\begin{aligned} 1 = P(X \in [0, 1]) &= P(X \in [0, 0.5]) + P(X \in [0.5, 1]) - P(X = 0.5) \\ &= P(X \in [0, 0.5]) + P(X \in [0.5, 1]). \end{aligned}$$

Combining these equations, we deduce that each event has probability $1/2$. This is an intuitively pleasing interpretation of "equally likely." Suitably generalized, it says that the probability of each subinterval equals the proportion of the entire interval occupied by the subinterval. In mathematical notation, our conclusion can be expressed as follows:

> *Suppose that $X(S) = [0,1]$ and each $x \in [0,1]$ is equally likely. If $0 \le a \le b \le 1$, then $P(X \in [a,b]) = b - a$.*

Notice that statements like $P(X \in [0, 0.5]) = 0.5$ cannot be deduced from knowledge that each $P(X = x) = 0$. To construct a probability distribution for this situation, it is necessary to assign probabilities to intervals, not just to individual points. This fact reveals the reason that, in Section 3.2, we introduced the concept of an event and insisted that probabilities be assigned to events rather than to outcomes.

The probability distribution that we have constructed is called the *continuous uniform distribution* on the interval $[0,1]$, denoted Uniform$[0,1]$. If $X \sim$ Uniform$[0,1]$, then the cdf of X is easily computed:

- If $y < 0$, then

$$\begin{aligned} F(y) &= P(X \le y) \\ &= P(X \in (-\infty, y]) \\ &= 0. \end{aligned}$$

- If $y \in [0,1]$, then

$$\begin{aligned} F(y) &= P(X \le y) \\ &= P(X \in (-\infty, 0)) + P(X \in [0, y]) \\ &= 0 + (y - 0) \\ &= y. \end{aligned}$$

- If $y > 1$, then

$$\begin{aligned} F(y) &= P(X \le y) \\ &= P(X \in (-\infty, 0)) + P(X \in [0, 1]) + P(X \in (1, y)) \\ &= 0 + (1 - 0) + 0 \\ &= 1. \end{aligned}$$

This function is plotted in Figure 5.1.

What about the pmf of X? In Section 4.1, we defined the pmf of a discrete random variable by $f(x) = P(X = x)$; we then used the pmf to calculate the probabilities of arbitrary events. In the present situation, $P(X = x) = 0$ for every x, so the pmf is not very useful. Instead of representing the probabilites of individual points, we need to represent the probabilities of intervals.

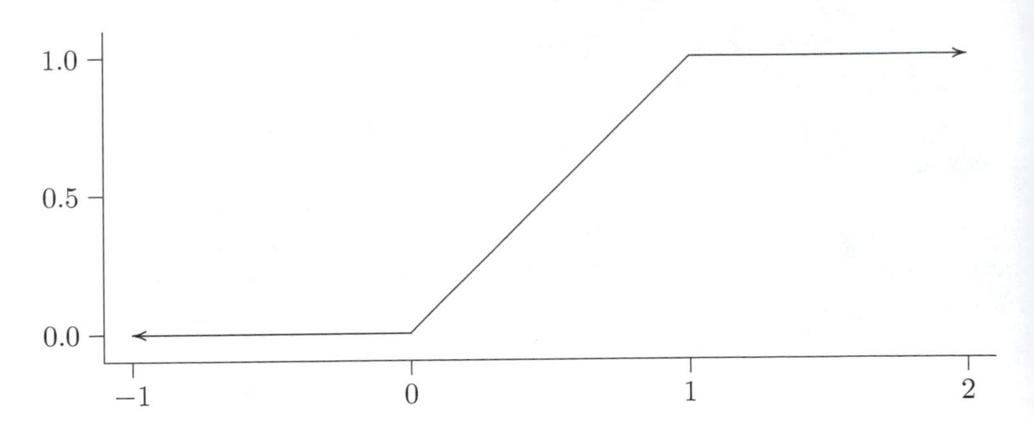

Figure 5.1: The cumulative distribution function of $X \sim \text{Uniform}(0,1)$.

Consider the function

$$f(x) = \left\{ \begin{array}{ll} 0 & x \in (-\infty, 0) \\ 1 & x \in [0,1] \\ 0 & x \in (1, \infty) \end{array} \right\}, \tag{5.3}$$

which is plotted in Figure 5.2. Notice that f is constant on $X(S) = [0,1]$, the set of equally likely possible values, and vanishes elsewhere. If $0 \leq a \leq b \leq 1$, then the area under the graph of f between a and b is the area of a rectangle with sides $b-a$ (horizontal direction) and 1 (vertical direction). Hence, the area in question is

$$(b-a) \cdot 1 = b - a = P(X \in [a,b]),$$

so that the probabilities of intervals can be determined from f. In the next section, we will base our definition of continuous random variables on this observation.

5.2 Basic Concepts

Consider the graph of a function $f : \Re \to \Re$, as depicted in Figure 5.3. Our interest is in the area of the shaded region. This region is bounded by the graph of f, the horizontal axis, and vertical lines at the specified endpoints a and b. We denote this area by $\text{Area}_{[a,b]}(f)$. Our intent is to identify such areas with the probabilities that random variables assume certain values.

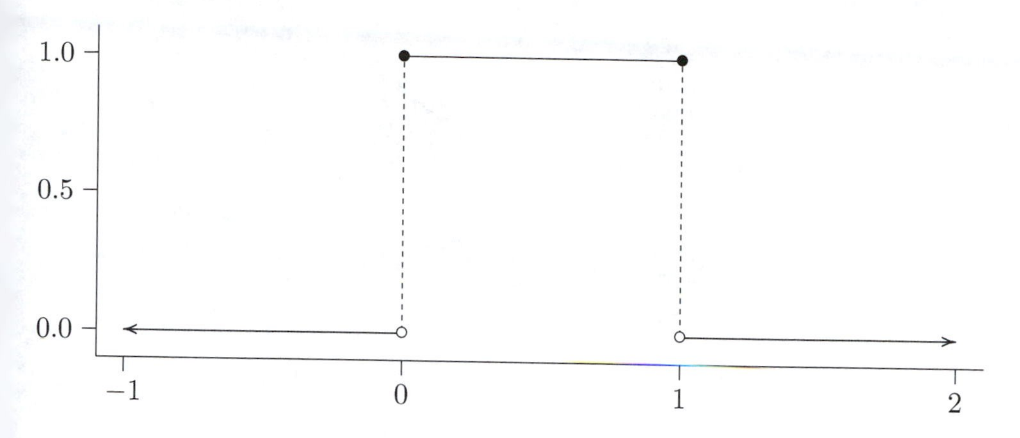

Figure 5.2: The probability density function of $X \sim \text{Uniform}(0,1)$.

For a very few functions, such as the one defined in (5.3), it is possible to determine $\text{Area}_{[a,b]}(f)$ by elementary geometric calculations. For most functions, some knowledge of calculus is required to determine $\text{Area}_{[a,b]}(f)$. Because we assume no previous knowledge of calculus, we will not be concerned with such calculations. Nevertheless, for the benefit of those readers who know some calculus, we find it helpful to borrow some notation and write

$$\text{Area}_{[a,b]}(f) = \int_a^b f(x)dx. \tag{5.4}$$

Readers who have no knowledge of calculus should interpret (5.4) as a definition of its right-hand side, which is pronounced "the integral of f from a to b". Readers who are familiar with the Riemann (or Lebesgue) integral should interpret this notation in its conventional sense.

We now introduce an alternative to the probability mass function.

Definition 5.1 *A probability density function (pdf) is a function $f : \Re \to \Re$ such that*

1. $f(x) \geq 0$ *for every* $x \in \Re$.
2. $\text{Area}_{(-\infty,\infty)}(f) = \int_{-\infty}^{\infty} f(x)dx = 1$.

Notice that the definition of a pdf is analogous to the definition of a pmf. Each is nonnegative and assigns unit probability to the set of possible values. The only difference is that summation in the definition of a pmf is replaced with integration in the case of a pdf.

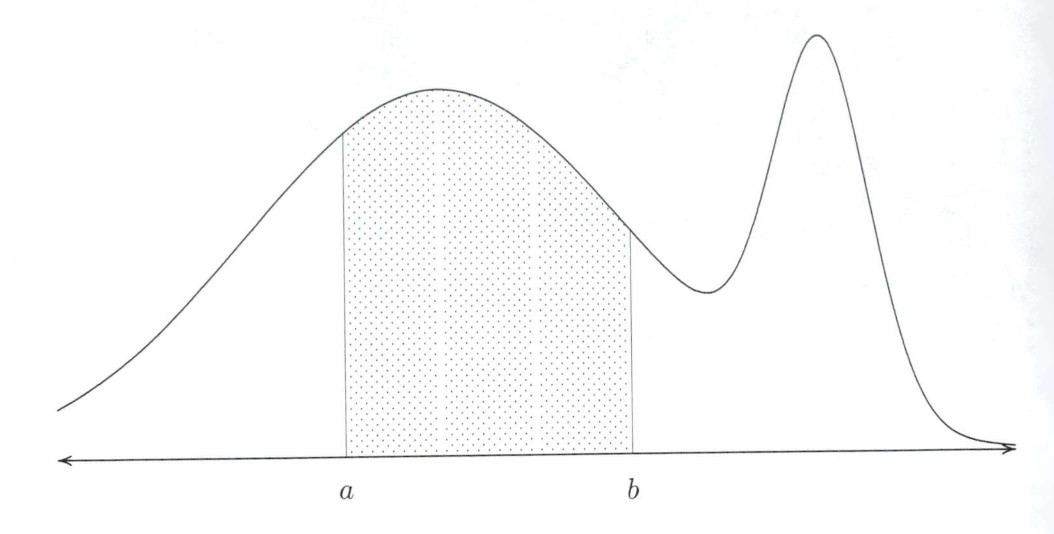

Figure 5.3: A continuous probability density function.

Definition 5.1 was made without reference to a random variable—we now use it to define a new class of random variables.

Definition 5.2 *A random variable X is continuous*[1] *if there exists a probability density function f such that*

$$P(X \in [a,b]) = \int_a^b f(x)dx.$$

It is immediately apparent from Definition 5.2 that the cdf of a continuous random variable X is

$$F(y) = P(X \le y) = P(X \in (-\infty, y]) = \int_{-\infty}^{y} f(x)dx. \tag{5.5}$$

Equation (5.5) should be compared to equation (4.1). In both cases, the value of the cdf at y is represented as the accumulation of values of the pmf/pdf at $x \le y$. The difference lies in the nature of the accumulating process: summation for the discrete case (pmf), integration for the continuous case (pdf).

[1]More precisely, a random variable is continuous if and only if its cdf is a continuous function. Definition 5.2 identifies a proper subset of the continuous random variables, those for which the cdf is an absolutely continuous function. For the concerns of statistical inference, this is a rather esoteric distinction. It is more convenient to define continuous random variables to be those that possess the property that we want to use.

Remark for Calculus Students By applying the Fundamental Theorem of Calculus to (5.5), we deduce that the pdf of a continuous random variable is the derivative of its cdf:

$$\frac{d}{dy}F(y) = \frac{d}{dy}\int_{-\infty}^{y} f(x)dx = f(y).$$

Remark on Notation It may strike the reader as curious that we have used f to denote both the pmf of a discrete random variable and the pdf of a continuous random variable. However, as our discussion of their relation to the cdf is intended to suggest, they play analogous roles. In advanced, *measure-theoretic* courses on probability, one learns that our pmf and pdf are actually two special cases of one general construction.

Likewise, the concept of expectation for continuous random variables is analogous to the concept of expectation for discrete random variables. Because $P(X = x) = 0$ if X is a continuous random variable, the notion of a probability-weighted average is not very useful in the continuous setting. However, if X is a discrete random variable, then $P(X = x) = f(x)$ and a probability-weighted average is identical to a pmf-weighted average. The notion of a pmf-weighted average *is* easily extended to the continuous setting: if X is a continuous random variable, then we introduce a pdf-weighted average of the possible values of X, where averaging is accomplished by replacing summation with integration.

Definition 5.3 *Suppose that X is a continuous random variable with probability density function f. Then the* expected value *of X is*

$$\mu = EX = \int_{-\infty}^{\infty} xf(x)dx,$$

assuming that this quantity exists.

If the function $g : \Re \to \Re$ is such that $Y = g(X)$ is a random variable, then it can be shown that

$$EY = Eg(X) = \int_{-\infty}^{\infty} g(x)f(x)dx,$$

assuming that this quantity exists. In particular,

Definition 5.4 *If $\mu = EX$ exists and is finite, then the* variance *of X is*

$$\sigma^2 = \text{Var}X = E(X-\mu)^2 = \int_{-\infty}^{\infty} (x-\mu)^2 f(x)dx.$$

Thus, for discrete *and* continuous random variables, the expected value is the pmf/pdf-weighted average of the possible values and the variance is the pmf/pdf-weighted average of the squared deviations of the possible values from the expected value.

Because calculus is required to compute the expected value and variance of most continuous random variables, our interest in these concepts lies not in computing them but in understanding what information they convey. We will return to this subject in Chapter 6.

5.3 Elementary Examples

In this section we consider some examples of continuous random variables for which probabilities can be calculated without recourse to calculus.

Example 5.1 *What is the probability that a battery-powered wristwatch will stop with its minute hand positioned between* 10 *and* 20 *minutes past the hour?*

To answer this question, let X denote the number of minutes past the hour to which the minute hand points when the watch stops. Then the possible values of X are $X(S) = [0, 60)$ and it is reasonable to assume that each value is equally likely. We must compute $P(X \in (10, 20))$. Because these values occupy one sixth of the possible values, it should be obvious that the answer is going to be $1/6$.

To obtain the answer using the formal methods of probability, we require a generalization of the Uniform$[0, 1]$ distribution that we studied in Section 5.1. The pdf that describes the notion of equally likely values in the interval $[0, 60)$ is

$$f(x) = \left\{ \begin{array}{cc} 0 & x \in (-\infty, 0) \\ 1/60 & x \in [0, 60) \\ 0 & x \in [60, \infty) \end{array} \right\}. \tag{5.6}$$

To check that f is really a pdf, observe that $f(x) \geq 0$ for every $x \in \Re$ and that

$$\text{Area}_{[0,60)}(f) = (60 - 0)\frac{1}{60} = 1.$$

Notice the analogy between the pdfs (5.6) and (5.3). The present pdf defines the continuous uniform distribution on the interval $[0, 60)$; thus, we describe the present situation by writing $X \sim \text{Uniform}[0, 60)$. To calculate the specified probability, we must determine the area of the shaded region in Figure 5.4, i.e.,

$$P(X \in (10,20)) = \text{Area}_{(10,20)}(f) = (20-10)\frac{1}{60} = \frac{1}{6}.$$

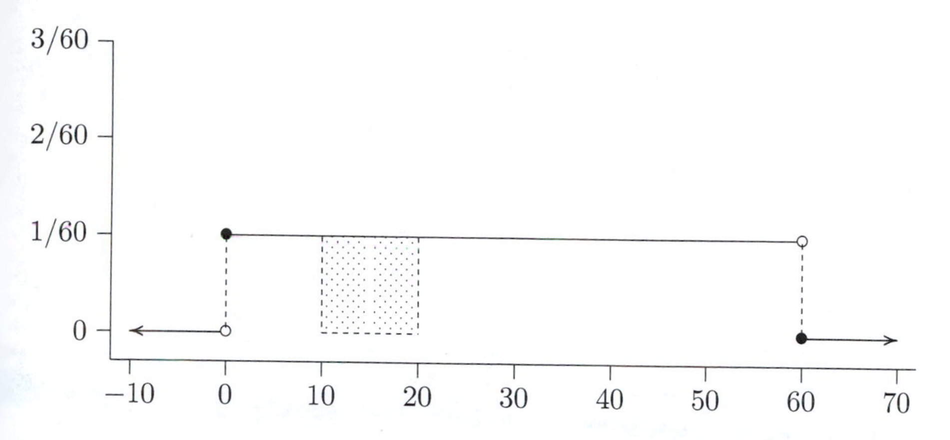

Figure 5.4: The probability density function of $X \sim \text{Uniform}[0,60)$.

Example 5.2 *Consider two battery-powered watches. Let X_1 denote the number of minutes past the hour at which the first watch stops and let X_2 denote the number of minutes past the hour at which the second watch stops. What is the probability that the larger of X_1 and X_2 will be between 30 and 50?*

Here we have two independent random variables, each distributed as Uniform$[0,60)$, and a third random variable,

$$Y = \max(X_1, X_2).$$

Let F denote the cdf of Y. We want to calculate

$$P(30 < Y < 50) = F(50) - F(30).$$

We proceed to derive the cdf of Y. It is evident that $Y(S) = [0,60)$, so $F(y) = 0$ if $y < 0$ and $F(y) = 1$ if $y \geq 1$. If $y \in [0,60)$, then (by the independence of X_1 and X_2)

$$\begin{aligned} F(y) = P(Y \leq y) &= P(\max(X_1,X_2) \leq y) = P(X_1 \leq y,\ X_2 \leq y) \\ &= P(X_1 \leq y) \cdot P(X_2 \leq y) = \frac{y-0}{60-0} \cdot \frac{y-0}{60-0} \\ &= \frac{y^2}{3600}. \end{aligned}$$

Thus, the desired probability is

$$P(30 < Y < 50) = F(50) - F(30) = \frac{50^2}{3600} - \frac{30^2}{3600} = \frac{4}{9}.$$

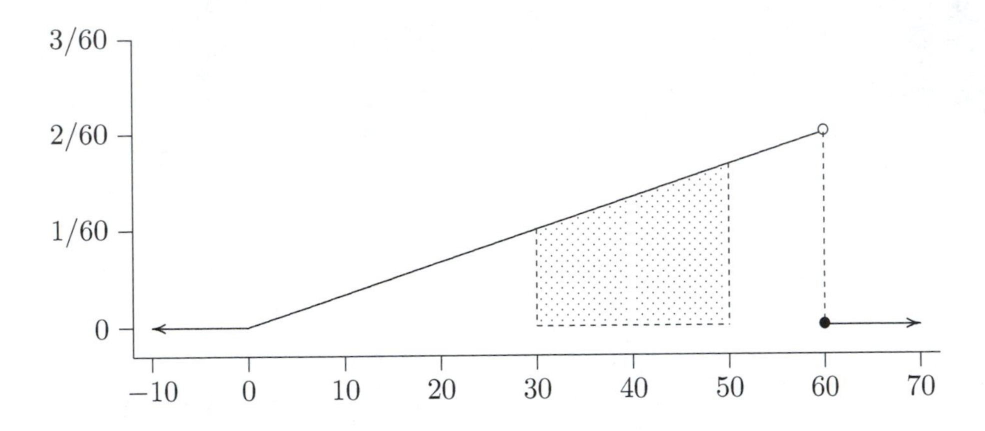

Figure 5.5: The probability density function for Example 5.2.

In preparation for Example 5.3, we claim that the pdf of Y is

$$f(y) = \left\{ \begin{array}{cl} 0 & y \in (-\infty, 0) \\ y/1800 & y \in [0, 60) \\ 0 & y \in [60, \infty) \end{array} \right\},$$

which is graphed in Figure 5.5. To check that f is really a pdf, observe that $f(y) \geq 0$ for every $y \in \Re$ and that

$$\text{Area}_{[0,60)}(f) = \frac{1}{2}(60 - 0)\frac{60}{1800} = 1.$$

To check that f is really the pdf of Y, observe that $f(y) = 0$ if $y \notin [0, 60)$ and that, if $y \in [0, 60)$, then

$$P(Y \in [0, y)) = P(Y \leq y) = F(y) = \frac{y^2}{3600} = \frac{1}{2}(y - 0)\frac{y}{1800} = \text{Area}_{[0,y)}(f).$$

If the pdf had been specified, then instead of deriving the cdf we would have simply calculated

$$P(30 < Y < 50) = \text{Area}_{(30,50)}(f)$$

by any of several convenient geometric arguments.

Example 5.3 *Consider two battery-powered watches. Let X_1 denote the number of minutes past the hour at which the first watch stops and let X_2 denote the number of minutes past the hour at which the second watch stops. What is the probability that the sum of X_1 and X_2 will be between 45 and 75?*

Again we have two independent random variables, each distributed as Uniform$[0, 60)$, and a third random variable,

$$Z = X_1 + X_2.$$

We want to calculate

$$P(45 < Z < 75) = P(Z \in (45, 75)).$$

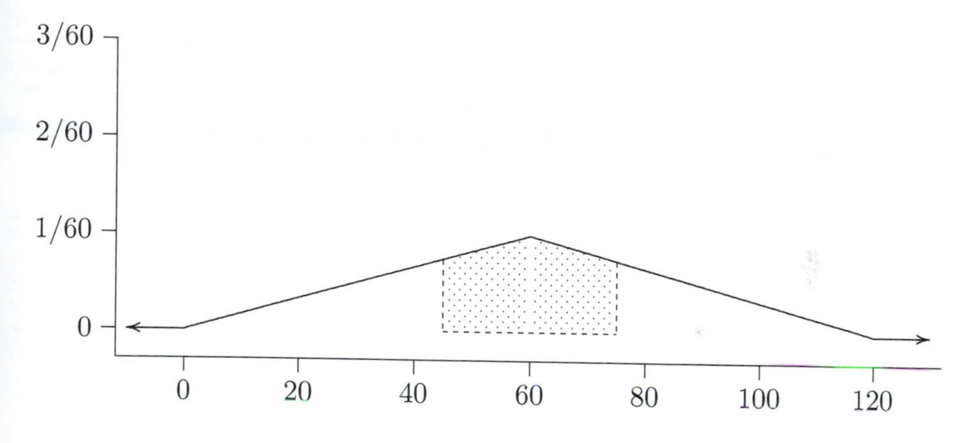

Figure 5.6: The probability density function for Example 5.3.

It is apparent that $Z(S) = [0, 120)$. Although we omit the derivation, it can be determined mathematically that the pdf of Z is

$$f(z) = \left\{ \begin{array}{rl} 0 & z \in (-\infty, 0) \\ z/3600 & z \in [0, 60) \\ (120 - z)/3600 & z \in [60, 120) \\ 0 & z \in [120, \infty) \end{array} \right\}.$$

This pdf is graphed in Figure 5.6, in which it is apparent that the area of the shaded region is

$$\begin{aligned} P(45 < Z < 75) &= P(Z \in (45, 75)) = \text{Area}_{(45,75)}(f) \\ &= 1 - \frac{1}{2}(45 - 0)\frac{45}{3600} - \frac{1}{2}(120 - 75)\frac{120 - 75}{3600} \\ &= 1 - \frac{45^2}{60^2} = \frac{7}{16}. \end{aligned}$$

5.4 Normal Distributions

We now introduce the most important family of distributions in probability and statistics, defined by the familiar *bell-shaped curve.*

Definition 5.5 *A continuous random variable X is normally distributed with mean μ and variance $\sigma^2 > 0$, denoted $X \sim Normal(\mu, \sigma^2)$, if and only if the pdf of X is*

$$f(x) = \frac{1}{\sqrt{2\pi}\sigma} \exp\left[-\frac{1}{2}\left(\frac{x-\mu}{\sigma}\right)^2\right]. \tag{5.7}$$

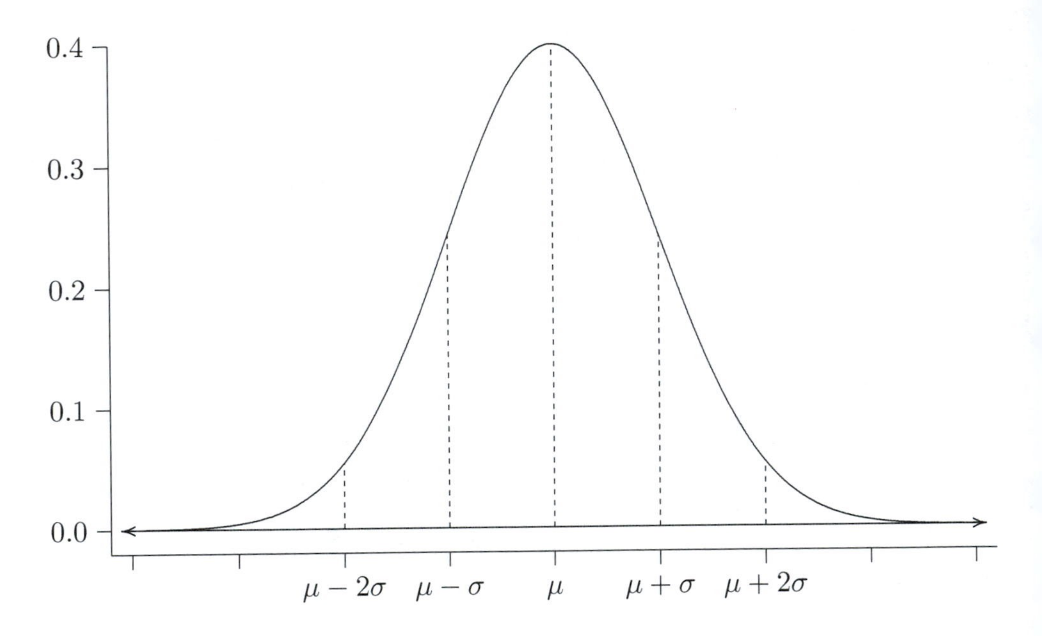

Figure 5.7: The probability density function of $X \sim \text{Normal}(\mu, \sigma^2)$.

Although we will not make extensive use of (5.7), a great many useful properties of normal distributions can be deduced directly from it. Most of the following properties can be discerned in Figure 5.7.

1. $f(x) > 0$. It follows that, for any nonempty interval (a, b),

$$P(X \in (a, b)) = \text{Area}_{(a,b)}(f) > 0,$$

and hence that $X(S) = (-\infty, +\infty)$.

2. f is symmetric about μ, i.e., $f(\mu + x) = f(\mu - x)$.

3. $f(x)$ decreases as $|x - \mu|$ increases. In fact, the decrease is very rapid. We express this by saying that f has very light tails.

4. $P(\mu - \sigma < X < \mu + \sigma) \doteq 0.683$.

5. $P(\mu - 2\sigma < X < \mu + 2\sigma) \doteq 0.954$.

6. $P(\mu - 3\sigma < X < \mu + 3\sigma) \doteq 0.997$.

Notice that there is no one normal distribution, but a 2-parameter family of uncountably many normal distributions. In fact, if we plot μ on a horizontal axis and $\sigma > 0$ on a vertical axis, then there is a distinct normal distribution for each point in the upper half-plane. However, Properties 4–6 above, which hold for *all* choices of μ and σ, suggest that there is a fundamental equivalence between different normal distributions. It turns out that, if one can compute probabilities for any one normal distribution, then one can compute probabilities for any other normal distribution. In anticipation of this fact, we distinguish one normal distribution to serve as a reference distribution:

Definition 5.6 *The standard normal distribution is Normal*$(0,1)$.

The following result is of enormous practical value:

Theorem 5.1 *If* $X \sim \text{Normal}(\mu, \sigma^2)$, *then*

$$Z = \frac{X - \mu}{\sigma} \sim \text{Normal}(0,1).$$

The transformation $Z = (X - \mu)/\sigma$ is called conversion to standard units.

Detailed tables of the standard normal cdf are widely available, as is computer software for calculating specified values. Combined with Theorem 5.1, this availability allows us to easily compute probabilities for arbitrary normal distributions. In the following examples, we let Φ denote the cdf of $Z \sim \text{Normal}(0,1)$ and we make use of the R function `pnorm`.

Example 5.4a *If* $X \sim \text{Normal}(1,4)$, *then what is the probability that* X *assumes a value no more than* 3*?*

Here, $\mu = 1$, $\sigma = 2$, and we want to calculate

$$P(X \le 3) = P\left(\frac{X - \mu}{\sigma} \le \frac{3 - \mu}{\sigma}\right) = P\left(Z \le \frac{3-1}{2} = 1\right) = \Phi(1).$$

We do so in R as follows:

```
> pnorm(1)
[1] 0.8413447
```

Remark The R function pnorm accepts optional arguments that specify a mean and standard deviation. Thus, in Example 5.4a, we could directly evaluate $P(X \le 3)$ as follows:

```
> pnorm(3,mean=1,sd=2)
[1] 0.8413447
```

This option, of course, is not available if one is using a table of the standard normal cdf. Because the transformation to standard units plays such a fundamental role in probability and statistics, we will emphasize computing normal probabilities via the standard normal distribution.

Example 5.4b *If $X \sim Normal(-1, 9)$, then what is the probability that X assumes a value of at least -7?*

Here, $\mu = -1$, $\sigma = 3$, and we want to calculate

$$\begin{aligned} P(X \ge -7) &= P\left(\frac{X-\mu}{\sigma} \ge \frac{-7-\mu}{\sigma}\right) \\ &= P\left(Z \ge \frac{-7+1}{3} = -2\right) \\ &= 1 - P(Z < -2) \\ &= 1 - \Phi(-2). \end{aligned}$$

We do so in R as follows:

```
> 1-pnorm(-2)
[1] 0.9772499
```

Example 5.4c *If $X \sim Normal(2, 16)$, then what is the probability that X assumes a value between 0 and 10?*

Here, $\mu = 2$, $\sigma = 4$, and we want to calculate

$$\begin{aligned} P(0 < X < 10) &= P\left(\frac{0-\mu}{\sigma} < \frac{X-\mu}{\sigma} < \frac{10-\mu}{\sigma}\right) \\ &= P\left(-0.5 = \frac{0-2}{4} < Z < \frac{10-2}{4} = 2\right) \\ &= P(Z < 2) - P(Z < -0.5) \\ &= \Phi(2) - \Phi(-0.5). \end{aligned}$$

We do so in R as follows:

```
> pnorm(2)-pnorm(-.5)
[1] 0.6687123
```

Example 5.4d *If $X \sim Normal(-3, 25)$, then what is the probability that $|X|$ assumes a value greater than 10?*

Here, $\mu = -3$, $\sigma = 5$, and we want to calculate

$$\begin{aligned}
P(|X| > 10) &= P(X > 10 \text{ or } X < -10) \\
&= P(X > 10) + P(X < -10) \\
&= P\left(\frac{X-\mu}{\sigma} > \frac{10-\mu}{\sigma}\right) + P\left(\frac{X-\mu}{\sigma} < \frac{-10-\mu}{\sigma}\right) \\
&= P\left(Z > \frac{10+3}{5} = 2.6\right) + P\left(Z < \frac{-10+3}{5} = -1.2\right) \\
&= 1 - \Phi(2.6) + \Phi(-1.2).
\end{aligned}$$

We do so in R as follows:

```
> 1-pnorm(2.6)+pnorm(-1.2)
[1] 0.1197309
```

Example 5.4e *If $X \sim Normal(4, 16)$, then what is the probability that X^2 assumes a value less than 36?*

Here, $\mu = 4$, $\sigma = 4$, and we want to calculate

$$\begin{aligned}
P(X^2 < 36) &= P(-6 < X < 6) \\
&= P\left(\frac{-6-\mu}{\sigma} < \frac{X-\mu}{\sigma} < \frac{6-\mu}{\sigma}\right) \\
&= P\left(-2.5 = \frac{-6-4}{4} < Z < \frac{6-4}{4} = 0.5\right) \\
&= P(Z < 0.5) - P(Z < -2.5) \\
&= \Phi(0.5) - \Phi(-2.5).
\end{aligned}$$

We do so in R as follows:

```
> pnorm(.5)-pnorm(-2.5)
[1] 0.6852528
```

We defer an explanation of why the family of normal distributions is so important until Section 8.3, concluding the present section with the following useful result:

Theorem 5.2 *If $X_1 \sim Normal(\mu_1, \sigma_1^2)$ and $X_2 \sim Normal(\mu_2, \sigma_2^2)$ are independent, then*

$$X_1 + X_2 \sim Normal(\mu_1 + \mu_2, \sigma_1^2 + \sigma_2^2).$$

5.5 Normal Sampling Distributions

A number of important probability distributions can be derived by considering various functions of normal random variables. These distributions play important roles in statistical inference. They are rarely used to describe data; rather, they arise when analyzing data that is sampled from a normal distribution. For this reason, they are sometimes called *sampling distributions.*

This section collects some definitions of and facts about several important sampling distributions. It is not important to read this section until you encounter these distributions in later chapters; however, it is convenient to collect this material in one easy-to-find place.

Chi-Squared Distributions Suppose that $Z_1, \ldots, Z_n \sim \text{Normal}(0, 1)$ and consider the continuous random variable

$$Y = Z_1^2 + \cdots + Z_n^2.$$

Because each $Z_i^2 \geq 0$, the set of possible values of Y is $Y(S) = [0, \infty)$. We are interested in the distribution of Y.

The distribution of Y belongs to a family of probability distributions called the *chi-squared* family. This family is indexed by a single real-valued parameter, $\nu \in [1, \infty)$, called the *degrees of freedom* parameter. We will denote a chi-squared distribution with ν degrees of freedom by $\chi^2(\nu)$. Figure 5.8 displays the pdfs of several chi-squared distributions.

Now we can state a theorem that characterizes the behavior of Y.

Theorem 5.3 *If $Z_1, \ldots, Z_n \sim Normal(0, 1)$ and $Y = Z_1^2 + \cdots + Z_n^2$, then $Y \sim \chi^2(n)$.*

In principle, Theorem 5.3 allows one to compute the probabilities of events defined by values of Y, e.g., $P(Y > 4.5)$. In practice, this requires evaluating the cdf of $\chi^2(\nu)$, a function for which there is no simple formula. Fortunately, there exist efficient algorithms for numerically evaluating these cdfs. The `R` function `pchisq` returns values of the cdf of any specified chi-squared distribution. For example, if $Y \sim \chi^2(2)$, then $P(Y > 4.5)$ is

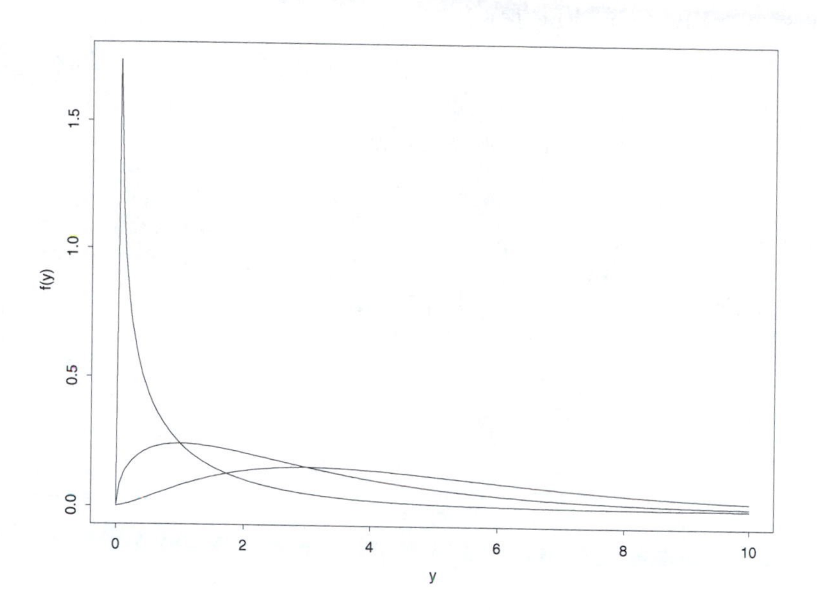

Figure 5.8: The probability density functions of $Y \sim \chi^2(\nu)$ for $\nu = 1, 3, 5$.

```
> 1-pchisq(4.5,df=2)
[1] 0.1053992
```

Finally, if $Z_i \sim \text{Normal}(0, 1)$, then

$$EZ_i^2 = \text{Var}\, Z_i + (EZ_i)^2 = 1.$$

It follows that

$$EY = E\left(\sum_{i=1}^{n} Z_i^2\right) = \sum_{i=1}^{n} EZ_i^2 = \sum_{i=1}^{n} 1 = n;$$

thus,

Corollary 5.1 *If $Y \sim \chi^2(n)$, then $EY = n$.*

Student's t Distributions Now let $Z \sim \text{Normal}(0, 1)$ and $Y \sim \chi^2(\nu)$ be independent random variables and consider the continuous random variable

$$T = \frac{Z}{\sqrt{Y/\nu}}.$$

The set of possible values of T is $T(S) = (-\infty, \infty)$. We are interested in the distribution of T.

Definition 5.7 *The distribution of T is called a t distribution with ν degrees of freedom. We will denote this distribution by $t(\nu)$.*

The standard normal distribution is symmetric about the origin; i.e., if $Z \sim \text{Normal}(0, 1)$, then $-Z \sim \text{Normal}(0, 1)$. It follows that $T = Z/\sqrt{Y/\nu}$ and $-T = -Z/\sqrt{Y/\nu}$ have the same distribution. Hence, if p is the pdf of T, then it must be that $p(t) = p(-t)$. Thus, t pdfs are symmetric about the origin, just like the standard normal pdf.

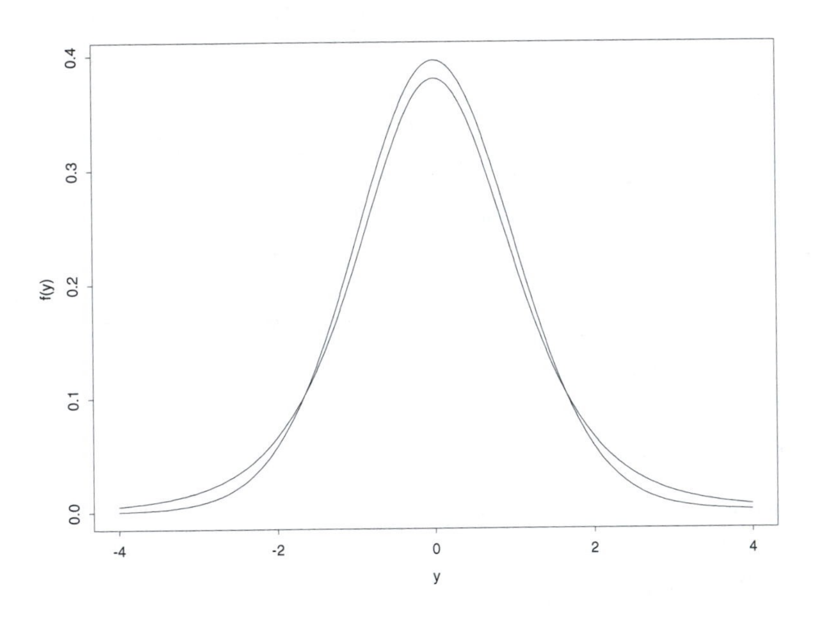

Figure 5.9: The probability density functions of $T \sim t(\nu)$ for $\nu = 5, 30$.

Figure 5.9 displays the pdfs of two t distributions. They can be distinguished by virtue of the fact that the variance of $t(\nu)$ decreases as ν increases. It may strike you that t pdfs closely resemble normal pdfs. In fact, the standard normal pdf is a limiting case of the t pdfs:

Theorem 5.4 *Let F_ν denote the cdf of $t(\nu)$ and let Φ denote the cdf of Normal$(0, 1)$. Then*

$$\lim_{\nu \to \infty} F_\nu(y) = \Phi(y)$$

for every $y \in (-\infty, \infty)$.

Thus, when ν is sufficiently large ($\nu > 40$ is a reasonable rule of thumb), $t(\nu)$ is approximately Normal$(0,1)$ and probabilities involving the former can be approximated by probabilities involving the latter.

In R, it is just as easy to calculate $t(\nu)$ probabilities as it is to calculate Normal$(0,1)$ probabilities. The R function pt returns values of the cdf of any specified t distribution. For example, if $T \sim t(14)$, then $P(T \leq -1.5)$ is

```
> pt(-1.5,df=14)
[1] 0.07791266
```

F Distributions Finally, let $Y_1 \sim \chi^2(\nu_1)$ and $Y_2 \sim \chi^2(\nu_2)$ be independent random variables and consider the continuous random variable

$$F = \frac{Y_1/\nu_1}{Y_2/\nu_2}.$$

Because $Y_i \geq 0$, the set of possible values of F is $F(S) = [0, \infty)$. We are interested in the distribution of F.

Definition 5.8 *The distribution of F is called an F distribution with ν_1 and ν_2 degrees of freedom. We will denote this distribution by $F(\nu_1, \nu_2)$. It is customary to call ν_1 the "numerator" degrees of freedom and ν_2 the "denominator" degrees of freedom.*

Figure 5.10 displays the pdfs of several F distributions.

There is an important relationship between t and F distributions. To anticipate it, suppose that $Z \sim$ Normal$(0,1)$ and $Y_2 \sim \chi^2(\nu_2)$ are independent random variables. Then $Y_1 = Z^2 \sim \chi^2(1)$, so

$$T = \frac{Z}{\sqrt{Y_2/\nu_2}} \sim t\left(\nu_2\right)$$

and

$$T^2 = \frac{Z^2}{Y_2/\nu_2} = \frac{Y_1/1}{Y_2/\nu_2} \sim F\left(1, \nu_2\right).$$

More generally,

Theorem 5.5 *If $T \sim t(\nu)$, then $T^2 \sim F(1, \nu)$.*

The R function pf returns values of the cdf of any specified F distribution. For example, if $F \sim F(2, 27)$, then $P(F > 2.5)$ is

```
> 1-pf(2.5,df1=2,df2=27)
[1] 0.1008988
```

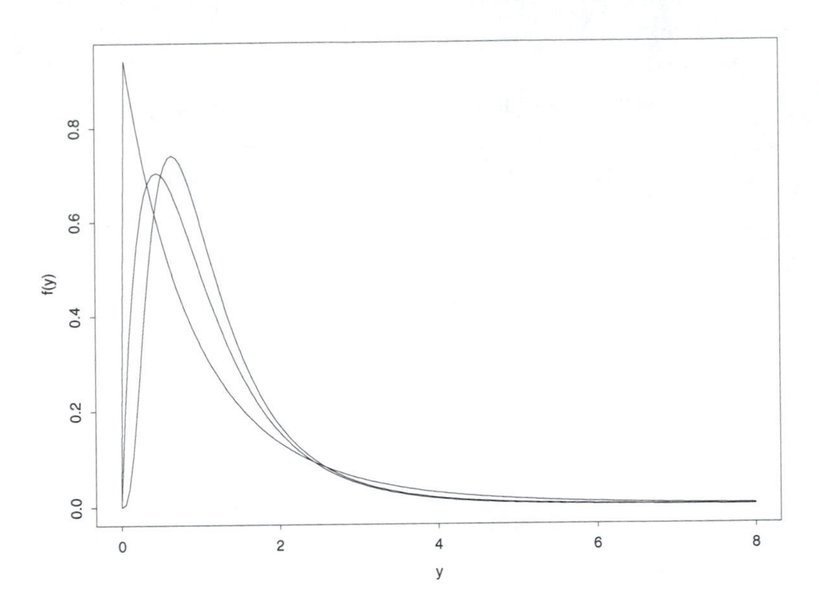

Figure 5.10: The probability density functions of $F \sim F(\nu_1, \nu_2)$ for $(\nu_1, \nu_2) = (2, 12), (4, 20), (9, 10)$.

5.6 Exercises

1. In this problem you will be asked to examine two equations. Several symbols from each equation will be identified. Your task will be to decide which symbols represent real numbers and which symbols represent functions. If a symbol represents a function, then you should state the domain and the type of labels that the function assigns to elements of the domain.

   Recall: A function is a rule of assignment. For example, when I grade your test, I assign a numeric value to your name. Grading is a function that assigns real numbers (the labels) to students (the domain).

   (a) In the equation $p = P(Z > 1.96)$, please identify each of the following symbols as a real number or a function:

      i. p
      ii. P
      iii. Z

(b) In the equation $\sigma^2 = E(X - \mu)^2$, please identify each of the following symbols as a real number or a function:

i. σ

ii. E

iii. X

iv. μ

2. Suppose that X is a continuous random variable with probability density function (pdf) f defined as follows:

$$f(x) = \left\{ \begin{array}{lll} 0 & \text{if} & x < 1 \\ 2(x-1) & \text{if} & 1 \le x \le 2 \\ 0 & \text{if} & x > 2 \end{array} \right\}.$$

(a) Graph f.

(b) Verify that f is a pdf.

(c) Compute $P(1.50 < X < 1.75)$.

3. Consider the function $f : \Re \to \Re$ defined by

$$f(x) = \left\{ \begin{array}{ll} 0 & x < 0 \\ cx & 0 < x < 1.5 \\ c(3-x) & 1.5 < x < 3 \\ 0 & x > 3 \end{array} \right\},$$

where c is an undetermined constant.

(a) For what value of c is f a probability density function?

(b) Suppose that a continuous random variable X has probability density function f. Compute EX. (Hint: Draw a picture of the pdf.)

(c) Compute $P(X > 2)$.

(d) Suppose that $Y \sim \text{Uniform}(0, 3)$. Which random variable has the larger variance, X or Y? (Hint: Draw a picture of the two pdfs.)

(e) Determine and graph the cumulative distribution function of X.

4. Imagine throwing darts at a circular dart board, B. Let us measure the dart board in units for which the radius of B is 1, so that the area of B is π. Suppose that the darts are thrown in such a way that they are certain to hit a point in B, and that each point in B is equally

likely to be hit. Thus, if $A \subset B$, then the probability of hitting a point in A is

$$P(A) = \frac{\text{area}(A)}{\text{area}(B)} = \frac{\text{area}(A)}{\pi}.$$

Define the random variable X to be the distance from the center of B to the point that is hit.

(a) What are the possible values of X?

(b) Compute $P(X \leq 0.5)$.

(c) Compute $P(0.5 < X \leq 0.7)$.

(d) Determine and graph the cumulative distribution function of X.

(e) [Optional—for those who know a little calculus.] Determine and graph the probability density function of X.

5. Imagine throwing darts at a triangular dart board,

$$B = \{(x, y) \, : \, 0 \leq y \leq x \leq 1\}.$$

Suppose that the darts are thrown in such a way that they are certain to hit a point in B, and that each point in B is equally likely to be hit. Define the random variable X to be the value of the x-coordinate of the point that is hit, and define the random variable Y to be the value of the y-coordinate of the point that is hit.

(a) Draw a picture of B.

(b) Compute $P(X \leq 0.5)$.

(c) Determine and graph the cumulative distribution function of X.

(d) Are X and Y independent?

6. Suppose that X has a normal distribution. Does $|X|$ have a normal distribution? Why or why not?

7. Let X be a normal random variable with mean $\mu = -5$ and standard deviation $\sigma = 10$. Compute the following:

(a) $P(X < 0)$

(b) $P(X > 5)$

(c) $P(-3 < X < 7)$

(d) $P(|X + 5| < 10)$

(e) $P(|X-3|>2)$

8. Suppose that $X_1 \sim \text{Normal}(1,9)$ and $X_2 \sim \text{Normal}(3,16)$ are independent. Determine the mean and variance of each of the following normal random variables:

   (a) $X_1 + X_2$

   (b) $-X_2$

   (c) $X_1 - X_2$

   (d) $2X_1$

   (e) $2X_1 - 2X_2$

Chapter 6

Quantifying Population Attributes

The distribution of a random variable is a mathematical abstraction of the possible outcomes of an experiment. Indeed, having identified a random variable of interest, we will often refer to its distribution as *the population.* If one's goal is to represent an entire population, then one can hardly do better than to display its entire probability mass or density function. Usually, however, one is interested in specific attributes of a population. This is true if only because it is through specific attributes that one comprehends the entire population, but it is also easier to draw inferences about a specific population attribute than about the entire population. Accordingly, this chapter examines several population attributes that are useful in statistics.

We will be especially concerned with measures of centrality and measures of dispersion. The former provide quantitative characterizations of where the "middle" of a population is located; the latter provide quantitative characterizations of how widely the population is spread. We have already introduced one important measure of centrality, the expected value of a random variable (the population mean, μ), and one important measure of dispersion, the standard deviation of a random variable (the population standard deviation, σ). This chapter discusses these measures in greater depth and introduces other, complementary measures.

6.1 Symmetry

We begin by considering the following question:

> *Where is the "middle" of a normal distribution?*

It is quite evident from Figure 5.7 that there is only one plausible answer to this question: if $X \sim \text{Normal}(\mu, \sigma^2)$, then the "middle" of the distribution of X is μ.

Let f denote the pdf of X. To understand why μ is the only plausible middle of f, recall a property of f that we noted in Section 5.4: for any x, $f(\mu+x) = f(\mu-x)$. This property states that f is *symmetric* about μ. It is the property of symmetry that restricts the plausible locations of "middle" to the central value μ.

To generalize the above example of a measure of centrality, we introduce an important qualitative property that a population may or may not possess:

Definition 6.1 *Let X be a continuous random variable with probability density function f. If there exists a value $\theta \in \Re$ such that*

$$f(\theta + x) = f(\theta - x)$$

for every $x \in \Re$, then X is a symmetric random variable and θ is its center of symmetry.

We have already noted that $X \sim \text{Normal}(\mu, \sigma^2)$ has center of symmetry μ. Another example of symmetry is illustrated in Figure 6.1: $X \sim \text{Uniform}[a, b]$ has center of symmetry $(a + b)/2$.

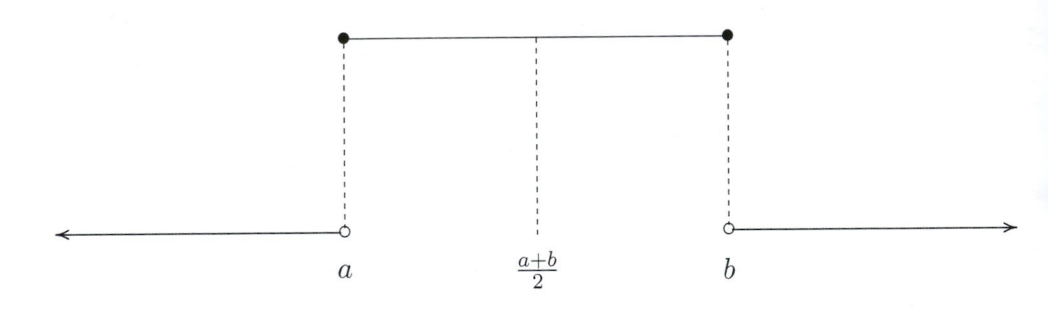

Figure 6.1: $X \sim \text{Uniform}[a, b]$ has center of symmetry $(a + b)/2$.

For symmetric random variables, the center of symmetry is the only plausible measure of centrality—of where the "middle" of the distribution is located. Symmetry will play an important role in our study of statistical inference. Our primary concern will be with continuous random variables, but the concept of symmetry can be used with other random variables as well. Here is a general definition:

Definition 6.2 *Let X be a random variable. If there exists a value $\theta \in \Re$ such that the random variables $X - \theta$ and $\theta - X$ have the same distribution, then X is a symmetric random variable and θ is its center of symmetry.*

Suppose that we attempt to compute the expected value of a symmetric random variable X with center of symmetry θ. Thinking of the expected value as a weighted average, we see that each $\theta + x$ will be weighted precisely as much as the corresponding $\theta - x$. Thus, if the expected value exists (there are a few pathological random variables for which the expected value is undefined), then it must equal the center of symmetry, i.e., $EX = \theta$. Of course, we have already seen that this is the case for $X \sim \text{Normal}(\mu, \sigma^2)$ and for $X \sim \text{Uniform}[a, b]$.

6.2 Quantiles

In this section we introduce population quantities that can be used for a variety of purposes. As in Section 6.1, these quantities are most easily understood in the case of continuous random variables.

Definition 6.3 *Let X be a continuous random variable and let $\alpha \in (0, 1)$. If $q = q(X; \alpha)$ is such that $P(X < q) = \alpha$ and $P(X > q) = 1 - \alpha$, then q is called an α* quantile *of X.*

If we express the probabilities in Definition 6.3 as percentages, then we see that q is the 100α percentile of the distribution of X.

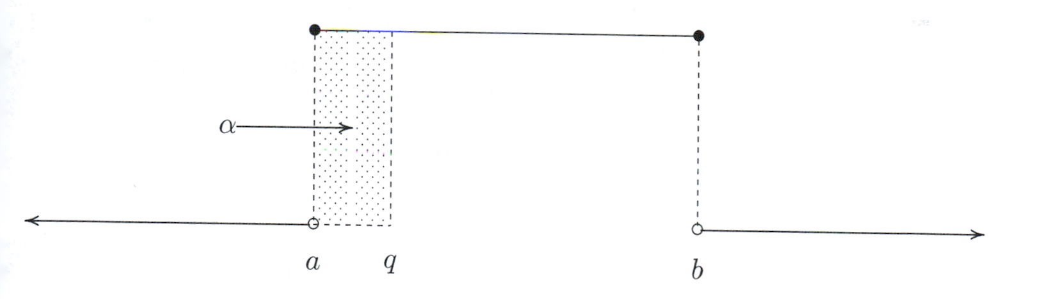

Figure 6.2: A quantile of a uniform distribution.

Example 6.1 Suppose that $X \sim \text{Uniform}[a,b]$ has pdf f, depicted in Figure 6.2. Then q is the value in (a,b) for which

$$\alpha = P(X < q) = \text{Area}_{[a,q]}(f) = (q-a) \cdot \frac{1}{b-a},$$

i.e., $q = a + \alpha(b-a)$. This expression is easily interpreted: to the lower endpoint a, add $100\alpha\%$ of the distance $b-a$ to obtain the 100α percentile.

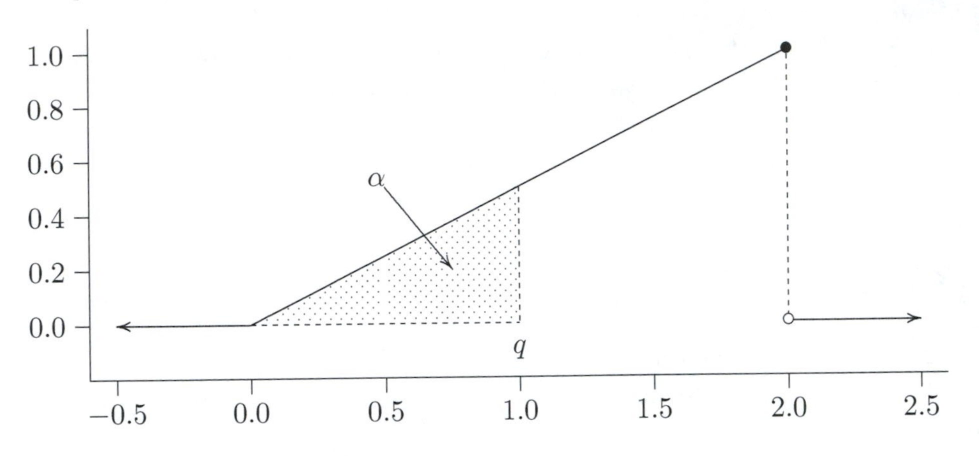

Figure 6.3: A quantile of another distribution.

Example 6.2 Suppose that X has pdf

$$f(x) = \left\{ \begin{array}{cc} x/2 & x \in [0,2] \\ 0 & \text{otherwise} \end{array} \right\},$$

depicted in Figure 6.3. Then q is the value in $(0,2)$ for which

$$\alpha = P(X < q) = \text{Area}_{[a,q]}(f) = \frac{1}{2} \cdot (q-0) \cdot \left(\frac{q}{2} - 0\right) = \frac{q^2}{4},$$

i.e., $q = 2\sqrt{\alpha}$.

Example 6.3 Suppose that $X \sim \text{Normal}(0,1)$ has cdf Φ. Then q is the value in $(-\infty, \infty)$ for which $\alpha = P(X < q) = \Phi(q)$, i.e., $q = \Phi^{-1}(\alpha)$. Unlike the previous examples, we cannot compute q by elementary calculations. Fortunately, the `R` function `qnorm` computes quantiles of normal distributions. For example, we compute the $\alpha = 0.95$ quantile of X as follows:

```
> qnorm(.95)
[1] 1.644854
```

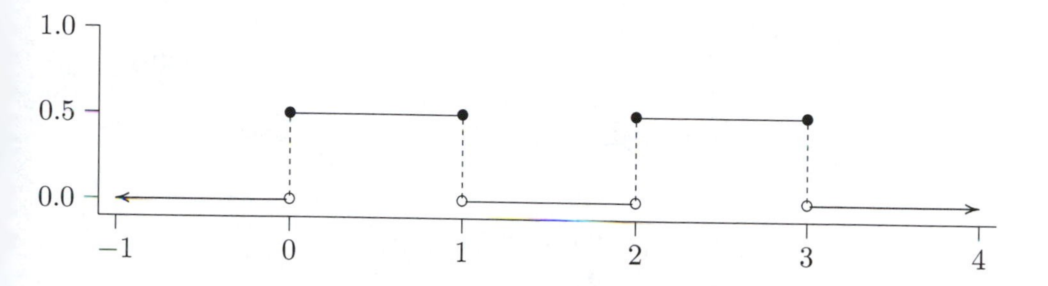

Figure 6.4: A distribution for which the $\alpha = 0.5$ quantile is not unique.

Example 6.4 Suppose that X has pdf

$$f(x) = \left\{ \begin{array}{cl} 1/2 & x \in [0,1] \cup [2,3] \\ 0 & \text{otherwise} \end{array} \right\},$$

depicted in Figure 6.4. Notice that $P(X \in [0,1]) = 0.5$ and $P(X \in [2,3]) = 0.5$. If $\alpha \in (0, 0.5)$, then we can use the same reasoning that we employed in Example 6.1 to deduce that $q = 2\alpha$. Similarly, if $\alpha \in (0.5, 1)$, then $q = 2 + 2(\alpha - 0.5) = 2\alpha + 1$. However, if $\alpha = 0.5$, then we encounter an ambiguity: the equalities $P(X < q) = 0.5$ and $P(X > q) = 0.5$ hold for *any* $q \in [1, 2]$. Accordingly, any $q \in [1,2]$ is an $\alpha = 0.5$ quantile of X. Thus, quantiles are not always unique.

To avoid confusion when a quantile is not unique, it is nice to have a convention for selecting one of the possible quantile values. In the case that $\alpha = 0.5$, there is a universal convention:

Definition 6.4 *The midpoint of the interval of all values of the* $\alpha = 0.5$ *quantile is called the* population median.

In Example 6.4, the population median is $q = 1.5$.

Working with the quantiles of a continuous random variable X is straightforward because $P(X = q) = 0$ for any choice of q. This means that $P(X < q) + P(X > q) = 1$; hence, if $P(X < q) = \alpha$, then $P(X > q) = 1 - \alpha$. Furthermore, it is always possible to find a q for which $P(X < q) = \alpha$. This is not the case if X is discrete.

Example 6.5 *Let X be a discrete random variable that assumes values in the set $\{1, 2, 3\}$ with probabilities $p(1) = 0.4$, $p(2) = 0.4$, and $p(3) = 0.2$. What is the median of X?*

Imagine accumulating probability as we move from $-\infty$ to ∞. At what point do we find that we have acquired half of the total probability? The answer is that we pass from having 40% of the probability to having 80% of the probability as we occupy the point $q = 2$. It makes sense to declare this value to be the median of X.

Here is another argument that appeals to Definition 6.3. If $q < 2$, then $P(X > q) = 0.6 > 0.5$. Hence, it would seem that the population median should not be less than 2. Similarly, if $q > 2$, then $P(X < q) = 0.8 > 0.5$. Hence, it would seem that the population median should not be greater than 2. We conclude that the population median should equal 2. But notice that $P(X < 2) = 0.4 < 0.5$ and $P(X > 2) = 0.2 < 0.5$! We conclude that Definition 6.3 will not suffice for discrete random variables. However, we can generalize the reasoning that we have just employed as follows:

Definition 6.5 *Let X be a random variable and let $\alpha \in (0, 1)$. If $q = q(X; \alpha)$ is such that $P(X < q) \leq \alpha$ and $P(X > q) \leq 1 - \alpha$, then q is called an α* quantile *of X.*

The remainder of this section describes how quantiles are often used to measure centrality and dispersion. The following three quantiles will be of particular interest:

Definition 6.6 *Let X be a random variable. The first, second, and third* quartiles *of X, denoted $q_1(X)$, $q_2(X)$, and $q_3(X)$, are the $\alpha = 0.25$, $\alpha = 0.50$, and $\alpha = 0.75$ quantiles of X. The second quartile is also called the* median *of X.*

6.2.1 The Median of a Population

If X is a symmetric random variable with center of symmetry θ, then

$$P(X < \theta) = P(X > \theta) = \frac{1 - P(X = \theta)}{2} \leq \frac{1}{2}$$

and $q_2(X) = \theta$. Even if X is not symmetric, the median of X is an excellent way to define the "middle" of the population. Many statistical procedures use the median as a measure of centrality.

Example 6.6 One useful property of the median is that it is rather insensitive to the influence of extreme values that occur with small probability. For example, let X_k denote a discrete random variable that assumes values in $\{-1, 0, 1, 10^k\}$ for $k = 1, 2, 3, \ldots$. Suppose that X_k has the following pmf:

x	$p_k(x)$
-1	0.19
0	0.60
1	0.19
10^k	0.02

Most of the probability (98%) is concentrated on the values $\{-1, 0, 1\}$. This probability is centered at $x = 0$. A small amount of probability is concentrated at a large value, $x = 10^k$. If we want to treat these large values as aberrations (perhaps our experiment produces a physically meaningful value $x \in \{-1, 0, 1\}$ with probability 0.98, but our equipment malfunctions and produces a physically meaningless value $x = 10^k$ with probability 0.02), then we might prefer to declare that $x = 0$ is the central value of X. In fact, no matter how large we choose k, the median refuses to be distracted by the aberrant value: $P(X < 0) = 0.19$ and $P(X > 0) = 0.21$, so the median of X is $q_2(X) = 0$.

6.2.2 The Interquartile Range of a Population

Now we turn our attention from the problem of measuring centrality to the problem of measuring dispersion. Can we use quantiles to quantify how widely spread are the values of a random variable? A natural approach is to choose two values of α and compute the corresponding quantiles. The distance between these quantiles is a measure of dispersion.

To avoid comparing apples and oranges, let us agree on which two values of α we will choose. Statisticians have developed a preference for $\alpha = 0.25$ and $\alpha = 0.75$, in which case the corresponding quantiles are the first and third quartiles.

Definition 6.7 *Let X be a random variable with first and third quartiles q_1 and q_3. The* interquartile range *of X is the quantity*

$$iqr(X) = q_3 - q_1.$$

If X is a continuous random variable, then $P(q_1 < X < q_3) = 0.5$, so the interquartile range is the interval of values on which is concentrated the central 50% of the probability.

Like the median, the interquartile range is rather insensitive to the influence of extreme values that occur with small probability. In Example 6.6, the central 50% of the probability is concentrated on the single value $x = 0$. Hence, the interquartile range is $0 - 0 = 0$, regardless of where the aberrant 2% of the probability is located.

6.3 The Method of Least Squares

Let us return to the case of a symmetric random variable X, in which case the "middle" of the distribution is unambiguously the center of symmetry θ. Given this measure of centrality, how might we construct a measure of dispersion? One possibility is to measure how far a "typical" value of X lies from its central value, i.e., to compute $E|X - \theta|$. This possibility leads to several remarkably fertile approaches to describing both dispersion and centrality.

Given a designated central value c and another value x, we say that the *absolute deviation* of x from c is $|x - c|$ and that the *squared deviation* of x from c is $(x - c)^2$. The magnitude of a typical absolute deviation is $E|X - c|$ and the magnitude of a typical squared deviation is $E(X - c)^2$. A natural approach to measuring centrality is to choose a value of c that typically results in small deviations, i.e., to choose c either to minimize $E|X - c|$ or to minimize $E(X - c)^2$. The second possibility is a simple example of the *method of least squares.*

Measuring centrality by minimizing the magnitude of a typical absolute or squared deviation results in two familiar quantities:

Theorem 6.1 *Let X be a random variable with population median q_2 and population mean $\mu = EX$. Then*

1. *The value of c that minimizes $E|X - c|$ is $c = q_2$.*

2. *The value of c that minimizes $E(X - c)^2$ is $c = \mu$.*

It follows that medians are naturally associated with absolute deviations and that means are naturally associated with squared deviations. Having discussed the former in Section 6.2.1, we now turn to the latter.

6.3.1 The Mean of a Population

Imagine creating a physical model of a probability distribution by distributing weights along the length of a board. The location of the weights are the values of the random variable and the weights represent the probabilities of those values. After gluing the weights in place, we position the board atop a fulcrum. How must the fulcrum be positioned in order that the board be perfectly balanced? It turns out that one should position the fulcrum at the mean of the probability distribution. For this reason, the expected value of a random variable is sometimes called its *center of mass.*

Thus, like the population median, the population mean has an appealing interpretation that commends its use as a measure of centrality. If X is a symmetric random variable with center of symmetry θ, then $\mu = EX = \theta$ and $q_2 = q_2(X) = \theta$, so the population mean and the population median agree. In general, this is not the case. If X is not symmetric, then one should think carefully about whether one is interested in the population mean or the population median. Of course, computing both measures and examining the discrepancy between them may be highly instructive. In particular, if $EX \neq q_2(X)$, then X is not a symmetric random variable.

In Section 6.2.1 we noted that the median is rather insensitive to the influence of extreme values that occur with small probability. The mean lacks this property. In Example 6,

$$EX_k = -0.19 + 0.00 + 0.19 + 10^k \cdot 0.02 = 2 \cdot 10^{k-2},$$

which equals 0.2 if $k = 1$, 2 if $k = 2$, 20 if $k = 3$, 200 if $k = 4$, and so on. The population mean follows the aberrant value toward infinity as k increases.

6.3.2 The Standard Deviation of a Population

Suppose that X is a random variable with $EX = \mu$ and $\operatorname{Var} X = \sigma^2$. If we adopt the method of least squares, then we obtain $c = \mu$ as our measure of centrality, in which case the magnitude of a typical squared deviation is $E(X - \mu)^2 = \sigma^2$, the population variance. The variance measures dispersion in squared units. For example, if X measures length in meters, then $\operatorname{Var} X$ is measured in meters squared. If, as in Section 6.2.2, we prefer to measure dispersion in the original units of measurement, then we must take the square root of the variance. Accordingly, we will emphasize the population standard deviation, σ, as a measure of dispersion.

Just as it is natural to use the median and the interquartile range together, so is it natural to use the mean and the standard deviation together. In the case of a symmetric random variable, the median and the mean agree. However, the interquartile range and the standard deviation measure dispersion in two fundamentally different ways. To gain insight into their relation to each other, suppose that $X \sim \text{Normal}(0,1)$, in which case the population standard deviation is $\sigma = 1$. We use R to compute $\text{iqr}(X)$:

```
> qnorm(.75)-qnorm(.25)
[1] 1.348980
```

We have derived a useful fact: *the interquartile range of a normal random variable is approximately* 1.35 *standard deviations.* If we encounter a random variable for which this is not the case, then that random variable is not normally distributed.

Like the mean, the standard deviation is sensitive to the influence of extreme values that occur with small probability. Consider Example 6. The variance of X_k is

$$\begin{aligned}\sigma_k^2 &= EX_k^2 - (EX_k)^2 = \left(0.19 + 0.00 + 0.19 + 100^k \cdot 0.02\right) - \left(2 \cdot 10^{k-2}\right)^2 \\ &= 0.38 + 2 \cdot 100^{k-1} - 4 \cdot 100^{k-2} = 0.38 + 196 \cdot 100^{k-2},\end{aligned}$$

so $\sigma_1 = \sqrt{2.34}$, $\sigma_2 = \sqrt{196.38}$, $\sigma_3 = \sqrt{19600.38}$, and so on. The population standard deviation tends toward infinity as the aberrant value tends toward infinity.

6.4 Exercises

1. Refer to the random variable X defined in Exercise 5.6.2. Compute the following two quantities: $q_2(X)$, the population median; and $\text{iqr}(X)$, the population interquartile range.

2. Consider the function $g : \Re \to \Re$ defined by

$$g(x) = \left\{ \begin{array}{cc} 0 & x < 0 \\ x & x \in [0,1] \\ 1 & x \in [1,2] \\ 3-x & x \in [2,3] \\ 0 & x > 3 \end{array} \right\}.$$

   Let $f(x) = cg(x)$, where c is an undetermined constant.

   (a) For what value of c is f a probability density function?

   (b) Suppose that a continuous random variable X has probability density function f. Compute $P(1.5 < X < 2.5)$.

   (c) Compute EX.

(d) Let F denote the cumulative distribution function of X. Compute $F(1)$.

(e) Determine the 0.90 quantile of f.

3. Suppose that X is a continuous random variable with probability density function

$$f(x) = \left\{ \begin{array}{cc} 0 & x < 0 \\ x & x \in (0,1) \\ (3-x)/4 & x \in (1,3) \\ 0 & x > 3 \end{array} \right\}.$$

(a) Compute $q_2(X)$, the population median.

(b) Which is greater, $q_2(X)$ or EX? Explain your reasoning.

(c) Compute $P(0.5 < X < 1.5)$.

(d) Compute $\text{iqr}(X)$, the population interquartile range.

4. Consider the dart-throwing experiment described in Exercise 5.6.5 and compute the following quantities:

(a) $q_2(X)$

(b) $q_2(Y)$

(c) $\text{iqr}(X)$

(d) $\text{iqr}(Y)$

5. Lynn claims that Lulu is the cutest dog in the world. Slightly more circumspect, Michael allows that Lulu is "one in a million." Seizing the opportunity to revel in Lulu's charm, Lynn devises a procedure for measuring CCQ (canine cuteness quotient), which she calibrates so that CCQ $\sim$ Normal$(100, 400)$. Assuming that Michael is correct, what is Lulu's CCQ score?

6. A random variable $X \sim \text{Uniform}(5, 15)$ has population mean $\mu = EX = 10$ and population variance $\sigma^2 = \text{Var}\, X = 225$. Let Y denote a normal random variable with the same mean and variance.

(a) Consider X. What is the ratio of its interquartile range to its standard deviation?

(b) Consider Y. What is the ratio of its interquartile range to its standard deviation?

7. Identify each of the following statements as *True* or *False*. Briefly explain each of your answers.

   (a) For every symmetric random variable X, the median of X equals the average of the first and third quartiles of X.

   (b) For every random variable X, the interquartile range of X is greater than the standard deviation of X.

   (c) For every random variable X, the expected value of X lies between the first and third quartile of X.

   (d) If the standard deviation of a random variable equals zero, then so does its interquartile range.

   (e) If the median of a random variable equals its expected value, then the random variable is symmetric.

8. For each of the following random variables, discuss whether the median or the mean would be a more useful measure of centrality.

   (a) The annual income of U.S. households.

   (b) The lifetime of 75-watt light bulbs.

   (c) The math SAT score of entering William & Mary freshmen.

   (d) The age of participants in the Bay to Breakers 12K road race.

9. The `R` function `qbinom` returns quantiles of the binomial distribution. For example, quartiles of $X \sim \text{Binomial}(n = 3; p = 0.5)$ can be computed as follows:

```
> alpha <- c(.25,.5,.75)
> qbinom(alpha,size=3,prob=.5)
[1] 1 1 2
```

   Notice that X is a symmetric random variable with center of symmetry $\theta = 1.5$, but `qbinom` computes $q_2(X) = 1$. This reveals that `R` may produce unexpected results when it computes the quantiles of discrete random variables. By experimenting with various choices of n and p, try to discover a rule according to which `qbinom` computes quartiles of the binomial distribution.

Chapter 7

Data

Chapters 3–6 developed mathematical tools for studying populations. Experiments are performed for the purpose of obtaining information about a population that is imperfectly understood. Experiments produce data, the raw material from which statistical procedures draw inferences about the population under investigation.

The probability distribution of a random variable X is a mathematical abstraction of an experimental procedure for sampling from a population. When we perform the experiment, we observe one of the possible values of X. To distinguish an observed value of a random variable from the random variable itself, we designate random variables by uppercase letters and observed values by corresponding lowercase letters.

Example 7.1 A coin is tossed and `Heads` is observed. The mathematical abstraction of this experiment is $X \sim \text{Bernoulli}(p)$ and the observed value of X is $x = 1$.

We will be concerned with experiments that are replicated a fixed number of times. By replication, we mean that each repetition of the experiment is performed under identical conditions and that the repetitions are mutually independent. Mathematically, we write $X_1, \ldots, X_n \sim P$. Let x_i denote the observed value of X_i. The set of observed values, $\vec{x} = \{x_1, \ldots, x_n\}$, is called a sample.

This chapter introduces several useful techniques for extracting information from samples. This information will be used to draw inferences about populations (for example, to guess the value of the population mean) and to assess assumptions about populations (for example, to decide whether or not the population can plausibly be modelled by a normal distribution).

Drawing inferences about population attributes (especially means) is the primary subject of subsequent chapters, which will describe specific procedures for drawing specific types of inferences. However, deciding which procedure is appropriate often involves assessing the validity of certain statistical assumptions. The methods described in this chapter will be our primary tools for making such assessments.

To assess whether or not an assumption is plausible, one must be able to investigate what happens when the assumption holds. For example, if a scientist needs to decide whether or not it is plausible that her sample was drawn from a normal distribution, then she needs to be able to recognize normally distributed data. For this reason, many of the samples studied in this chapter were generated under carefully controlled conditions, by computer simulation. Simulation allows us to investigate how samples drawn from specified distributions *should* behave, thereby providing a standard against which to compare experimental data for which the true distribution can never be known. Fortunately, R provides several convenient functions for simulating random sampling.

Example 7.2 Consider the experiment of tossing a fair die $n = 20$ times. We can simulate this experiment as follows:

```
> SampleSpace <- c(1,2,3,4,5,6)
> sample(x=SampleSpace,size=20,replace=TRUE)
[1]  1 6 3 2 2 3 5 3 6 4 3 2 5 3 2 2 3 2 4 2
```

Example 7.3 Consider the experiment of drawing a sample of size $n = 5$ from Normal$(2, 3)$. We can simulate this experiment as follows:

```
> rnorm(5,mean=2,sd=sqrt(3))
[1] 1.3274812 0.5901923 2.5881013 1.2222812 3.4748139
```

7.1 The Plug-In Principle

We will employ a general methodology for relating samples to populations. In Chapters 3–6 we developed a formidable apparatus for studying populations (probability distributions). We would like to exploit this apparatus fully. Given a sample, we will pretend that the sample is a finite population (discrete probability distribution) and then we will use methods for studying finite populations to learn about the sample. This approach is sometimes called the Plug-In Principle.

The Plug-In Principle employs a fundamental construction:

Definition 7.1 *Let $\vec{x} = (x_1, \ldots, x_n)$ be a sample. The* empirical probability distribution *associated with $\vec{x}$, denoted $\hat{P}_n$, is the discrete probability distribution defined by assigning probability $1/n$ to each $\{x_i\}$.*

Notice that, if a sample contains several copies of the same numerical value, then *each copy* is assigned probability $1/n$. This is illustrated in the following example.

Example 7.2 (continued) A fair die is rolled $n = 20$ times, resulting in the sample

$$\vec{x} = \{1, 6, 3, 2, 2, 3, 5, 3, 6, 4, 3, 2, 5, 3, 2, 2, 3, 2, 4, 2\}. \tag{7.1}$$

The empirical distribution $\hat{P}_{20}$ is the discrete distribution that assigns the following probabilities:

x_i	$\#\{x_i\}$	$\hat{P}_{20}(\{x_i\})$
1	1	0.05
2	7	0.35
3	6	0.30
4	2	0.10
5	2	0.10
6	2	0.10

Notice that, although the true probabilities are $P(\{x_i\}) = 1/6$, the empirical probabilities range from 0.05 to 0.35. The fact that $\hat{P}_{20}$ differs from P is an example of sampling variation. Statistical inference is concerned with determining what the empirical distribution (the sample) tells us about the true distribution (the population).

The empirical distribution, $\hat{P}_n$, is an intuitively appealing approximation of the actual probability distribution, P, from which the sample was drawn. Notice that the empirical probability of any event A is just

$$\hat{P}_n(A) = \#\{x_i \in A\} \cdot \frac{1}{n},$$

the observed frequency with which A occurs in the sample. Because the empirical distribution is an authentic probability distribution, all of the methods that we developed for studying (discrete) distributions are available for studying samples. For example,

Definition 7.2 *The empirical cdf, usually denoted $\hat{F}_n$, is the cdf associated with $\hat{P}_n$, i.e.*

$$\hat{F}_n(y) = \hat{P}_n(X \le y) = \frac{\#\{x_i \le y\}}{n}.$$

The empirical cdf of sample (7.1) is graphed in Figure 7.1.

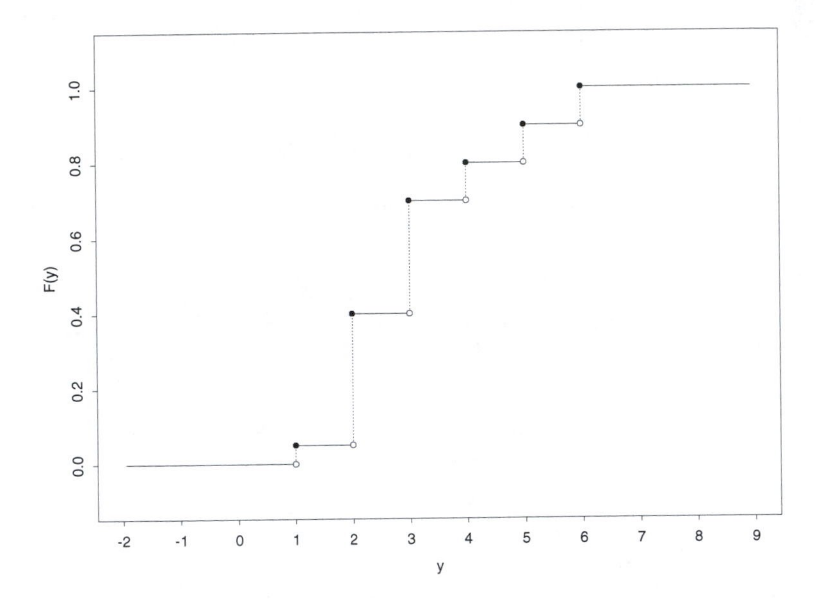

Figure 7.1: An empirical cdf.

In R, one can graph the empirical cdf of a sample x with the following command:

```
> plot.ecdf(x)
```

7.2 Plug-In Estimates of Mean and Variance

Population quantities defined by expected values are easily estimated by the plug-in principle. For example, suppose that $X_1, \ldots, X_n \sim P$ and that we observe a sample $\vec{x} = \{x_1, \ldots, x_n\}$. Let $\mu = EX_i$ denote the population mean. Then

Definition 7.3 *The plug-in estimate of μ, denoted $\hat{\mu}_n$, is the mean of the empirical distibution:*

$$\hat{\mu}_n = \sum_{i=1}^{n} x_i \cdot \frac{1}{n} = \frac{1}{n} \sum_{i=1}^{n} x_i = \bar{x}_n.$$

This quantity is called the sample mean.

Example 7.2 (continued) The population mean is

$$\mu = EX_i = 1 \cdot \frac{1}{6} + 2 \cdot \frac{1}{6} + 3 \cdot \frac{1}{6} + 4 \cdot \frac{1}{6} + 5 \cdot \frac{1}{6} + 6 \cdot \frac{1}{6} = \frac{1+2+3+4+5+6}{6} = 3.5.$$

The sample mean of sample (7.1) is

$$\begin{aligned} \hat{\mu}_{20} = \bar{x}_{20} &= 1 \cdot \frac{1}{20} + 6 \cdot \frac{1}{20} + \cdots + 4 \cdot \frac{1}{20} + 2 \cdot \frac{1}{20} \\ &= 1 \times 0.05 + 2 \times 0.35 + 3 \times 0.30 + 4 \times 0.10 + \\ &\quad 5 \times 0.10 + 6 \times 0.10 \\ &= 3.15. \end{aligned}$$

Notice that $\hat{\mu}_{20} \neq \mu$. This is another example of sampling variation.

The variance can be estimated in the same way. Let $\sigma^2 = \text{Var}\, X_i$ denote the population variance; then

Definition 7.4 *The plug-in estimate of σ^2, denoted $\widehat{\sigma_n^2}$, is the variance of the empirical distribution:*

$$\widehat{\sigma_n^2} = \sum_{i=1}^{n} (x_i - \hat{\mu}_n)^2 \cdot \frac{1}{n} = \frac{1}{n} \sum_{i=1}^{n} (x_i - \bar{x}_n)^2 = \frac{1}{n} \sum_{i=1}^{n} x_i^2 - \left(\frac{1}{n} \sum_{i=1}^{n} x_i \right)^2.$$

Notice that we do not refer to $\widehat{\sigma_n^2}$ as the sample variance. As will be discussed in Section 9.2.2, most authors designate another, equally plausible estimate of the population variance as *the* sample variance.

Example 7.2 (continued) The population variance is

$$\sigma^2 = EX_i^2 - (EX_i)^2 = \frac{1^2 + 2^2 + 3^2 + 4^2 + 5^2 + 6^2}{6} - 3.5^2 = \frac{35}{12} \doteq 2.9167.$$

The plug-in estimate of the variance is

$$\begin{aligned}\widehat{\sigma^2_{20}} &= \Big(1^2 \times 0.05 + 2^2 \times 0.35 + 3^2 \times 0.30 + \\ &\quad 4^2 \times 0.10 + 5^2 \times 0.10 + 6^2 \times 0.10\Big) - 3.15^2 \\ &= 1.9275.\end{aligned}$$

Again, notice that $\widehat{\sigma^2_{20}} \neq \sigma^2$, yet another example of sampling variation.

There are many ways to compute the preceding plug-in estimates using `R`. Assuming that `x` contains the sample, here are two possibilities:

```
> n <- length(x)
> plug.mean <- sum(x)/n
> plug.var <- sum(x^2)/n - plug.mean^2

> plug.mean <- mean(x)
> plug.var <- mean(x^2) - plug.mean^2
```

7.3 Plug-In Estimates of Quantiles

Population quantities defined by quantiles can also be estimated by the plug-in principle. Again, suppose that $X_1, \ldots, X_n \sim P$ and that we observe a sample $\vec{x} = \{x_1, \ldots, x_n\}$. Then

Definition 7.5 *The plug-in estimate of a population quantile is the corresponding quantile of the empirical distribution. In particular, the* sample median *is the median of the empirical distribution. The* sample interquartile range *is the interquartile range of the empirical distribution.*

Example 7.4 Consider the experiment of drawing a sample of size $n = 20$ from Uniform$(1, 5)$. This probability distribution has a population median of 3 and a population interquartile range of $4 - 2 = 2$. I simulated this experiment (and listed the sample in increasing order) with the following `R` command:

```
> x <- sort(runif(20,min=1,max=5))
```

This resulted in the following sample:

1.124600	1.161286	1.445538	1.828181	1.853359
1.934939	1.943951	2.107977	2.372500	2.448152
2.708874	3.297806	3.418913	3.437485	3.474940
3.698471	3.740666	4.039637	4.073617	4.195613

The sample median is

$$\frac{2.448152 + 2.708874}{2} = 2.578513,$$

which also can be computed with the following R command:

```
> median(x)
[1] 2.578513
```

Notice that the sample median does not exactly equal the population median. This is another example of sampling variation.

To compute the sample interquartile range, we require the first and third sample quartiles, i.e., the $\alpha = 0.25$ and $\alpha = 0.75$ sample quantiles. We must now confront the fact that Definition 6.5 may not specify unique quantile values. For the empirical distribution of the sample above, any number in $[1.853359, 1.934939]$ is a sample first quartile and any number in $[3.474940, 3.698471]$ is a sample third quartile.

The statistical community has not agreed on a convention for resolving the ambiguity in the definition of quartiles. One natural and popular possibility is to use the central value in each interval of possible quartiles. If we adopt that convention here, then the sample interquartile range is

$$\frac{3.474940 + 3.698471}{2} - \frac{1.853359 + 1.934939}{2} = 1.692556.$$

R adopts a slightly different convention, illustrated below. The following command computes the 0.25 and 0.75 quantiles:

```
> quantile(x,probs=c(.25,.75))
     25%      75%
1.914544 3.530823
```

The following command computes the 0.25 and 0.75 quantiles, plus several other useful sample quantities:

```
> summary(x)
    Min.  1st Qu.   Median     Mean  3rd Qu.     Max.
1.124600 1.914544 2.578513 2.715325 3.530823 4.195613
```

If we use the R definition of quantile, then the sample interquartile range is $3.530823 - 1.914544 = 1.616279$. Rather than typing the quartiles into R, we can compute the sample interquartile range as follows:

```
> q <- as.vector(quantile(x,probs=c(.25,.75)))
> q[2]-q[1]
[1] 1.616279
```

This is sufficiently complicated that we might prefer to create a function that computes the interquartile range of a sample:

```
> iqr <- function(x) {
+ q <- as.vector(quantile(x,probs=c(.25,.75)))
+ return(q[2]-q[1])
+ }
> iqr(x)
[1] 1.616279
```

Notice that the sample quantities do not exactly equal the population quantities that they estimate, regardless of which convention we adopt for defining quartiles. This is another example of sampling variation.

Used judiciously, sample quantiles can be extremely useful when trying to discern various features of the population from which the sample was drawn. The remainder of this section describes two graphical techniques for assimilating and displaying sample quantile information.

7.3.1 Box Plots

Information about sample quartiles is often displayed visually, in the form of a *box plot*. A box plot of a sample consists of a rectangle that extends from the first to the third sample quartile, thereby drawing attention to the central 50% of the data. Thus, the length of the rectangle equals the sample interquartile range. The location of the sample median is also identified, and its location within the rectangle often provides insight into whether or not the population from which the sample was drawn is symmetric. Whiskers extend from the ends of the rectangle, either to the extreme values of the data or to 1.5 times the sample interquartile range, whichever is less. Values that lie beyond the whiskers are called *outliers* and are individually identified.

Example 7.5 The pdf of the asymmetric distribution $\chi^2(3)$ is plotted in Figure 5.8. The following `R` commands draw a random sample of $n = 100$ observed values from this population, then construct a box plot of the sample:

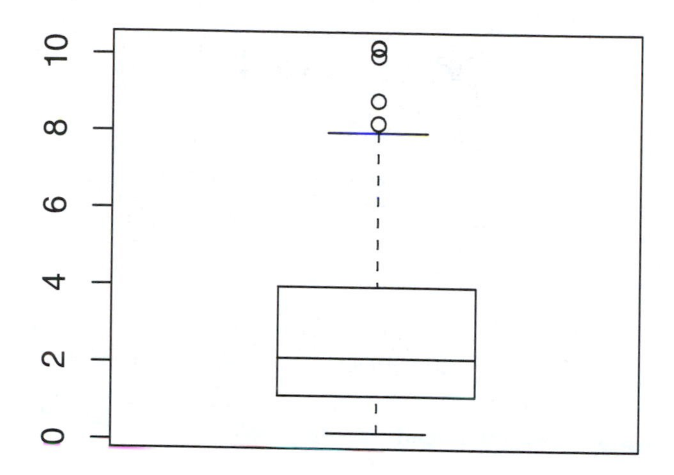

Figure 7.2: A box plot of a sample from $\chi^2(3)$.

```
> x <- rchisq(100,df=3)
> boxplot(x)
```

An example of a box plot produced by these commands is displayed in Figure 7.2. In this box plot, the numerical values in the sample are represented by the vertical axis.

The third quartile of the box plot in Figure 7.2 is farther above the median than the first quartile is below it. The short lower whisker extends from the first quartile to the minimal value in the sample, whereas the long upper whisker extends 1.5 interquartile ranges beyond the third quartile. Furthermore, there are 4 outliers beyond the upper whisker. Once we learn to discern these key features of the box plot, we can easily recognize that the population from which the sample was drawn is not symmetric.

The frequency of outliers in a sample often provides useful diagnostic information. Recall that, in Section 6.3, we computed that the interquartile range of a normal distribution is 1.34898 standard deviations. A value is an outlier if it lies more than

$$z = \frac{1.34898}{2} + 1.5 \cdot 1.34898 = 2.69796$$

standard deviations from the mean. Hence, the probability that an observation drawn from a normal distribution is an outlier is

```
> 2*pnorm(-2.69796)
[1] 0.006976582
```

and we would expect a sample drawn from a normal distribution to contain approximately 7 outliers per 1000 observations. A sample that contains a dramatically different proportion of outliers, as in Example 7.5, is not likely to have been drawn from a normal distribution.

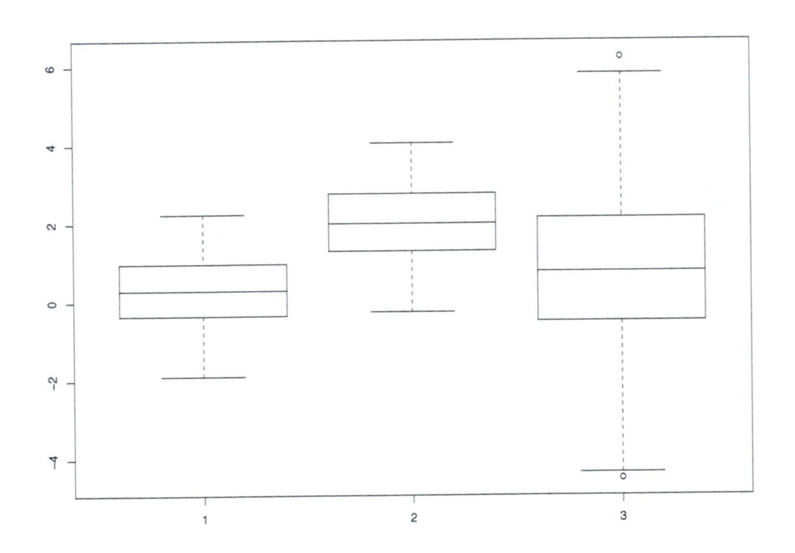

Figure 7.3: Box plots of samples from three normal distributions.

Box plots are especially useful for comparing several populations.

Example 7.6 I drew samples of 100 observations from three normal populations: Normal(0, 1), Normal(2, 1), and Normal(1, 4). To attempt to discern in the samples the various differences in population mean and standard deviation, we examine side-by-side box plots. These are constructed by the following R commands:

```
> z1 <- rnorm(100)
> z2 <- rnorm(100,mean=2,sd=1)
> z3 <- rnorm(100,mean=1,sd=2)
> boxplot(z1,z2,z3)
```

An example of the output of these commands is displayed in Figure 7.3.

7.3.2 Normal Probability Plots

Another powerful graphical technique that relies on quantiles are quantile-quantile (QQ) plots, which plot the quantiles of one distribution against the quantiles of another. QQ plots are used to compare the shapes of two distributions, most commonly by plotting the observed quantiles of an empirical distribution against the corresponding quantiles of a theoretical normal distribution. In this case, a QQ plot is often called a *normal probability plot*. If the shape of the empirical distribution resembles a normal distribution, then the points in a normal probability plot should tend to fall on a straight line. If they do not, then we should be skeptical that the sample was drawn from a normal distribution. Extracting useful information from normal probability plots requires some practice, but the patient data analyst will be richly rewarded.

Example 7.4 (continued) A normal probability plot of the sample generated in Example 7.5 against a theoretical normal distribution is displayed in Figure 7.4. This plot was created using the following R command:

```
> qqnorm(x)
```

Notice the systematic and asymmetric bending away from linearity in this plot. In particular, the smaller quantiles are much closer to the central values than should be the case for a normal distribution. This suggests that this sample was drawn from a nonnormal distribution that is skewed to the right. Of course, we know that this sample was drawn from $\chi^2(3)$, which is in fact skewed to the right.

When using normal probability plots, one must guard against overinterpreting slight departures from linearity. Remember: *some departures from linearity will result from sampling variation.* Consequently, before drawing

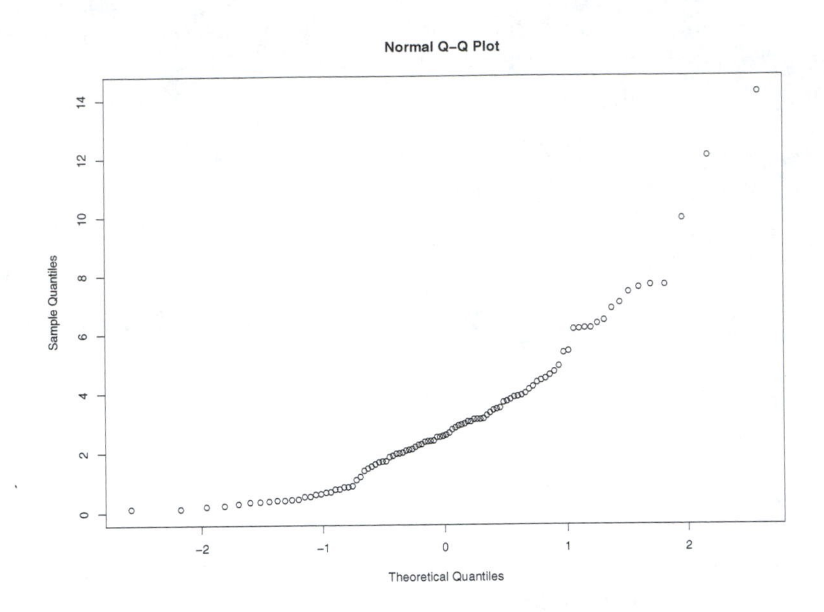

Figure 7.4: A normal probability plot of a sample from $\chi^2(3)$.

definitive conclusions, the wise data analyst will generate several random samples from the theoretical distribution of interest in order to learn how much sampling variation is to be expected. Before dismissing the possibility that the sample in Example 7.5 was drawn from a normal distribution, one should generate several normal samples of the same size for comparison. The normal probability plots of four such samples are displayed in Figure 7.5. In none of these plots did the points fall exactly on a straight line. However, upon comparing the normal probability plot in Figure 7.4 to the normal probability plots in Figure 7.5, it is abundantly clear that the sample in Example 7.5 was not drawn from a normal distribution.

7.4 Kernel Density Estimates

Suppose that $\vec{x} = \{x_1, \ldots, x_n\}$ is a sample drawn from an unknown pdf f. Box plots and normal probability plots are extremely useful graphical techniques for discerning in $\vec{x}$ certain important attributes of f, e.g., centrality, dispersion, asymmetry, nonnormality. To discern more subtle features of f, we now ask if it is possible to construct from $\vec{x}$ a pdf $\hat{f}_n$ that approximates f. This is a difficult problem, one that remains a vibrant topic of research

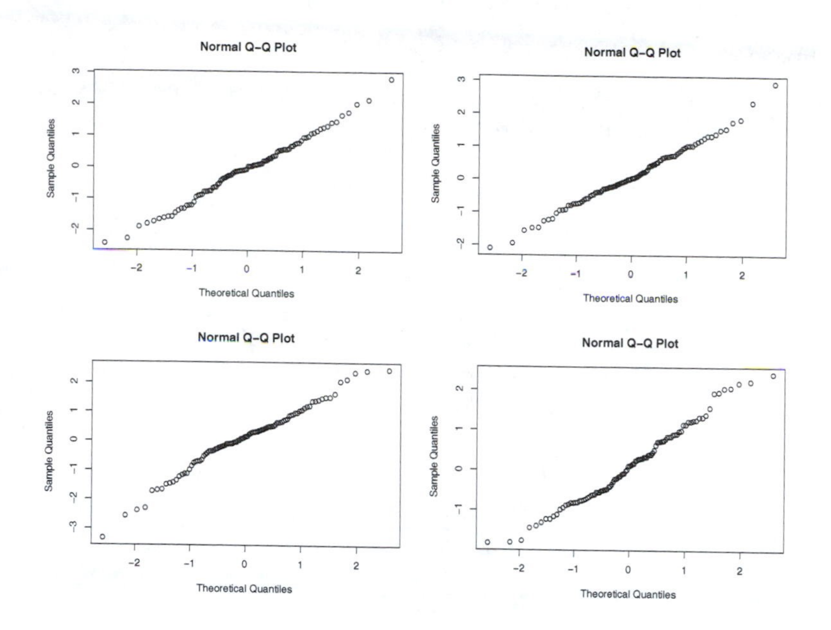

Figure 7.5: Normal probability plots of four samples from Normal(0, 1).

and about which little is said in introductory courses. However, using the concept of the empirical distribution, one can easily motivate one of the most popular techniques for *nonparametric probability density estimation.*

The logic of the empirical distribution is this: by assigning probability $1/n$ to each x_i, one accumulates more probability in the regions that produced more observed values. However, because the entire amount $1/n$ is placed exactly on the value x_i, the resulting empirical distribution is necessarily discrete. If the population from which the sample was drawn is discrete, then the empirical distribution estimates the probability mass function. However, if the population from which the sample was drawn is continuous, then *all* possible values occur with zero probability. In this case, there is nothing special about the precise values that were observed—what is important are the regions in which they occurred.

Instead of placing all of the probability $1/n$ assigned to x_i exactly on the value x_i, we now imagine distributing it in a neighborhood of x_i according to some probability density function. This construction will also result in more probability accumulating in regions that produced more values, but it will produce a pdf instead of a pmf. Here is a general description of this approach, usually called *kernel density estimation*:

1. Choose a probability density function K, the *kernel*. Typically, K is a symmetric pdf centered at the origin. Common choices of K include the Normal$(0, 1)$ and Uniform$[-0.5, 0.5]$ pdfs.

2. At each x_i, center a rescaled copy of the kernel. This pdf,

$$\frac{1}{h} K \left(\frac{x - x_i}{h} \right), \tag{7.2}$$

will control the distribution of the $1/n$ probability assigned to x_i. The parameter h is variously called the *smoothing parameter*, the *window width*, or the *bandwidth*.

3. The difficult decision in constructing a kernel density estimate is the choice of h. The technical details of this issue are beyond the scope of this book, but the underlying principles are quite simple:

   - Small values of h mean that the standard deviation of (7.2) will be small, so that the $1/n$ probability assigned to x_i will be distributed close to x_i. This is appropriate when n is large and the x_i are tightly packed.
   - Large values of h mean that the standard deviation of (7.2) will be large, so that the $1/n$ probability assigned to x_i will be widely distributed in the general vicinity of x_i. This is appropriate when n is small and the x_i are sparse.

4. After choosing K and h, the kernel density estimate of f is

$$\hat{f}_n(x) = \sum_{i=1}^{n} \frac{1}{n} \frac{1}{h} K \left(\frac{x - x_i}{h} \right) = \frac{1}{nh} \sum_{i=1}^{n} K \left(\frac{x - x_i}{h} \right).$$

   Such estimates are easily computed and graphed using the `R` functions `density` and `plot`.

Example 7.7 *Consider the probability density function f displayed in Figure 7.6. The most striking feature of f is that it is* bimodal. *Can we detect this feature using a sample drawn from f?*

We drew a sample of size $n = 100$ from f. A box plot and a normal probability plot of this sample are displayed in Figure 7.7. It is difficult to discern anything unusual from the box plot. The normal probability plot contains all of the information in the sample, but it is encoded in such a way that the feature of interest is not easily extracted. In contrast, the kernel

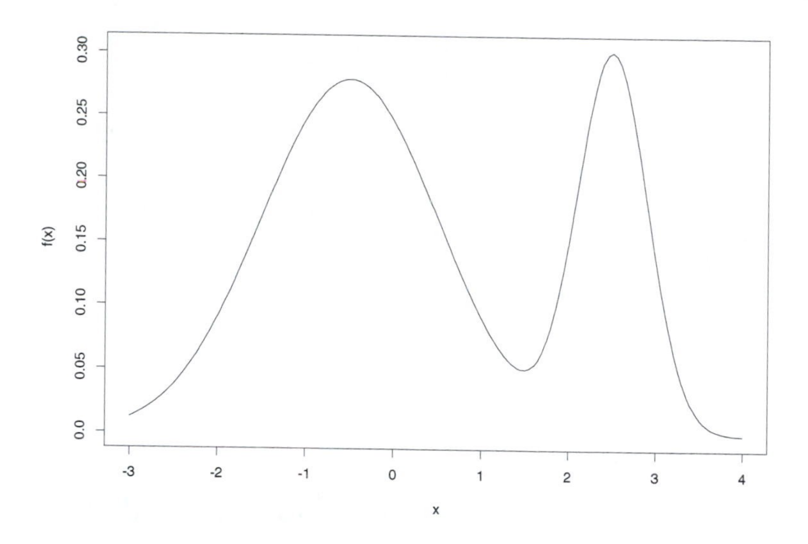

Figure 7.6: A bimodal probability density function.

density estimate displayed in Figure 7.8 clearly reveals that the sample was drawn from a bimodal population. After storing the sample in the vector `x`, this estimate was computed and plotted using the following R command:

```
> plot(density(x))
```

7.5 Case Study: Are Forearm Lengths Normally Distributed?

Many of the inferential procedures that statisticians have developed assume that the data to be analyzed were drawn from one (or more) normal distributions. These procedures are often the most elegant and powerful methods available to the data analyst, but they can easily mislead when applied to nonnormal data. Conveniently, the normality assumption is often quite plausible; just as often, however, it is not. It is therefore essential that the data analyst be able to make informed decisions about whether or not to assume normality. One of our primary uses for the methods introduced in the present chapter will be to assist us in making such decisions. Be warned: because we cannot *know* the true distributions of (most) random variables

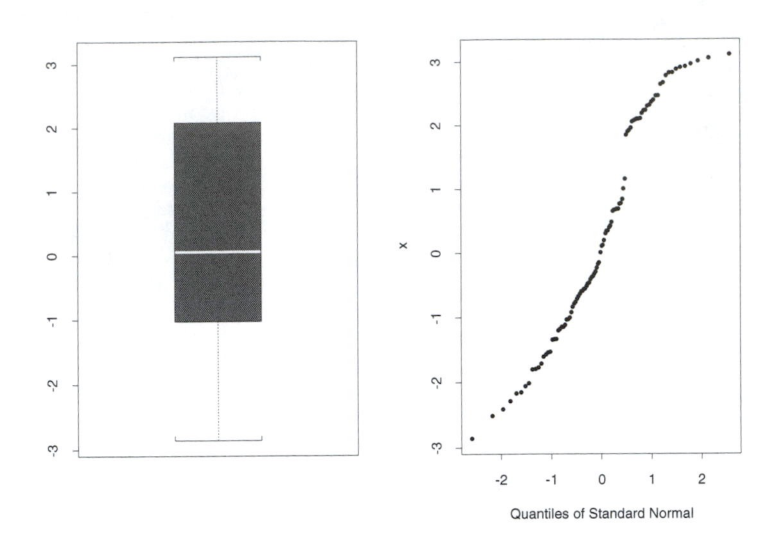

Figure 7.7: A box plot and a normal probability plot for Example 7.7.

encountered in scientific experimentation, we cannot *know* whether or not they are in fact normally distributed. Such ignorance should humble and can intimidate, but it should not paralyze. To analyze data, one must proceed somehow, and it is best to do so with as much information as possible.

It is often (but not universally) the case that measurements of linear dimension (height, length, width, depth, breadth, etc.) are normally distributed. To further illustrate the methods introduced in the present chapter, we apply them to a famous data set, measurements of forearm length made on $n = 140$ adult males, inquiring whether or not it appears plausible to assume that forearm lengths are normally distributed. These data, displayed in Table 7.1, were studied by K. Pearson and A. Lee[1] and subsequently reproduced as Data Set 139 in *A Handbook of Small Data Sets*.

Examining the numbers in Table 7.1, we note that the measurements were made with a precision of 0.1 inches and that many values occur several times. For example, 9 of the 140 men had forearms with a measured length of 18.5 inches. Because the probability that any two continuous random variables will be equal is zero, the existence of equal values in the sample should cause one to consider whether or not these measurements should

[1]K. Pearson and A. Lee (1903). On the laws of inheritance in man. I. Inheritance of physical characters. *Biometrika*, 2:357–462.

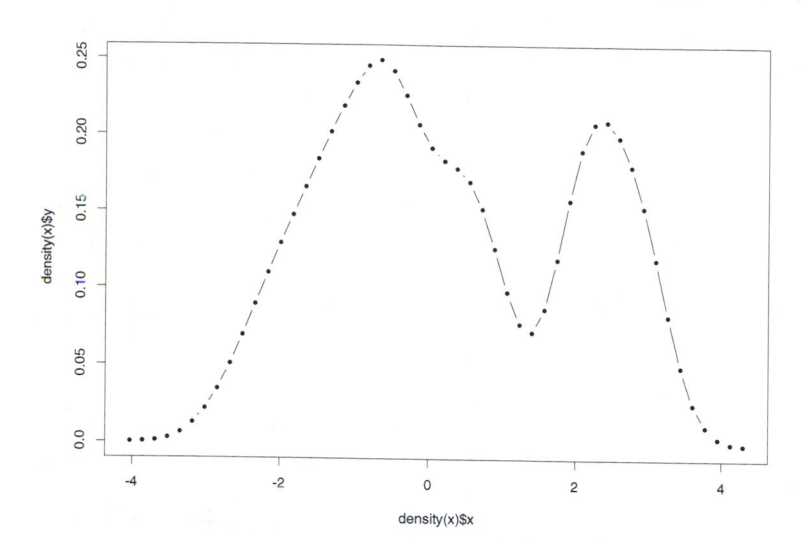

Figure 7.8: A kernel density estimate for Example 7.7.

be modelled as observed values of continuous random variables. In this case, it makes sense to proceed. Actual (as opposed to measured) forearm length is surely continuous, and there are 47 distinct values in Table 7.1. To preserve important numerical relations, e.g., $19.5 - 18.5 = 2(18.5 - 18)$, we can accomplish far more with continuous random variables than we might with discrete random variables. We proceed to investigate the plausibility of assuming that the continuous random variables are normal random variables.

Figure 7.9 displays a box plot, a normal probability plot, and a kernel density estimate, constructed from the 140 forearm measurements in Table 7.1 by the following R commands:

```
> par(mfrow=c(1,3))
> boxplot(forearms,main="Box Plot")
> qqnorm(forearms)
> plot(density(forearms),type="l",main="PDF Estimate")
```

Examining the box plot, we first note that the sample median lies roughly halfway between the first and third sample quartiles, and that the whiskers are of roughly equal length. This is precisely what we would expect to observe if the data were drawn from a symmetric distribution. We also note

17.3	18.4	20.9	16.8	18.7	20.5	17.9	20.4	18.3	20.5
19.0	17.5	18.1	17.1	18.8	20.0	19.1	19.1	17.9	18.3
18.2	18.9	19.4	18.9	19.4	20.8	17.3	18.5	18.3	19.4
19.0	19.0	20.5	19.7	18.5	17.7	19.4	18.3	19.6	21.4
19.0	20.5	20.4	19.7	18.6	19.9	18.3	19.8	19.6	19.0
20.4	17.3	16.1	19.2	19.6	18.8	19.3	19.1	21.0	18.6
18.3	18.3	18.7	20.6	18.5	16.4	17.2	17.5	18.0	19.5
19.9	18.4	18.8	20.1	20.0	18.5	17.5	18.5	17.9	17.4
18.7	18.6	17.3	18.8	17.8	19.0	19.6	19.3	18.1	18.5
20.9	19.8	18.1	17.1	19.8	20.6	17.6	19.1	19.5	18.4
17.7	20.2	19.9	18.6	16.6	19.2	20.0	17.4	17.1	18.3
19.1	18.5	19.6	18.0	19.4	17.1	19.9	16.3	18.9	20.7
19.7	18.5	18.4	18.7	19.3	16.3	16.9	18.2	18.5	19.3
18.1	18.0	19.5	20.3	20.1	17.2	19.5	18.8	19.2	17.7

Table 7.1: Forearm lengths (in inches) of $n = 140$ adult males, studied by K. Pearson and A. Lee (1903).

that these data contain no outliers. These features are consistent with the possibility that these data were drawn from a normal distribution, but they do not preclude other symmetric distributions.

Both normal probability plots and kernel density estimates reveal far more about the data than do box plots. More information is generally desirable, but seeing too much creates the danger that patterns created by chance variation will be overinterpreted by the too-eager data analyst. Key to the proper use of normal probability plots and kernel density estimates is mature judgment about which features reflect on the population and which features are due to chance variation.

The normal probability plot of the forearm data is generally straight, but should we worry about the kink at the lower end? The kernel density estimate of the forearm data is unimodal and nearly symmetric, but should we be concerned by its apparent lack of inflection points at ± 1 standard deviations? The best way to investigate such concerns is to generate pseudorandom normal samples, each of the same size as the observed sample (here $n = 140$), and consider what—if anything—distinguishes the observed sample from the normal samples. I generated three pseudorandom normal samples using the `rnorm` function. The four normal probability plots are displayed in Figure 7.10 and the four kernel density estimates are displayed

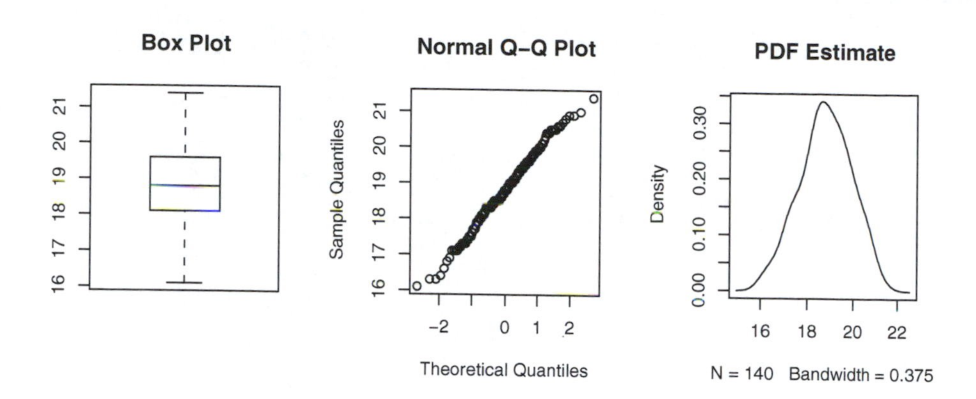

Figure 7.9: Three displays of 140 forearm measurements.

in Figure 7.11. I am unable to advance a credible argument that the forearm sample looks any less normal than the three normal samples.

In addition to the admittedly subjective comparison of normal probability plots and kernel density estimates, it may be helpful to compare certain quantitative attributes of the sample to known quantitative attributes of normal distributions. In Section 6.3, for example, we noted that the ratio of population interquartile range to population standard deviation is $1.34898 \doteq 1.35$ for a normal distribution. The analogous ratio of sample interquartile range to sample standard deviation can be quite helpful in deciding whether or not the sample was drawn from a normal distribution. It should be noted, however, that not all distributions with this ratio are normal; thus, although a ratio substantially different from 1.35 may suggest that the sample was not drawn from a normal distribution, a ratio close to 1.35 does not prove that it was.

To facilitate the calculation of iqr:sd ratios, we define an R function that performs the necessary operations. Here are the R commands that define our new function, iqrsd:

```
> iqrsd <- function(x) {
+ x.mean <- mean(x)
+ x.var <- mean(x^2)-x.mean^2
+ q <- as.vector(quantile(x,probs=c(.25,.75)))
+ x.iqr <- q[2]-q[1]
+ return(x.iqr/sqrt(x.var))
+ }
```

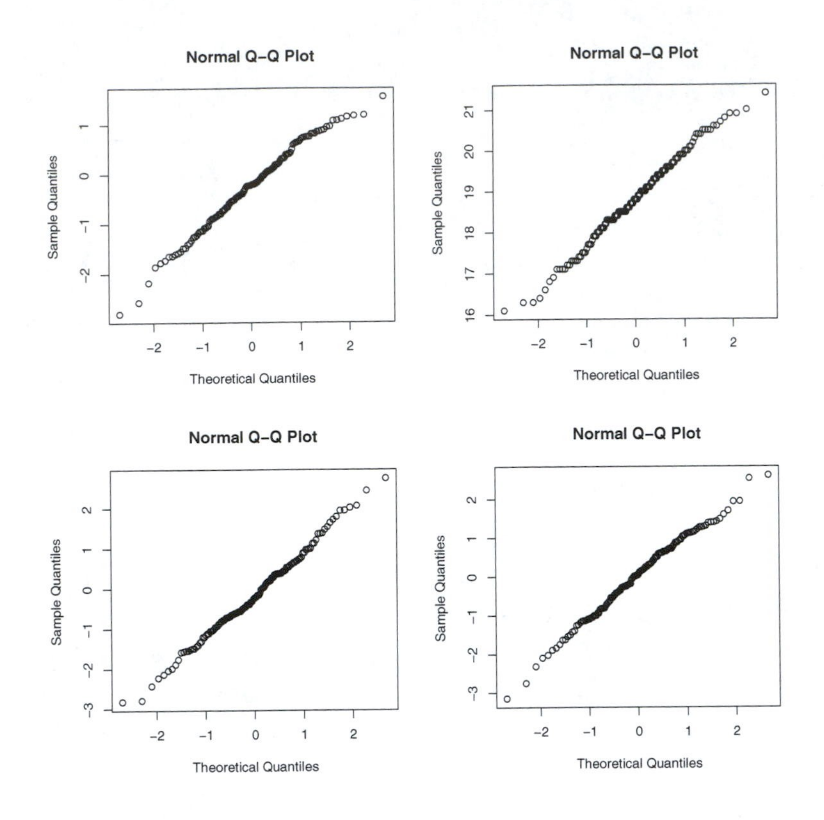

Figure 7.10: Normal probability plots of the forearm data and three pseudorandom samples from normal(0, 1).

I generated 10 pseudorandom normal samples, each of size $n = 140$, using the `rnorm` function, then applied the new `iqrsd` function to each sample. The resulting ratios ranged from a minimum of 1.178 to a maximum of 1.545. The ratio for the forearm data is 1.344, so one can hardly object to assuming normality on the basis of this quantity.

Overall, the forearm data look about as normal as one could ever hope to encounter with actual experimental data. If one is hoping to use an inferential procedure that assumes normality, then this is the ideal case. Unfortunately, one rarely encounters situations in which one can so comfortably assume normality.

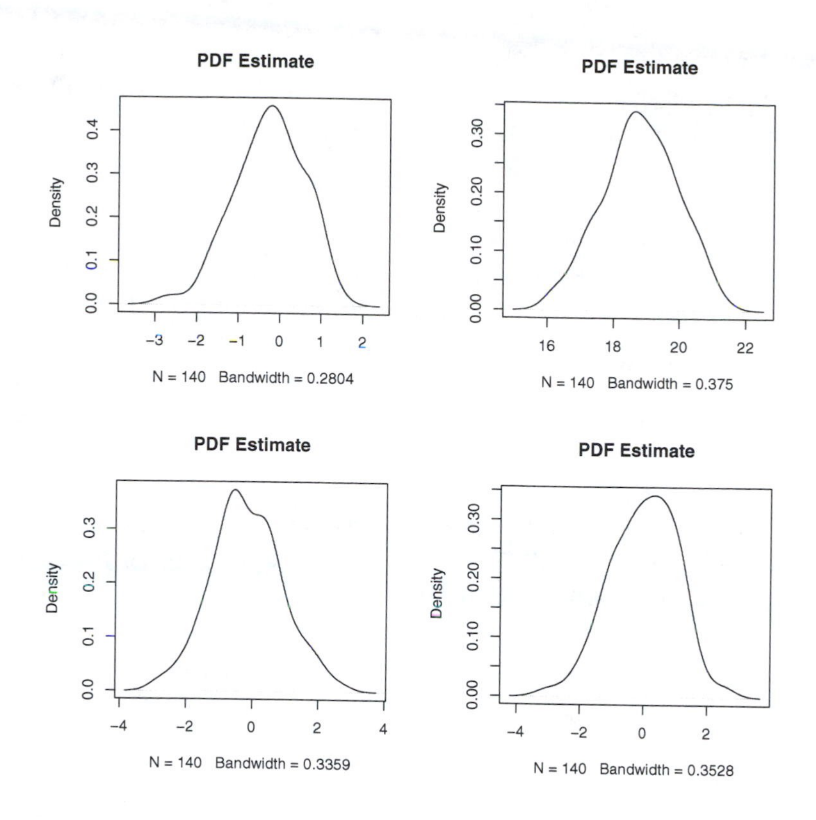

Figure 7.11: Kernel density estimates constructed from the forearm data and three pseudorandom samples from normal$(0, 1)$.

7.6 Transformations

When a scientist decides to measure a quantity, she usually chooses a familiar unit of measurement that facilitates interpretation of the phenomenon under investigation. For example, a speech scientist may feel that she will have better intuition about lung volume if she records X_i, lung volume in cubic centimeters, than if she records $Y_i = \log(X_i)$. Unfortunately, Nature may not arrange matters so that the quantities that are most easily interpreted are the quantities that are most easily analyzed. What if the scientist records X_i, but $Y_i \sim$ Normal? If Y_i is normally distributed, then $X_i = \exp(Y_i)$ is *not* normally distributed. The distribution of X_i, a *lognormal* distribution, is skewed to the right and a statistician might prefer to analyze $\vec{y} = \{\log(x_1), \ldots, \log(x_n)\}$ rather than $\vec{x} = \{x_1, \ldots, x_n\}$.

Replacing $\vec{x}$ with $\vec{y}$ illustrates the process of data transformation. One should not transform data frivolously, but there may be compelling reasons to do so. For example, a great many statistical procedures rely on averaging. Means average values; variances average squared deviations. To use these procedures, one must adopt units with respect to which averaging is reasonable.

Suppose that a psychologist, interested in the effect of a new drug on cognitive function in Alzheimer's patients, administers a neuropsychological test to $n = 5$ patients before and after the study period. Let B_i denote the score of patient i before the study and let A_i denote the the score of patient i after the study. The values of the B_i provide baseline measurements against which the values of the A_i will be compared. But *how* should they be compared?

Two obvious comparisons are $A_i - B_i$ and A_i/B_i. Which makes more sense may require considerable reflection. Should we interpret $b_1 = 2$ and $a_1 = 4$ as a doubling of cognitive function, the equivalent of $b_2 = 20$ and $a_2 = 40$? Or should we interpret it as an increase of 2 points, the equivalent of $b_3 = 20$ and $a_3 = 22$?

Suppose that the psychologist prefers the former interpretation and opts to measure $X_i = A_i/B_i$, so that $X_i = 1$ indicates no effect, $X_i > 1$ indicates a beneficial effect, and $X_i < 1$ indicates an adverse affect. Suppose that he observes the following data:

i	B_i	A_i	X_i
1	16	32	2.00
2	16	24	1.50
3	16	16	1.00
4	16	12	0.75
5	16	8	0.50

Overall, there is no evidence in this sample of an effect. Patient 3 scored the same after treatment as before. Patient 1 scored twice as high after, but patient 5 scored twice as high before. Patient 2 scored 1.5 times as high after, but patient 4 scored 1.5 times as high before. However, if we average the x_i, we obtain $\bar{x} = 1.15$, which seems to indicate that the drug had a slightly beneficial effect on cognitive function. What happened?

There really is no evidence of an effect. Suppose that we had computed B_i/A_i instead of A_i/B_i, in which case values greater than 1 indicate an adverse effect. Instead of averaging $\{2.00, 1.50, 1.00, 0.75, 0.50\}$ to obtain 1.15 and inferring a benefical effect, we would average $\{0.50, 0.75, 1.00, 1.50, 2.00\}$ to obtain 1.15 and infer an adverse effect! Clearly, something is wrong.

The problem is that ratios are measured asymmetrically. For X_i, the neutral ratio is 1, the beneficial ratios are $(1, \infty)$, and the adverse ratios are $(0, 1)$. But consider what happens if we transform the ratios logarithmically, by setting $Y_i = \log(X_i)$. The neutral value of Y_i is 0, the beneficial values are $(0, \infty)$, and the adverse values are $(-\infty, 0)$. Because $\log(1/x_i) = -\log(x_i)$, the average of the five transformed values is precisely 0, the neutral value that correctly represents the overall absence of an effect.

Transformations are often applied to data sampled from skewed distributions. The goal is to analyze data that are more normally, or at least more symmetrically, distributed. Taking logarithms ($y_i = \log(x_i)$) and taking square roots ($y_i = \sqrt{x_i}$) are two especially popular transformations. Although statisticians have devised entire families of transformations, together with ways of estimating which transformation is best suited to a given sample, the choice of transformation remains something of an art. When contemplating data transformation, always check to see if the transformed data really do behave better than the observed data, always consider how easily an analysis of the transformed data can be interpreted, and never transform unless there is a good reason to do so.

7.7 Exercises

1. Let $\vec{x}$ denote the following sample from an unknown population:

462	425	164	784	625
472	658	658	663	928
92	230	96	626	1277
225	150	320	496	157
458	933	861	174	431

   (a) Graph the empirical cdf of $\vec{x}$.

   (b) Compute the plug-in estimates of the population mean and variance.

   (c) Compute the plug-in estimates of the population median and interquartile range.

   (d) Compute the ratio of the plug-in estimate of the interquartile range to the square root of the plug-in estimate of the variance.

   (e) Construct a boxplot

   (f) Construct a normal probability plot.

(g) Construct a kernel density estimate.

(h) Do you think that this sample was drawn from a normal distribution? Why or why not?

2. Let $\vec{x}$ denote the following sample of pulse rates of Peruvian Indians.[2]

88	76	84	64	60	64	60	64	68	74
68	68	72	76	72	52	72	64	60	56
72	88	80	76	64	72	60	76	88	72
64	60	60	72	92	80	72	64	68	

(a) Graph the empirical cdf of $\vec{x}$.

(b) Compute the plug-in estimates of the population mean and variance.

(c) Compute the plug-in estimates of the population median and interquartile range.

(d) Compute the ratio of the plug-in estimate of the interquartile range to the square root of the plug-in estimate of the variance.

(e) Construct a boxplot.

(f) Construct a normal probability plot.

(g) Construct a kernel density estimate.

(h) Do you think that this sample was drawn from a normal distribution? Why or why not?

3. The following independent samples were drawn from four populations:

Sample 1	Sample 2	Sample 3	Sample 4
5.098	4.627	3.021	7.390
2.739	5.061	6.173	5.666
2.146	2.787	7.602	6.616
5.006	4.181	6.250	7.868
4.016	3.617	1.875	2.428
9.026	3.605	6.996	6.740
4.965	6.036	4.850	7.605
5.016	4.745	6.661	10.868
6.195	2.340	6.360	1.739
4.523	6.934	7.052	1.996

[2]T. A. Ryan, Jr., B. L. Joiner, and B. F. Ryan (1985). *The Minitab Student Handbook*. Duxbury Press, Boston, pp. 317–318. These data appear as Data Set 345 in *A Handbook of Small Data Sets*.

(a) Use the `boxplot` function to create side-by-side box plots of these samples. Does it appear that these samples were all drawn from the same population? Why or why not?

(b) Use the `rnorm` function to draw four independent samples, each of size $n = 10$, from one normal distribution. Examine box plots of these samples. Is it possible that Samples 1–4 were all drawn from the same normal distribution?

4. The following sample, $\vec{x}$, was observed and sorted:

0.246	0.327	0.423	0.425	0.434
0.530	0.583	0.613	0.641	1.054
1.098	1.158	1.163	1.439	1.464
2.063	2.105	2.106	4.363	7.517

(a) Graph the empirical cdf of $\vec{x}$.

(b) Calculate the plug-in estimates of the mean, the variance, the median, and the interquartile range.

(c) Take the square root of the plug-in estimate of the variance and compare it to the plug-in estimate of the interquartile range. Do you think that $\vec{x}$ was drawn from a normal distribution? Why or why not?

(d) Use the `qqnorm` function to create a normal probability plot. Do you think that $\vec{x}$ was drawn from a normal distribution? Why or why not?

(e) Now consider the transformed sample $\vec{y}$ produced by replacing each x_i with its natural logarithm. If $\vec{x}$ is stored in the vector `x`, then $\vec{y}$ can be computed by the following `R` command:

```
> y <- log(x)
```

Do you think that $\vec{y}$ was drawn from a normal distribution? Why or why not?

5. In January 2002, twelve students enrolled in Math 351 (Applied Statistics) at the College of William & Mary reported the following results for the experiment described in Exercise 1.5.2.[3]

[3]Two students reported more than one measurement, but only one measurement per student is reported here.

$143\frac{3}{16}$	$144\frac{4}{16}$	$140\frac{14}{16}$	$144\frac{7}{16}$	$143\frac{12}{16}$	$153\frac{13}{16}$
$119\frac{10}{16}$	$143\frac{1}{16}$	$143\frac{14}{16}$	$144\frac{3}{16}$	$144\frac{7}{16}$	$148\frac{3}{16}$

(a) Do these measurements appear to be a sample from a normal distribution? Why or why not?

(b) Suggest possible explanations for the surprising amount of variation in these measurements.

(c) Use these measurements to estimate the true length of the table. Justify your estimation procedure.

6. Forty-one students taking Math 351 (Applied Statistics) at the College of William & Mary were administered a test. The following test scores were observed and sorted:

90	90	89	88	85	85	84	82	82	82	
81	81	81	80	79	79	78	76	75	74	
72	71	70	66	65	63	62	62	61	59	
58	58	57	56	56	53	48	44	40	35	33

(a) Do these numbers appear to be a random sample from a normal distribution?

(b) Does this list of numbers have any interesting anomalies?

7. Consider an urn that contains 10 tickets, labelled

$$\{1, 1, 1, 1, 2, 5, 5, 10, 10, 10\}.$$

From this urn, I propose to draw (with replacement) $n = 40$ tickets. I am interested in the sum, Y, of the 40 ticket values that I draw.

(a) Write an R function named `urn.model` that simulates this experiment, i.e., evaluating `urn.model` is like observing a value, y, of the random variable Y.

(b) Use `urn.model` to generate a sample, $\vec{y} = \{y_1, \ldots, y_{25}\}$, of $n = 25$ observed sums. The random variable Y is discrete. Does it appear that the distribution of Y can be approximated by a normal distribution? Why or why not?

8. Experiment with using `R` to generate simulated random samples of various sizes. Use the `summary` function to compute the quartiles of these samples. Try to discern the convention that this function uses to define sample quartiles.

Chapter 8

Lots of Data

Throughout Chapter 7 we emphasized that, because of sampling variation, the plug-in estimate of a population quantity rarely equals the actual value of the population quantity. The present chapter explores this phenomenon in greater depth.

Suppose that $X_1, \ldots, X_n \sim P$ and that an experimental scientist wants to estimate the population mean, $\mu = EX_i$. To do so, she observes values $x_1, \ldots, x_n$ of $X_1, \ldots, X_n$, then computes

$$\bar{x}_n = \frac{1}{n} \sum_{i=1}^{n} x_i,$$

the plug-in estimate of μ. Mathematically, this is equivalent to first defining a new random variable,

$$\bar{X}_n = \frac{1}{n} \sum_{i=1}^{n} X_i,$$

then observing the value $\bar{x}_n$ of $\bar{X}_n$. The random variable $\bar{X}_n$ is the average of the random variables $X_1, \ldots, X_n$. Both the random variable $\bar{X}_n$ and the observed value $\bar{x}_n$ are called the *sample mean*. This is potentially confusing, but the convention of using uppercase letters for random variables and lowercase letters for observed values allows us to be clear about which concept we have in mind when we use the phrase "sample mean." In this chapter, we study the behavior of $\bar{X}_n$.

We begin with an example. Suppose that, unbeknownst to the scientist, P is the asymmetric probability distribution $\chi^2(3)$, with pdf depicted in Figure 5.8. Because of Corollary 5.1, it follows that $\mu = 3$. Hence, we can assess the quality of the scientist's estimates of μ by comparing the estimates to the correct value, $\mu = 3$. We will use simulation to explore what might occur in this situation.

First, consider drawing a small sample of $n = 5$ observations. Here is what happened when I performed that experiment three times:

```
> x <- rchisq(5,df=3)
> mean(x)
[1] 3.650077

> x <- rchisq(5,df=3)
> mean(x)
[1] 2.963841

> x <- rchisq(5,df=3)
> mean(x)
[1] 2.063129
```

Due to sampling variation, the first estimate is too high, the second estimate is just about right, and the third estimate is too low. These results suggest that small samples may be unreliable. Of course, if we admit the possibility that small samples are unreliable, then it might be wise to perform the simulation more than three times! So, I performed the same simulation 1000 times, each time observing values of $X_1, \ldots, X_5 \sim \chi^2(3)$ and then computing $\bar{x}_5$, the observed value of $\bar{X}_5$. To display the results, I applied the method described in Section 7.4 to the 1000 observed values of $\bar{X}_5$. This produced a kernel density estimate, displayed in Figure 8.1, of the pdf of $\bar{X}_5$. Notice the considerable variation in the observed values of $\bar{X}_5$.

Next, consider drawing a moderate sample of $n = 20$ observations. I did this 1000 times, each time observing values of $X_1, \ldots, X_{20} \sim \chi^2(3)$ and then computing $\bar{x}_{20}$, the observed value of $\bar{X}_{20}$. From these 1000 observed values of $\bar{X}_{20}$, I constructed a kernel density estimate of the pdf of $\bar{X}_{20}$. This estimated pdf is displayed in Figure 8.2. Notice that the observed values of $\bar{X}_{20}$ tend to be more tightly clustered around $\mu = 3$ than do the observed values of $\bar{X}_5$, suggesting that moderate samples are more reliable than small samples.

Finally, consider drawing a large sample of $n = 80$ observations. I did this 1000 times, each time observing values of $X_1, \ldots, X_{80} \sim \chi^2(3)$ and then computing $\bar{x}_{80}$, the observed value of $\bar{X}_{80}$. From these 1000 observed values of $\bar{X}_{80}$, I constructed a kernel density estimate of the pdf of $\bar{X}_{80}$. This estimated pdf is displayed in Figure 8.3. Notice that the observed values of $\bar{X}_{80}$ tend to be more tightly clustered around $\mu = 3$ than do the observed

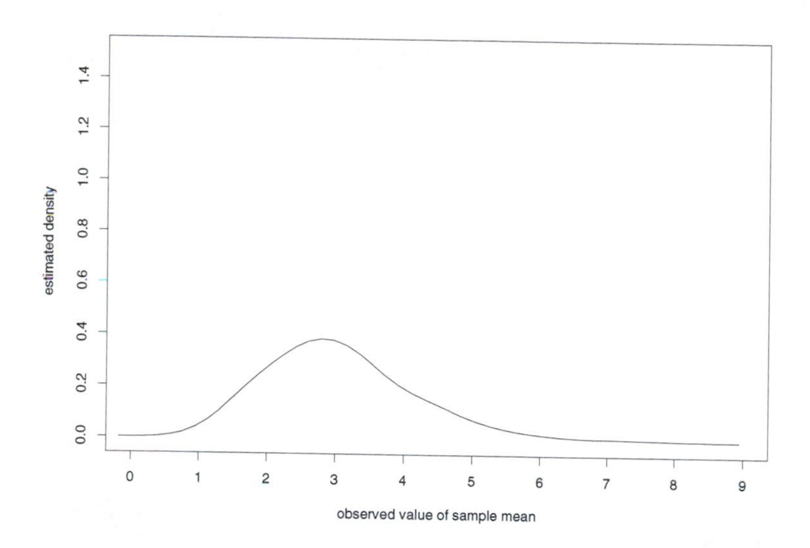

Figure 8.1: Kernel density estimate constructed from 1000 observed values of $\bar{X}_n$ for $n = 5$. $X_1, \ldots, X_n \sim \chi^2(3)$ and $\mu = EX_i = 3$.

values of $\bar{X}_{20}$, suggesting that large samples are more reliable than moderate samples.

The sections in this chapter generalize the preceding observations. We consider any experiment that can be performed, independently and identically, as many times as we please. We describe this situation by supposing the existence of a sequence of independent and identically distributed random variables, $X_1, X_2, \ldots$, and we assume that these random variables have a finite mean $\mu = EX_i$ and a finite variance $\sigma^2 = \operatorname{Var} X_i$. Under these assumptions, we study the behavior of the sample mean, $\bar{X}_n$, as n increases.

8.1 Averaging Decreases Variation

By definition, $EX_i = \mu$. Thus, the population mean is the average value assumed by the random variable X_i. This statement is also true of the sample mean:

$$E\bar{X}_n = \frac{1}{n}\sum_{i=1}^{n} EX_i = \frac{1}{n}\sum_{i=1}^{n} \mu = \mu;$$

however, there is a crucial distinction between X_i and $\bar{X}_n$.

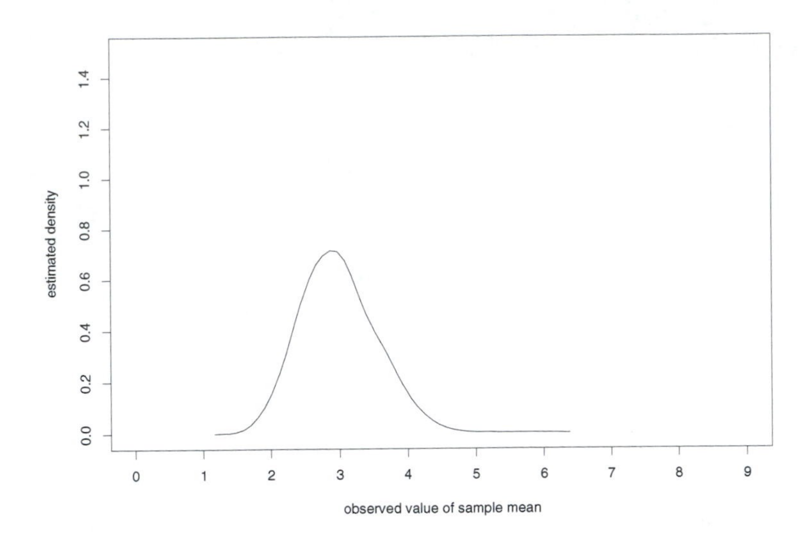

Figure 8.2: Kernel density estimate constructed from 1000 observed values of $\bar{X}_n$ for $n = 20$. $X_1, \ldots, X_n \sim \chi^2(3)$ and $\mu = EX_i = 3$.

The tendency of a random variable to assume a value that is close to its expected value is quantified by computing its variance. By definition, $\operatorname{Var} X_i = \sigma^2$, but

$$\operatorname{Var} \bar{X}_n = \operatorname{Var}\left(\frac{1}{n}\sum_{i=1}^{n} X_i\right) = \frac{1}{n^2}\sum_{i=1}^{n} \operatorname{Var} X_i = \frac{1}{n^2}\sum_{i=1}^{n} \sigma^2 = \frac{\sigma^2}{n}.$$

Hence, the sample mean has less variability than any of the individual random variables that are being averaged. *Averaging decreases variation.* Furthermore, as $n \to \infty$, $\operatorname{Var} \bar{X}_n \to 0$. Thus, by repeating our experiment enough times, we can make the variation in the sample mean as small as we please.

The preceding remarks suggest that, if the population mean is unknown, then we can draw inferences about it by observing the behavior of the sample mean. This fundamental insight is the basis for a considerable portion of this book. The remainder of this chapter refines the relation between the population mean and the behavior of the sample mean.

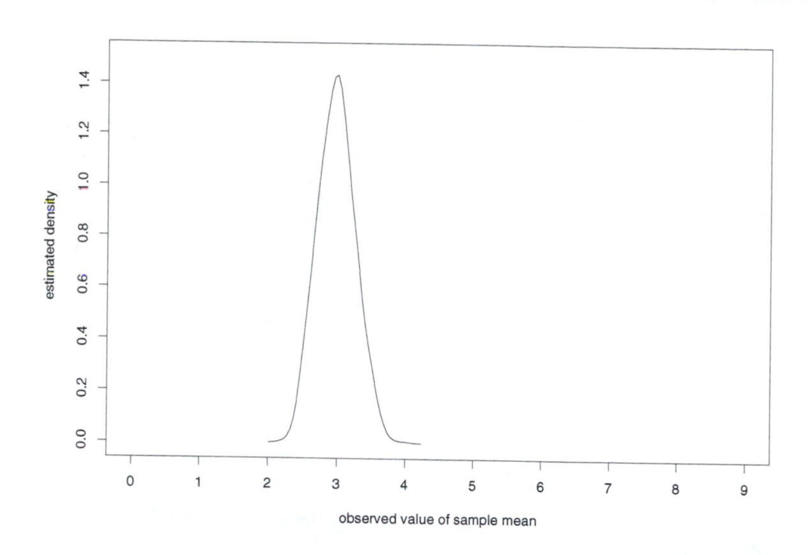

Figure 8.3: Kernel density estimate constructed from 1000 observed values of $\bar{X}_n$ for $n = 80$. $X_1, \ldots, X_n \sim \chi^2(3)$ and $\mu = EX_i = 3$.

8.2 The Weak Law of Large Numbers

Recall Definition 2.12 from Section 2.4: a sequence of real numbers $\{y_n\}$ converges to a limit $c \in \Re$ if and only if, for every $\epsilon > 0$, there exists a natural number N such that $y_n \in (c - \epsilon, c + \epsilon)$ for each $n \geq N$. Our first task is to generalize from convergence of a sequence of real numbers to convergence of a sequence of random variables.

If we replace $\{y_n\}$, a sequence of real numbers, with $\{Y_n\}$, a sequence of random variables, then the event that $Y_n \in (c-\epsilon, c+\epsilon)$ is uncertain. Rather than demand that this event *must* occur for n sufficiently large, we ask only that the probability of this event tend to unity as n tends to infinity. This results in

Definition 8.1 *A sequence of random variables $\{Y_n\}$ converges in probability to a constant c, written $Y_n \xrightarrow{P} c$, if and only if, for every $\epsilon > 0$,*

$$\lim_{n\to\infty} P\left(Y_n \in (c - \epsilon, c + \epsilon)\right) = 1.$$

Convergence in probability is depicted in Figure 8.4 using the pdfs f_n of continuous random variables Y_n. (One could also use the pmfs of discrete

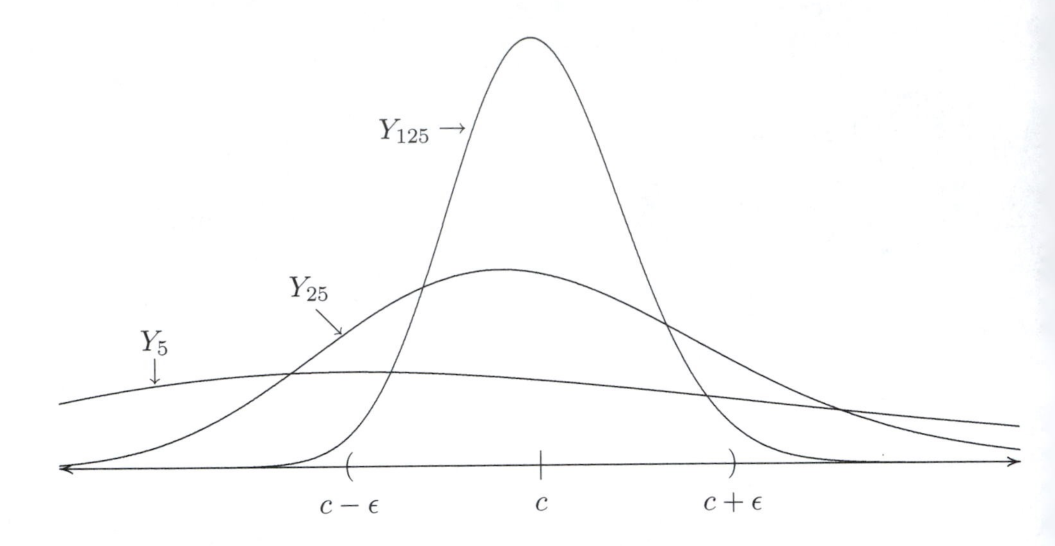

Figure 8.4: An example of convergence in probability.

random variables.) We see that

$$p_n = P\left(Y_n \in (c - \epsilon, c + \epsilon)\right) = \int_{c-\epsilon}^{c+\epsilon} f_n(x)\, dx$$

is tending to unity as n increases. Notice, however, that each $p_n < 1$.

The concept of convergence in probability allows us to state an important result.

Theorem 8.1 *(Weak Law of Large Numbers) Let $X_1, X_2, \ldots$ be any sequence of independent and identically distributed random variables having finite mean μ and finite variance σ^2. Then*

$$\bar{X}_n \xrightarrow{P} \mu.$$

This result is of considerable consequence. It states that, as we average more and more X_i, the average values that we observe tend to be distributed closer and closer to the theoretical average of the X_i. This property of the sample mean strengthens our contention that the behavior of $\bar{X}_n$ provides more and more information about the value of μ as n increases.

The Weak Law of Large Numbers (WLLN) has an important special case.

Corollary 8.1 *(Law of Averages) Let A be any event and consider a sequence of independent and identical experiments in which we observe whether or not A occurs. Let $p = P(A)$ and define independent and identically distributed random variables by*

$$X_i = \left\{ \begin{array}{ll} 1 & A \text{ occurs} \\ 0 & A^c \text{ occurs} \end{array} \right\}.$$

Then $X_i \sim$ Bernoulli(p), $\bar{X}_n$ is the observed frequency with which A occurs in n trials, and $\mu = EX_i = p = P(A)$ is the theoretical probability of A. The WLLN states that the former tends to the latter as the number of trials increases.

The Law of Averages formalizes our common experience that "things tend to average out in the long run." For example, we might be surprised if we tossed a fair coin $n = 10$ times and observed $\bar{X}_{10} = 0.9$; however, if we knew that the coin was indeed fair ($p = 0.5$), then we would remain confident that, as n increased, $\bar{X}_n$ would eventually tend to 0.5.

Notice that the *conclusion* of the Law of Averages is the frequentist *interpretation* of probability. Instead of defining probability via the notion of long-run frequency, we defined probability via the Kolmogorov axioms. Although our approach does not require us to interpret probabilities in any one way, the Law of Averages states that probability necessarily behaves in the manner specified by frequentists.

Finally, recall from Section 7.1 that the empirical probability of an event A is the observed frequency with which A occurs in the sample:

$$\hat{P}_n(A) = \# \{x_i \in A\} \cdot \frac{1}{n},$$

By the Law of Averages, this quantity tends to the true probability of A as the size of the sample increases. Thus, the theory of probability provides a mathematical justification for approximating P with $\hat{P}_n$ when P is unknown.

8.3 The Central Limit Theorem

The Weak Law of Large Numbers states a precise sense in which the distribution of values of the sample mean collapses to the population mean as the size of the sample increases. As interesting and useful as this fact is, it leaves several obvious questions unanswered:

1. How rapidly does the sample mean tend toward the population mean?

2. How does the shape of the sample mean's distribution change as the sample mean tends toward the population mean?

To answer these questions, we convert the random variables in which we are interested to standard units.

We have supposed the existence of a sequence of independent and identically distributed random variables, $X_1, X_2, \ldots$, with finite mean $\mu = EX_i$ and finite variance $\sigma^2 = \text{Var}\, X_i$. We are interested in the sum and/or the average of $X_1, \ldots, X_n$. It will be helpful to identify several crucial pieces of information for each random variable of interest:

random variable	expected value	standard deviation	standard units
X_i	μ	σ	$(X_i - \mu)/\sigma$
$\sum_{i=1}^n X_i$	$n\mu$	$\sqrt{n}\,\sigma$	$(\sum_{i=1}^n X_i - n\mu) \div (\sqrt{n}\,\sigma)$
$\bar{X}_n$	μ	$\sigma/\sqrt{n}$	$(\bar{X}_n - \mu) \div (\sigma/\sqrt{n})$

First we consider X_i. Notice that converting to standard units does *not* change the *shape* of the distribution of X_i. For example, if $X_i \sim$ Bernoulli(0.5), then the distribution of X_i assigns equal probability to each of two values, $x = 0$ and $x = 1$. If we convert to standard units, then the distribution of

$$Z_1 = \frac{X_i - \mu}{\sigma} = \frac{X_i - 0.5}{0.5}$$

also assigns equal probability to each of two values, $z_1 = -1$ and $z_1 = 1$. In particular, notice that converting X_i to standard units does *not* automatically result in a normally distributed random variable.

Next we consider the sum and the average of $X_1, \ldots, X_n$. Notice that, after converting to standard units, these quantities are identical:

$$Z_n = \frac{\sum_{i=1}^n X_i - n\mu}{\sqrt{n}\sigma} = \frac{(1/n)}{(1/n)} \frac{\sum_{i=1}^n X_i - n\mu}{\sqrt{n}\,\sigma} = \frac{\bar{X}_n - \mu}{\sigma/\sqrt{n}}.$$

It is this new random variable on which we shall focus our attention.

We begin by observing that

$$\text{Var}\left[\sqrt{n}\left(\bar{X}_n - \mu\right)\right] = \text{Var}\left(\sigma Z_n\right) = \sigma^2\, \text{Var}\left(Z_n\right) = \sigma^2$$

is constant. The WLLN states that

$$(\bar{X}_n - \mu) \xrightarrow{P} 0,$$

so $\sqrt{n}$ is a "magnification factor" that maintains random variables with a constant positive variance. We conclude that $1/\sqrt{n}$ measures how rapidly the sample mean tends toward the population mean.

Now we turn to the more refined question of how the distribution of the sample mean changes as the sample mean tends toward the population mean. By converting to standard units, we are able to distinguish changes in the shape of the distribution from changes in its mean and variance. Despite our inability to make general statements about the behavior of Z_1, it turns out that we can say quite a bit about the behavior of Z_n as n becomes large. The following theorem is one of the most remarkable and useful results in all of mathematics. It is fundamental to the study of both probability and statistics.

Theorem 8.2 *(Central Limit Theorem) Let $X_1, X_2, \ldots$ be any sequence of independent and identically distributed random variables having finite mean μ and finite variance σ^2. Let*

$$Z_n = \frac{\bar{X}_n - \mu}{\sigma/\sqrt{n}},$$

let F_n denote the cdf of Z_n, and let Φ denote the cdf of the standard normal distribution. Then, for any fixed value $z \in \Re$,

$$P\left(Z_n \leq z\right) = F_n(z) \to \Phi(z)$$

as $n \to \infty$.

The Central Limit Theorem (CLT) states that the behavior of the average (or, equivalently, the sum) of a large number of independent and identically distributed random variables will resemble the behavior of a standard normal random variable. *This is true regardless of the distribution of the random variables that are being averaged.* Thus, the CLT allows us to approximate a variety of probabilities that otherwise would be intractable. Of course, we require some sense of how many random variables must be averaged in order for the normal approximation to be reasonably accurate. This *does* depend on the distribution of the random variables, but a popular rule of thumb is that the normal approximation can be used if $n \geq 30$. Often, the normal approximation works quite well with even smaller n.

Notation We use the symbol $\approx$ to indicate asymptotic approximation, writing $Z_n \approx \text{Normal}(0,1)$ and $F_n(z) \approx \Phi(z)$.

Example 8.1 *A chemistry professor is attempting to determine the conformation of a certain molecule. To measure the distance between a pair of nearby hydrogen atoms, she uses NMR spectroscopy. She knows that this measurement procedure has an expected value equal to the actual distance and a standard deviation of* 0.5 *angstroms. If she replicates the experiment* 36 *times, then what is the probability that the average measured value will fall within* 0.1 *angstroms of the true value?*

Let X_i denote the measurement obtained from replication i, for $i = 1, \ldots, 36$. We are told that $\mu = EX_i$ is the actual distance between the atoms and that $\sigma^2 = \text{Var}\, X_i = 0.5^2$. Let $Z \sim \text{Normal}(0,1)$. Then, applying the CLT,

$$\begin{aligned} P\left(\mu - 0.1 < \bar{X}_{36} < \mu + 0.1\right) &= P\left(\mu - 0.1 - \mu < \bar{X}_{36} - \mu < \mu + 0.1 - \mu\right) \\ &= P\left(\frac{-0.1}{0.5/6} < \frac{\bar{X}_{36} - \mu}{0.5/6} < \frac{0.1}{0.5/6}\right) \\ &= P\left(-1.2 < Z_n < 1.2\right) \\ &\approx P\left(-1.2 < Z < 1.2\right) \\ &= \Phi(1.2) - \Phi(-1.2). \end{aligned}$$

Now we use R:

```
> pnorm(1.2)-pnorm(-1.2)
[1] 0.7698607
```

We conclude that there is a chance of approximately 77% that the average of the measured values will fall with 0.1 angstroms of the true value.

Notice that it is not possible to compute the exact probability. To do so would require knowledge of the distribution of the X_i.

It is sometimes useful to rewrite the normal approximations derived from the CLT as statements of the approximate distributions of the sum and the average. For the sum we obtain

$$\sum_{i=1}^{n} X_i \approx \text{Normal}\left(n\mu, n\sigma^2\right) \tag{8.1}$$

and for the average we obtain

$$\bar{X}_n \approx \text{Normal}\left(\mu, \frac{\sigma^2}{n}\right). \tag{8.2}$$

These approximations are especially useful when combined with Theorem 5.2.

Example 8.2 *The chemistry professor in Example 8.1 asks her graduate student to replicate her experiment. The student does so* 64 *times. What is the probability that the averages of their respective measured values will fall within* 0.1 *angstroms of each other?*

The professor's measurements are

$$X_1, \ldots, X_{36} \sim \left(\mu, 0.5^2\right).$$

Applying (8.2), we obtain

$$\bar{X}_{36} \approx \text{Normal}\left(\mu, \frac{0.25}{36}\right).$$

Similarly, the student's measurements are

$$Y_1, \ldots, Y_{64} \sim \left(\mu, 0.5^2\right).$$

Applying (8.2), we obtain

$$\bar{Y}_{64} \approx \text{Normal}\left(\mu, \frac{0.25}{64}\right) \quad \text{or} \quad -\bar{Y}_{64} \approx \text{Normal}\left(-\mu, \frac{0.25}{64}\right).$$

Now we apply Theorem 5.2 to conclude that

$$\bar{X}_{36} - \bar{Y}_{64} = \bar{X}_{36} + (-\bar{Y}_{64}) \approx \text{Normal}\left(0, \frac{0.25}{36} + \frac{0.25}{64} = \frac{5^2}{48^2}\right).$$

Converting to standard units, it follows that

$$\begin{aligned} P\left(-0.1 < \bar{X}_{36} - \bar{Y}_{64} < 0.1\right) &= P\left(\frac{-0.1}{5/48} < \frac{\bar{X}_{36} - \bar{Y}_{64}}{5/48} < \frac{0.1}{5/48}\right) \\ &\approx P\left(-0.96 < Z < 0.96\right) \\ &= \Phi(0.96) - \Phi(-0.96). \end{aligned}$$

Now we use R:

```
> pnorm(.96)-pnorm(-.96)
[1] 0.6629448
```

We conclude that there is a chance of approximately 66% that the two averages will fall within 0.1 angstroms of each other.

The CLT has a long history. For the special case of $X_i \sim \text{Bernoulli}(p)$, a version of the CLT was obtained by De Moivre in the 1730s. The first attempt at a more general CLT was made by Laplace in 1810, but definitive results were not obtained until the second quarter of the 20th century. Theorem 8.2 is actually a very special case of far more general results established during that period. However, with one exception to which we now turn, it is sufficiently general for our purposes.

The astute reader may have noted that, in Examples 8.1 and 8.2, we assumed that the population mean μ was unknown but that the population variance σ^2 was known. Is this plausible? In Examples 8.1 and 8.2, it might be that the nature of the instrumentation is sufficiently well understood that the population variance may be considered known. In general, however, it seems somewhat implausible that we would know the population variance and not know the population mean.

The normal approximations employed in Examples 8.1 and 8.2 require knowledge of the population variance. If the variance is not known, then it must be estimated from the measured values. Chapters 7 and 9 describe procedures for doing so. For use with those procedures, we state the following generalization of Theorem 8.2:

Theorem 8.3 *Let $X_1, X_2, \ldots$ be any sequence of independent and identically distributed random variables having finite mean μ and finite variance σ^2. Suppose that $D_1, D_2, \ldots$ is a sequence of random variables with the property that $D_n^2 \xrightarrow{P} \sigma^2$ and let*

$$T_n = \frac{\bar{X}_n - \mu}{D_n/\sqrt{n}}.$$

Let F_n denote the cdf of T_n, and let Φ denote the cdf of the standard normal distribution. Then, for any fixed value $t \in \Re$,

$$P\left(T_n \leq t\right) = F_n(t) \rightarrow \Phi(t)$$

as $n \rightarrow \infty$.

We conclude this section with a warning. Statisticians usually invoke the CLT in order to approximate the distribution of a sum or an average of random variables $X_1, \ldots, X_n$ that are observed in the course of an experiment. The X_i need not be normally distributed themselves—indeed, the grandeur of the CLT is that it does *not* assume normality of the X_i. Nevertheless, we will discover that many important statistical procedures do assume that the X_i are normally distributed. Researchers who hope to use these procedures naturally want to believe that their X_i are normally distributed. Some look to the CLT for reassurance, but misconstrue its meaning. They hope that if only they replicate their experiment enough times, then somehow their observations will be drawn from a normal distribution. This is absurd! Suppose that a fair coin is tossed once. Let X_1 denote the number of `Heads`, so that $X_1 \sim \text{Bernoulli}(0.5)$. The Bernoulli distribution is not at all like a normal distribution. If we toss the coin one million times, then each $X_i \sim \text{Bernoulli}(0.5)$. The Bernoulli distribution does not miraculously become a normal distribution. Remember,

> *The Central Limit Theorem does* not *say that a large sample was necessarily drawn from a normal distribution!*

On some occasions, it is possible to invoke the CLT to anticipate that the random variable to be observed will behave like a normal random variable. This involves recognizing that the observed random variable is the sum or the average of lots of independent and identically distributed random variables that are not observed.

Example 8.3 *To study the effect of an insect growth regulator (IGR) on termite appetite, an entomologist plans an experiment. Each replication of the experiment will involve placing* 100 *ravenous termites in a container with a dried block of wood. The block of wood will be weighed before the experiment begins and after a fixed number of days. The random variable of interest is the decrease in weight, the amount of wood consumed by the termites. Can we anticipate the distribution of this random variable?*

The total amount of wood consumed is the sum of the amounts consumed by each termite. Assuming that the termites behave independently and identically, the CLT suggests that this sum should be approximately normally distributed.

When reasoning as in Example 8.3, one should construe the CLT as no more than suggestive. Most natural processes are far too complicated to be modelled so simplistically with any guarantee of accuracy. One should *always* examine the observed values to see if they are consistent with one's theorizing.

8.4 Exercises

1. Suppose that I toss a fair coin 100 times and observe 60 `Heads`. Now I decide to toss the same coin another 100 times. Does the Law of Averages imply that I should expect to observe another 40 `Heads`?

2. In Example 7.7, we observed a sample of size $n = 100$. A normal probability plot and kernel density estimate constructed from this sample suggested that the observations had been drawn from a nonnormal distribution. True or False: *It follows from the Central Limit Theorem that a kernel density estimate constructed from a much larger sample would more closely resemble a normal distribution.*

3. Suppose that an astragalus has the following probabilities of producing the four possible uppermost faces: $P(1) = P(6) = 0.1$, $P(3) = P(4) = 0.4$. This astragalus is to be thrown 100 times. Let X_i denote the value of the uppermost face that results from throw i.

   (a) Compute the expected value and the variance of X_i.

   (b) Compute the probability that the average value of the 100 throws will exceed 3.6.

4. Chris owns a laser pointer that is powered by two `AAAA` batteries. A pair of batteries will power the pointer for an average of five hours use, with a standard deviation of 30 minutes. Chris decides to take advantage of a sale and buys 20 2-packs of `AAAA` batteries. What is the probability that he will get to use his laser pointer for at least 105 hours before he needs to buy more batteries?

5. Consider an urn that contains 10 tickets, labelled

$$\{1, 1, 1, 1, 2, 5, 5, 10, 10, 10\}.$$

   From this urn, I propose to draw (with replacement) $n = 40$ tickets. Let Y denote the sum of the values on the tickets that are drawn.

   (a) To approximate $P(170.5 < Y < 199.5)$, one Math 351 student writes an `R` function `urn.model` that simulates the proposed experiment. Evaluating `urn.model` is like observing a value, y, of the random variable Y. Then she writes a loop that repeatedly evaluates `urn.model` and computes the proportion of times that `urn.model` produces $y \in (170.5, 199.5)$. She reasons that, if she evaluates `urn.model` a large number of times, then the observed proportion of $y \in (170.5, 199.5)$ should approximate $P(170.5 < Y < 199.5)$. Is her reasoning justified? Why or why not?

(b) Another student suggests that $P(170.5 < Y < 199.5)$ can be approximated by performing the following R commands:

```
> se <- sqrt(585.6)
> pnorm(199.5,mean=184,sd=se)-
+ pnorm(170.5,mean=184,sd=se)
```

Do you agree? Why or why not?

(c) Which approach will produce the more accurate approximation of $P(170.5 < Y < 199.5)$? Explain your reasoning.

6. A certain financial theory posits that daily fluctuations in stock prices are independent random variables. Suppose that the daily price fluctuations (in dollars) of a certain value stock are independent and identically distributed random variables $X_1, X_2, X_3, \ldots$, with $EX_i = 0.01$ and $\text{Var}\, X_i = 0.01$. (Thus, if today's price of this stock is \$50, then tomorrow's price is $\$50 + X_1$, etc.) Suppose that the daily price fluctuations (in dollars) of a certain growth stock are independent and identically distributed random variables $Y_1, Y_2, Y_3, \ldots$, with $EY_j = 0$ and $\text{Var}\, Y_j = 0.25$.

   Now suppose that both stocks are currently selling for \$50 per share and you wish to invest \$50 in one of these two stocks for a period of 400 market days. Assume that the costs of purchasing and selling a share of either stock are zero.

   (a) Approximate the probability that you will make a profit on your investment if you purchase a share of the value stock.

   (b) Approximate the probability that you will make a profit on your investment if you purchase a share of the growth stock.

   (c) Approximate the probability that you will make a profit of at least \$20 if you purchase a share of the value stock.

   (d) Approximate the probability that you will make a profit of at least \$20 if you purchase a share of the growth stock.

   (e) Assuming that the growth stock fluctuations and the value stock fluctuations are independent, approximate the probability that, after 400 days, the price of the growth stock will exceed the price of the value stock.

Chapter 9

Inference

Given a specific probability distribution, we can calculate the probabilities of various events. For example, knowing that $Y \sim \text{Binomial}(n = 100; p = 0.5)$, we can calculate $P(40 \leq Y \leq 60)$. Roughly speaking, statistics is concerned with the opposite sort of problem. For example, knowing that $Y \sim \text{Binomial}(n = 100; p)$, *where the value of* p *is unknown*, and having observed $Y = y$ (say $y = 32$), what can we say about p? The phrase *statistical inference* describes any procedure for extracting information about a probability distribution from an observed sample.

The present chapter introduces fundamental principles of statistical inference. We will discuss three types of statistical inference—point estimation, hypothesis testing, and set estimation—in the context of drawing inferences about a single population mean. More precisely, we will consider the following situation:

1. $X_1, \ldots, X_n$ are independent and identically distributed random variables. We observe a sample, $\vec{x} = \{x_1, \ldots, x_n\}$.

2. Both $EX_i = \mu$ and $\text{Var}\, X_i = \sigma^2$ exist and are finite. We are interested in drawing inferences about the population mean μ, a quantity that is fixed but unknown.

3. The sample size, n, is sufficiently large that we can use the normal approximation provided by the Central Limit Theorem.

We begin, in Section 9.1, by examining a narrative that is sufficiently nuanced to motivate each type of inferential technique. We then proceed to discuss point estimation (Section 9.2), hypothesis testing (Sections 9.3 and 9.4), and set estimation (Section 9.5). Although we are concerned exclusively with large-sample inferences about a single population mean, it should be appreciated that this concern often arises in practice. More importantly, the fundamental concepts that we introduce in this context are common to virtually all problems that involve statistical inference.

9.1 A Motivating Example

We consider an artificial example that permits us to scrutinize the precise nature of statistical reasoning. Two siblings, a magician (Arlen) and an attorney (Robynne) agree to resolve their disputed ownership of an Erté painting by tossing a penny. Arlen produces a penny and, just as Robynne is about to toss it in the air, Arlen smoothly suggests that spinning the penny on a table might ensure better randomization. Robynne assents and spins the penny. As it spins, Arlen calls "Tails!" The penny comes to rest with `Tails` facing up and Arlen takes possession of the Erté. Robynne is left with the penny.

That evening, Robynne wonders if she has been had. She decides to perform an experiment. She spins the same penny on the same table 100 times and observes 68 `Tails`. It occurs to Robynne that perhaps spinning this penny was not entirely fair, but she is reluctant to accuse her brother of impropriety until she is convinced that the results of her experiment cannot be dismissed as coincidence. How should she proceed?

It is easy to devise a mathematical model of Robynne's experiment: each spin of the penny is a Bernoulli trial and the experiment is a sequence of $n = 100$ trials. Let X_i denote the outcome of spin i, where $X_i = 1$ if `Heads` is observed and $X_i = 0$ if `Tails` is observed. Then $X_1, \ldots, X_{100} \sim \text{Bernoulli}(p)$, where p is the fixed but unknown (to Robynne!) probability that a single spin will result in `Heads`. The probability distribution Bernoulli(p) is our mathematical abstraction of a population and the population parameter of interest is $\mu = EX_i = p$, the population mean.

Let

$$Y = \sum_{i=1}^{100} X_i,$$

the total number of `Heads` obtained in $n = 100$ spins. Under the mathematical model that we have proposed, $Y \sim \text{Binomial}(p)$. In performing her experiment, Robynne observes a sample $\vec{x} = \{x_1, \ldots, x_{100}\}$ and computes

$$y = \sum_{i=1}^{100} x_i,$$

the total number of **Heads** in her sample. In our narrative, $y = 32$.

We emphasize that $p \in [0, 1]$ is fixed but unknown. Robynne's goal is to draw inferences about this fixed but unknown quantity. We consider three sets of questions that she might ask:

1. What is the true value of p? More precisely, what is a reasonable guess as to the true value of p?

2. Is $p = 0.5$? Specifically, is the evidence that $p \neq 0.5$ so compelling that Robynne can comfortably accuse Arlen of impropriety?

3. What are plausible values of p? In particular, is there a subset of $[0, 1]$ that Robynne can confidently claim contains the true value of p?

The first set of questions introduces a type of inference that statisticians call *point estimation*. We have already encountered (in Chapter 7) a natural approach to point estimation, the plug-in principle. In the present case, the plug-in principle suggests estimating the theoretical probability of success, p, by computing the observed proportion of successes,

$$\hat{p} = \frac{y}{n} = \frac{32}{100} = 0.32.$$

The second set of questions introduces a type of inference that statisticians call *hypothesis testing*. Having calculated $\hat{p} = 0.32 \neq 0.5$, Robynne is inclined to guess that $p \neq 0.5$. But how compelling is the evidence that $p \neq 0.5$? Let us play devil's advocate: perhaps $p = 0.5$, but chance produced "only" $y = 32$ instead of a value nearer $EY = np = 100 \times 0.5 = 50$. This is a possibility that we can quantify. If $Y \sim \text{Binomial}(n = 100; p = 0.5)$, then the probability that Y will deviate from its expected value by at least $|50 - 32| = 18$ is

$$\begin{aligned}
\mathbf{p} &= P(|Y - 50| \geq 18) \\
&= P(Y \leq 32 \text{ or } Y \geq 68) \\
&= P(Y \leq 32) + P(Y \geq 68) \\
&= P(Y \leq 32) + 1 - P(Y \leq 67) \\
&= \texttt{pbinom(32,100,.5)+1-pbinom(67,100,.5)} \\
&= 0.0004087772.
\end{aligned}$$

This *significance probability* seems fairly small—perhaps small enough to convince Robynne that in fact $p \neq 0.5$.

The third set of questions introduces a type of inference that statisticians call *set estimation*. We have just tested the possibility that $p = p_0$ in the special case $p_0 = 0.5$. Now, imagine testing the possibility that $p = p_0$ for each $p_0 \in [0, 1]$. Those p_0 that are not rejected as inconsistent with the observed data, $y = 32$, will constitute a set of plausible values of p.

To implement this procedure, Robynne will have to adopt a standard of implausibility. Perhaps she decides to reject p_0 as implausible when the corresponding significance probability,

$$\begin{aligned} \mathbf{p} &= P\left(|Y - 100p_0| \geq |32 - 100p_0|\right) \\ &= P\left(Y - 100p_0 \geq |32 - 100p_0|\right) + P\left(Y - 100p_0 \leq -|32 - 100p_0|\right) \\ &= P\left(Y \geq 100p_0 + |32 - 100p_0|\right) + P\left(Y \leq 100p_0 - |32 - 100p_0|\right), \end{aligned}$$

satisfies $\mathbf{p} \leq 0.1$. Recalling that $Y \sim \text{Binomial}(100; p_0)$ and using the R function `pbinom`, some trial and error reveals that $\mathbf{p} > 0.1$ if p_0 lies in the interval $[0.245, 0.404]$. (The endpoints of this interval are included.) Notice that this interval does *not* contain $p_0 = 0.5$, which we had already rejected as implausible.

9.2 Point Estimation

The goal of point estimation is to make a reasonable guess of the unknown value of a designated population quantity, e.g., the population mean. The quantity that we hope to guess is called the *estimand*.

9.2.1 Estimating a Population Mean

Suppose that the estimand is μ, the population mean. The plug-in principle suggests estimating μ by computing the mean of the empirical distribution. This leads to the plug-in estimate of μ, $\hat{\mu} = \bar{x}_n$. Thus, we estimate the mean of the population by computing the mean of the sample, which is certainly a natural thing to do.

We will distinguish between

$$\bar{x}_n = \frac{1}{n} \sum_{i=1}^{n} x_i,$$

a real number that is calculated from the sample $\vec{x} = \{x_1, \ldots, x_n\}$, and

$$\bar{X}_n = \frac{1}{n} \sum_{i=1}^{n} X_i,$$

a random variable that is a function of the random variables $X_1, \ldots, X_n$. (Such a random variable is called a *statistic*.) The latter is our rule for guessing, an *estimation procedure* or *estimator*. The former is the guess itself, the result of applying our rule for guessing to the sample that we observed, an *estimate*. An estimate is an observed value of an estimator.

The quality of an individual estimate depends on the individual sample from which it was computed and therefore is affected by chance variation. Furthermore, it is rarely possible to assess how close to correct an individual estimate may be. For these reasons, we study estimation procedures and identify the statistical properties that these random variables possess. In the present case, two properties are worth noting:

1. We know that $E\bar{X}_n = \mu$. Thus, on the average, our procedure for guessing the population mean produces the correct value. We express this property by saying that $\bar{X}_n$ is an *unbiased* estimator of μ.

   The property of unbiasedness is intuitively appealing and sometimes is quite useful. However, many excellent estimation procedures are biased and some unbiased estimators are unattractive. For example, $EX_1 = \mu$ by definition, so X_1 is also an unbiased estimator of μ; but most researchers would find the prospect of estimating a population mean with a single observation to be rather unappetizing. Indeed,

   $$\operatorname{Var} \bar{X}_n = \frac{\sigma^2}{n} < \sigma^2 = \operatorname{Var} X_1,$$

   so the unbiased estimator $\bar{X}_n$ has smaller variance than the unbiased estimator X_1.

2. The Weak Law of Large Numbers states that $\bar{X}_n \xrightarrow{P} \mu$. Thus, as the sample size increases, the estimator $\bar{X}_n$ converges in probability to the estimand μ. We express this property by saying that $\bar{X}_n$ is a *consistent* estimator of μ.

   The property of consistency is essential—it is difficult to conceive a circumstance in which one would be willing to use an estimation procedure that might fail regardless of how much data one collected. Notice that the unbiased estimator X_1 is not consistent.

9.2.2 Estimating a Population Variance

Now suppose that the estimand is σ^2, the population variance. Although we are concerned with drawing inferences about the population mean, we will discover that hypothesis testing and set estimation require knowing the population variance. If the population variance is not known, then it must be estimated from the sample.

The plug-in principle suggests estimating σ^2 by computing the variance of the empirical distribution. This leads to the plug-in estimate of σ^2,

$$\widehat{\sigma^2} = \frac{1}{n} \sum_{i=1}^{n} (x_i - \bar{x}_n)^2 .$$

The plug-in estimator of σ^2 is *biased*; in fact,

$$E\left[\frac{1}{n} \sum_{i=1}^{n} (X_i - \bar{X}_n)^2\right] = \frac{n-1}{n}\sigma^2 < \sigma^2.$$

This does not present any particular difficulties; however, if we desire an unbiased estimator, then we simply multiply the plug-in estimator by the factor $n/(n-1)$, obtaining

$$S_n^2 = \frac{n}{n-1}\left[\frac{1}{n} \sum_{i=1}^{n} (X_i - \bar{X}_n)^2\right] = \frac{1}{n-1} \sum_{i=1}^{n} (X_i - \bar{X}_n)^2 . \tag{9.1}$$

The statistic S_n^2 is the most popular estimator of σ^2 and many books refer to the estimate

$$s_n^2 = \frac{1}{n-1} \sum_{i=1}^{n} (x_i - \bar{x}_n)^2$$

as *the* sample variance. (For example, the `R` command `var` computes s_n^2.) In fact, both estimators are perfectly reasonable, consistent estimators of σ^2. We prefer S_n^2 for the rather mundane reason that using it will simplify some of the formulas that we will encounter.

9.3 Heuristics of Hypothesis Testing

Hypothesis testing is appropriate for situations in which one wants to guess which of two possible statements about a population is correct. For example, in Section 9.1 we considered the possibility that spinning a penny is fair ($p = 0.5$) versus the possibility that spinning a penny is not fair ($p \neq 0.5$). The logic of hypothesis testing is of a familar sort:

> *If an alleged coincidence seems too implausible, then we tend to believe that it wasn't really a coincidence.*

Man has engaged in this kind of reasoning for millenia. In Cicero's *De Divinatione*, Quintus exclaims:

> They are entirely fortuitous you say? Come! Come! Do you really mean that? ... When the four dice [astragali] produce the venus-throw you may talk of accident: but suppose you made a hundred casts and the venus-throw appeared a hundred times; could you call that accidental?[1]

The essence of hypothesis testing is captured by the familiar saying, "Where there's smoke, there's fire." In this section we formalize such reasoning, appealing to three prototypical examples:

1. Assessing circumstantial evidence in a criminal trial.

   For simplicity, suppose that the defendant has been charged with a single count of premeditated murder and that the jury has been instructed that it should either convict of murder in the first degree or acquit. The defendant had motive, means, and opportunity. Furthermore, two types of blood were found at the crime scene. One type was evidently the victim's. Laboratory tests demonstrated that the other type was not the victim's, but failed to demonstrate that it was not the defendant's. What should the jury do?

   The evidence used by the prosecution to try to establish a connection between the blood of the defendant and blood found at the crime scene is probabilistic, i.e., circumstantial. It will likely be presented to the jury in the language of mathematics, e.g., "Both blood samples have characteristics x, y and z; yet only 0.5% of the population has such blood." The defense will argue that this is merely an unfortunate coincidence. The jury must evaluate the evidence and decide whether or not such a coincidence is too extraordinary to be believed, i.e., they must decide if their assent to the proposition that the defendant committed the murder rises to a level of certainty sufficient to convict. If the combined weight of the evidence against the defendant is a chance of one in ten, then the jury is likely to acquit; if it is a chance of one in a million, then the jury is likely to convict.

[1]Quoted by F. N. David (1962). *Games, Gods and Gambling: A History of Probability and Statistical Ideas.* Dover Publications, New York, p. 24. Cicero rejected the conclusion that a run of one hundred venus-throws is so improbable that it must have been caused by divine intervention; however, Cicero was castigating the practice of divination. Quintus was entirely correct in suggesting that a run of one hundred venus-throws should not be rationalized as "entirely fortuitous." A modern scientist might conclude that an unusual set of astragali had been used to produce this remarkable result.

2. Assessing data from a scientific experiment.

   A study of termite foraging behavior reached the controversial conclusion that two species of termites compete for scarce food resources.[2] In this study, a site in the Sonoran desert was cleared of dead wood and toilet paper rolls were set out as food sources. The rolls were examined regularly over a period of many weeks and it was observed that only very rarely was a roll infested with both species of termites. Was this just a coincidence or were the two species competing for food?

   The scientists constructed a mathematical model of termite foraging behavior under the assumption that the two species forage independently of each other. This model was then used to quantify the probability that infestation patterns such as the one observed arise due to chance. This probability turned out to be just one in many billions—a coincidence far too extraordinary to be dismissed as such—and the researchers concluded that the two species were competing.

3. Assessing the results of Robynne's penny-spinning experiment.

   In Section 9.1 we noted that Robynne observed only $y = 32$ `Heads` when she would expect $EY = 50$ `Heads` if indeed $p = 0.5$. This is a discrepancy of $|32 - 50| = 18$, and we considered the possibility that such a large discrepancy might have been produced by chance. More precisely, we calculated $\mathbf{p} = P(|Y - EY| \geq 18)$ under the assumption that $p = 0.5$, obtaining $\mathbf{p} \doteq 0.0004$. On this basis, we speculated that Robynne might be persuaded to accuse her brother of cheating.

In each of the preceding examples, a binary decision was based on a level of assent to probabilistic evidence. At least conceptually, this level can be quantified as a *significance probability*, which we loosely interpret to mean the probability that chance would produce a coincidence at least as extraordinary as the phenomenon observed. This begs an obvious question, which we pose now for subsequent consideration: how small should a significance probability be for one to conclude that a phenomenon is not a coincidence?

We now proceed to explicate a formal model for statistical hypothesis testing that was proposed by J. Neyman and E. S. Pearson in the late 1920s and 1930s. Our presentation relies heavily on drawing simple analogies to criminal law, which we suppose is a more familiar topic than statistics to most students.

---

[2]S.C. Jones and M.W. Trosset (1991). Interference competition in desert subterranean termites. *Entomologia Experimentalis et Applicata*, 61:83–90.

The States of Nature

The states of nature are the possible mechanisms that might have produced the observed phenomenon. Mathematically, they are the possible probability distributions under consideration. Thus, in the penny-spinning example, the states of nature are the Bernoulli trials indexed by $p \in [0, 1]$. In hypothesis testing, the states of nature are partitioned into two sets or *hypotheses.* In the penny-spinning example, the hypotheses that we formulated were $p = 0.5$ (penny-spinning is fair) and $p \neq 0.5$ (penny-spinning is not fair); in the legal example, the hypotheses are that the defendant did commit the murder (the defendant is factually guilty) and that the defendant did not commit the murder (the defendant is factually innocent).

The goal of hypothesis testing is to decide which hypothesis is correct, i.e., which hypothesis contains the true state of nature. In the penny-spinning example, Robynne wants to determine whether or not spinning a penny is fair. In the termite example, Jones and Trosset wanted to determine whether or not termites were foraging independently. More generally, scientists usually partition the states of nature into a hypothesis that corresponds to a theory that the experiment is designed to investigate and a hypothesis that corresponds to a chance explanation; the goal of hypothesis testing is to decide which explanation is correct. In a criminal trial, the jury would like to determine whether the defendant is factually innocent or factually guilty. In the words of the United States Supreme Court:

> Underlying the question of guilt or innocence is an objective truth: the defendant, in fact, did or did not commit the acts constituting the crime charged. From the time an accused is first suspected to the time the decision on guilt or innocence is made, our criminal justice system is designed to enable the trier of fact to discover that truth according to law.[3]

Formulating appropriate hypotheses can be a delicate business. In the penny-spinning example, we formulated hypotheses $p = 0.5$ and $p \neq 0.5$. These hypotheses are appropriate if Robynne wants to determine whether or not penny-spinning is fair. However, one can easily imagine that Robynne is not interested in whether or not penny-spinning is fair, but rather in whether or not her brother gained an advantage by using the procedure. If so, then appropriate hypotheses would be $p < 0.5$ (penny-spinning favored Arlen) and $p \geq 0.5$ (penny-spinning did not favor Arlen).

---

[3] *Bullington v. Missouri*, 451 U. S. 430 (1981).

The Actor

The states of nature having been partitioned into two hypotheses, it is necessary for a decision maker (the actor) to choose between them. In the penny-spinning example, the actor is Robynne; in the termite example, the actor is the team of researchers; in the legal example, the actor is the jury.

Statisticians often describe hypothesis testing as a game that they play against Nature. To study this game in greater detail, it becomes necessary to distinguish between the two hypotheses under consideration. In each example, we declare one hypothesis to be the *null hypothesis* (H_0) and the other to be the *alternative hypothesis* (H_1). Roughly speaking, the logic for determining which hypothesis is H_0 and which is H_1 is the following: H_0 should be the hypothesis to which one defaults if the evidence is equivocal and H_1 should be the hypothesis that one requires compelling evidence to embrace.

We shall have a great deal more to say about distinguishing null and alternative hypotheses, but for now suppose that we have declared the following: (1) H_0: the defendant did not commit the murder, (2) H_0: the termites are foraging independently, and (3) H_0: spinning the penny is fair. Having done so, the game takes the following form:

		State of Nature	
		H_0	H_1
Actor's Choice	H_0		Type II error
	H_1	Type I error	

There are four possible outcomes to this game, two of which are favorable and two of which are unfavorable. If the actor chooses H_1 when in fact H_0 is true, then we say that a Type I error has been committed. If the actor chooses H_0 when in fact H_1 is true, then we say that a Type II error has been committed. In a criminal trial, a Type I error occurs when a jury convicts a factually innocent defendant and a Type II error occurs when a jury acquits a factually guilty defendant.

Innocent Until Proven Guilty

Because we are concerned with probabilistic evidence, any decision procedure that we devise will occasionally result in error. Obviously, we would like to devise procedures that minimize the probabilities of committing errors. Unfortunately, there is an inevitable tradeoff between Type I and Type

II error that precludes simultaneously minimizing the probabilities of both types. To appreciate this, consider two juries. The first jury always acquits and the second jury always convicts. Then the first jury *never* commits a Type I error and the second jury *never* commits a Type II error. The only way to simultaneously better both juries is to never commit an error of either type, which is impossible with probabilistic evidence.

The distinguishing feature of hypothesis testing (and Anglo-American criminal law) is the manner in which it addresses the tradeoff between Type I and Type II error. The Neyman-Pearson formulation of hypothesis testing accords the null hypothesis a privileged status: H_0 will be maintained unless there is compelling evidence against it. It is instructive to contrast the asymmetry of this formulation with situations in which neither hypothesis is privileged. In statistics, this is the problem of determining which hypothesis better explains the data. This is *discrimination*, not hypothesis testing. In law, this is the problem of determining whether the defendant or the plaintiff has the stronger case. This is the criterion in civil suits, not in criminal trials.

In the penny-spinning example, Robynne required compelling evidence against the privileged null hypothesis that penny-spinning is fair to overcome her scruples about accusing her brother of impropriety. In the termite example, Jones and Trosset required compelling evidence against the privileged null hypothesis that two termite species forage independently in order to write a credible article claiming that two species were competing with each other. In a criminal trial, the principle of according the null hypothesis a privileged status has a familiar characterization: the defendant is "innocent until proven guilty."

According the null hypothesis a privileged status is equivalent to declaring Type I errors to be more egregious than Type II errors. This connection was eloquently articulated by Supreme Court Justice John Harlan:

> The standard of proof influences the relative frequency of these two types of erroneous outcomes. If, for example, the standard of proof for a criminal trial were a preponderance of the evidence, rather than proof beyond a reasonable doubt, there would be a smaller risk of factual errors that result in freeing guilty persons, but a far greater risk of factual errors that result in convicting the innocent. Because the standard of proof affects the comparative frequency of these two types of erroneous outcomes, the choice of the standard to be applied in a particular kind of litigation should, in a rational world, reflect an assessment of the comparative social disutility of each.[4]

[4] *In re Winship*, 397 U. S. 358 (1970).

A preference for Type II errors instead of Type I errors can often be glimpsed in scientific applications. For example, because science is conservative, it is generally considered better to wrongly accept than to wrongly reject the prevailing wisdom that termite species forage independently. Moreover, just as this preference is the foundation of statistical hypothesis testing, so is it a fundamental principle of criminal law. In his famous *Commentaries*, William Blackstone opined that "it is better that ten guilty persons escape, than that one innocent man suffer;" and in his influential *Practical Treatise on the Law of Evidence* (1824), Thomas Starkie suggested that "The maxim of the law... is that it is better that ninety-nine... offenders shall escape than that one innocent man be condemned." In *Reasonable Doubts*, Alan Dershowitz quotes both maxims and notes anecdotal evidence that jurors actually do prefer committing Type II to Type I errors: on *Prime Time Live* (October 4, 1995), O.J. Simpson juror Anise Aschenbach stated, "If we made a mistake, I would rather it be a mistake on the side of a person's innocence than the other way."[5]

Beyond a Reasonable Doubt

To operationalize an aversion to Type I errors, the Neyman-Pearson formulation imposes an upper bound on the maximal probability of Type I error that will be tolerated. This bound is the *significance level*, conventionally denoted α. The significance level is specified (prior to examining the data) and only decision rules for which the probability of Type I error is no greater than α are considered. Such tests are called *level α tests.*

To fix ideas, we consider the penny-spinning example and specify a significance level of α. Let $\mathbf{p}$ denote the significance probability that results from performing the analysis in Section 9.1 and consider a rule that rejects the null hypothesis $H_0 : p = 0.5$ if and only if $\mathbf{p} \leq \alpha$. Then a Type I error occurs if and only if $p = 0.5$ and we observe y such that $\mathbf{p} = P(|Y-50| \geq |y-50|) \leq \alpha$. We claim that the probability of observing such a y is just α, in which case we have constructed a level α test.

To see why this is the case, let $W = |Y-50|$ denote the *test statistic.* The decision to accept or reject the null hypothesis H_0 depends on the observed value, w, of this random variable. Let

$$\mathbf{p}(w) = P_{H_0}(W \geq w)$$

[5] A. M. Dershowitz (1996). *Reasonable Doubts: The O. J. Simpson Case and the Criminal Justice System.* Simon & Schuster, New York, pp. 38, 212, 85.

denote the significance probability associated with w. Notice that w is the $1 - \mathbf{p}(w)$ quantile of the random variable W under H_0. Let q denote the $1 - \alpha$ quantile of W under H_0, i.e.,

$$\alpha = P_{H_0}(W \geq q).$$

We reject H_0 if and only if we observe

$$P_{H_0}(W \geq w) = \mathbf{p}(w) \leq \alpha = P_{H_0}(W \geq q),$$

i.e., if and only $w \geq q$. If H_0 is true, then the probability of committing a Type I error is precisely

$$P_{H_0}(W \geq q) = \alpha,$$

as claimed above. We conclude that α quantifies the level of assent that we require to risk rejecting H_0, i.e., the significance level specifies how small a significance probability is required in order to conclude that a phenomenon is not a coincidence.

In statistics, the significance level α is a number in the interval $[0, 1]$. It is not possible to quantitatively specify the level of assent required for a jury to risk convicting an innocent defendant, but the legal principle is identical: in a criminal trial, the operative significance level is *beyond a reasonable doubt.* Starkie (1824) described the possible interpretations of this phrase in language derived from British empirical philosopher John Locke:

> Evidence which satisfied the minds of the jury of the truth of the fact in dispute, to the entire exclusion of every reasonable doubt, constitute full proof of the fact. ... Even the most direct evidence can produce nothing more than such a high degree of probability as amounts to moral certainty. From the highest it may decline, by an infinite number of gradations, until it produces in the mind nothing more than a preponderance of assent in favour of the particular fact.[6]

The gradations that Starkie described are not intrinsically numeric, but it is evident that the problem of defining reasonable doubt in criminal law is the problem of specifying a significance level in statistical hypothesis testing.

In both criminal law and statistical hypothesis testing, actions typically are described in language that acknowledges the privileged status of the null

[6] T. Starkie (1824). *Practical Treatise on the Law of Evidence*, 2 volumes. London, p. 478 of the 1833 edition. Quoted by B. J. Shapiro (1991). *"Beyond Reasonable Doubt" and "Probable Cause": Historical Perspectives on the Anglo-American Law of Evidence.* University of California Press, Berkeley, p. 35.

hypothesis and emphasizes that the decision criterion is based on the probability of committing a Type I error. In describing the action of choosing H_0, many statisticians prefer the phrase "fail to reject the null hypothesis" to the less awkward "accept the null hypothesis" because choosing H_0 does *not* imply an affirmation that H_0 is correct, only that the level of evidence against H_0 is not sufficiently compelling to warrant its rejection at significance level α. In precise analogy, juries render verdicts of "not guilty" rather than "innocent" because acquital does not imply an affirmation that the defendant did not commit the crime, only that the level of evidence against the defendant's innocence was not beyond a reasonable doubt.[7]

And To a Moral Certainty

The Neyman-Pearson formulation of statistical hypothesis testing is a mathematical abstraction. Part of its generality derives from its ability to accommodate *any* specified significance level. As a practical matter, however, α must be specified and we now ask how to do so.

In the penny-spinning example, Robynne is making a personal decision and is free to choose α as she pleases. In the termite example, the researchers were guided by decades of scientific convention. In 1925, in his extremely influential *Statistical Methods for Research Workers*, Ronald Fisher[8] suggested that $\alpha = 0.10$, $\alpha = 0.05$, and $\alpha = 0.02$ are often appropriate significance levels. These suggestions were intended as practical guidelines, but they have become enshrined (especially $\alpha = 0.05$) in the minds of many scientists as a sort of Delphic determination of whether or not a hypothesized theory is true.[9] While some degree of conformity is desirable (it inhibits a researcher from choosing—after the fact—a significance level that will permit rejecting

[7]In contrast, Scottish law permits a jury to return a verdict of "not proven," thereby reserving a verdict of "not guilty" to affirm a defendant's innocence.

[8]R. A. Fisher (1925). *Statistical Methods for Research Workers.* Re-issued in *Statistical Methods, Experimental Design, and Scientific Inference*, Oxford University Press, Oxford, 1995. Sir Ronald Fisher is properly regarded as the single most important figure in the history of statistics. It should be noted that he did not subscribe to all of the particulars of the Neyman-Pearson formulation of hypothesis testing. His fundamental objection to it, that it may not be possible to fully specify the alternative hypothesis, does not impact our development, as we are concerned with situations in which both hypotheses are fully specified. Fisher, Neyman, and Pearson all accepted the fundamental principle that the null hypothesis should be accorded a privileged status and maintained unless there is compelling evidence against it.

[9]Perhaps this development was inevitable. For decades, one could perform a test only by comparing the value of one's test statistic to a table of critical values. Fisher (1925) was obliged to choose a small number of significance levels in constructing his tables of critical values; researchers who used those tables were obliged to adopt one of Fisher's

the null hypothesis in favor of the alternative in which s/he may be invested), many statisticians are disturbed by the scientific community's slavish devotion to a single standard and by its often uncritical interpretation of the resulting conclusions.[10]

The imposition of an arbitrary standard like $\alpha = 0.05$ is possible because of the precision with which mathematics allows hypothesis testing to be formulated. Applying this precision to legal paradigms reveals the issues with great clarity, but is of little practical value when specifying a significance level, i.e., when trying to define the meaning of "beyond a reasonable doubt." Nevertheless, legal scholars have endeavored for centuries to position "beyond a reasonable doubt" along the infinite gradations of assent that correspond to the continuum $[0, 1]$ from which α is selected. The phrase "beyond a reasonable doubt" is still often connected to the archaic phrase "to a moral certainty." This connection survived because moral certainty was actually a significance level, intended to invoke an enormous body of scholarly writings and specify a level of assent:

> Throughout this development two ideas to be conveyed to the jury have been central. The first idea is that there are two realms of human knowledge. In one it is possible to obtain the absolute certainty of mathematical demonstration, as when we say that the square of the hypotenuse is equal to the sum of the squares of the other two sides of a right triangle. In the other, which is the empirical realm of events, absolute certainty of this kind is not possible. The second idea is that, in this realm of events, just because absolute certainty is not possible, we ought not to treat everything as merely a guess or a matter of opinion. Instead, in this realm there are levels of certainty, and we reach higher levels of certainty as the quantity and quality of the evidence available to us increase. The highest level of certainty in this empirical realm in which no absolute certainty is possible is what traditionally was called "moral certainty," a certainty which there was no reason to doubt.[11]

Although it is rarely (if ever) possible to quantify a juror's level of assent, those comfortable with statistical hypothesis testing may be inclined

---

signficance levels. Our use of `R` instead of tables affords us considerably greater freedom.

[10]See, for example, J. Cohen (1994). The world is round ($p < .05$). *American Psychologist*, 49:997–1003.

[11]B. J. Shapiro (1991). *"Beyond Reasonable Doubt" and "Probable Cause": Historical Perspectives on the Anglo-American Law of Evidence*, University of California Press, Berkeley, p. 41. I am greatly indebted to Shapiro's fascinating study.

to wonder what values of α correspond to conventional interpretations of reasonable doubt. If a juror believes that there is a 5 percent probability that chance alone could have produced the circumstantial evidence presented against a defendant accused of premeditated murder, is the juror's level of assent beyond a reasonable doubt and to a moral certainty? We hope not. We may be willing to tolerate a 5 percent probability of a Type I error when studying termite foraging behavior, but the analogous prospect of a 5 percent probability of wrongly convicting a factually innocent defendant is abhorrent.[12]

In fact, little is known about how anyone in the legal system quantifies reasonable doubt. Mary Gray cites a 1962 Swedish case in which a judge trying an overtime parking case explicitly ruled that a significance probability of 1/20736 was beyond reasonable doubt but that a significance probability of 1/144 was not.[13] In contrast, Alan Dershowitz relates a provocative classroom exercise in which his students preferred to acquit in one scenario with a significance probability of 10 percent and to convict in an analogous scenario with a significance probability of 15 percent.[14]

9.4 Testing Hypotheses about a Population Mean

We now apply the heuristic reasoning described in Section 9.3 to the problem of testing hypotheses about a population mean. Initially, we consider testing $H_0 : \mu = \mu_0$ versus $H_1 : \mu \neq \mu_0$.

The intuition that we are seeking to formalize is fairly straightfoward. By virtue of the Weak Law of Large Numbers, the observed sample mean ought to be fairly close to the true population mean. Hence, if the null hypothesis is true, then $\bar{x}_n$ ought to be fairly close to the hypothesized mean, μ_0. If we observe $\bar{X}_n = \bar{x}_n$ far from μ_0, then we guess that $\mu \neq \mu_0$, i.e., we reject H_0.

Given a significance level α, we want to calculate a significance probability $\mathbf{p}$. The significance level is a real number that is fixed by and known to

[12]This discrepancy illustrates that the consequences of committing a Type I error influence the choice of a significance level. The consequences of Jones and Trosset wrongly concluding that termite species compete are not commensurate with the consequences of wrongly imprisoning a factually innocent citizen.

[13]M.W. Gray (1983). Statistics and the law. *Mathematics Magazine*, 56:67–81. As a graduate of Rice University, I cannot resist quoting another of Gray's examples of statistics-as-evidence: "In another case, that of millionaire W. M. Rice, the signature on his will was disputed, and the will was declared a forgery on the basis of probability evidence. As a result, the fortune of Rice went to found Rice Institute."

[14]A. M. Dershowitz (1996). *Reasonable Doubts: The O. J. Simpson Case and the Criminal Justice System.* Simon & Schuster, New York, p. 40.

the researcher, e.g., $\alpha = 0.05$. The significance probability is a real number that is determined by the sample, e.g., $\mathbf{p} \doteq 0.0004$ in Section 9.1. We will reject H_0 if and only if $\mathbf{p} \leq \alpha$.

In Section 9.3 we interpreted the significance probability as the probability that chance would produce a coincidence at least as extraordinary as the phenomenon observed. Our first challenge is to make this notion mathematically precise; how we do so depends on the hypotheses that we want to test. In the present situation, we submit that a natural significance probability is

$$\mathbf{p} = P_{\mu_0}\left(|\bar{X}_n - \mu_0| \geq |\bar{x}_n - \mu_0|\right). \tag{9.2}$$

To understand why, it is essential to appreciate the following details:

1. The hypothesized mean, μ_0, is a real number that is fixed by and known to the researcher.

2. The estimated mean, $\bar{x}_n$, is a real number that is calculated from the observed sample and known to the researcher; hence, the quantity $|\bar{x}_n - \mu_0|$ is a fixed real number.

3. The estimator, $\bar{X}_n$, is a random variable. Hence, the inequality

$$|\bar{X}_n - \mu_0| \geq |\bar{x}_n - \mu_0| \tag{9.3}$$

   defines an event that may or may not occur each time the experiment is performed. Specifically, (9.3) is the event that the sample mean assumes a value at least as far from the hypothesized mean as the researcher observed.

4. The significance probability, $\mathbf{p}$, is the probability that (9.3) occurs. The notation P_{μ_0} reminds us that we are interested in the probability that this event occurs *under the assumption that the null hypothesis is true*, i.e., under the assumption that $\mu = \mu_0$.

Having formulated an appropriate significance probability for testing $H_0 : \mu = \mu_0$ versus $H_1 : \mu \neq \mu_0$, our second challenge is to find a way to compute $\mathbf{p}$. We remind the reader that we have assumed that n is large.

Case 1: The population variance is known or specified by the null hypothesis.

We define two new quantities, the random variable

$$Z_n = \frac{\bar{X}_n - \mu_0}{\sigma/\sqrt{n}}$$

and the real number

$$z = \frac{\bar{x}_n - \mu_0}{\sigma/\sqrt{n}}.$$

Under the null hypothesis that $\mu = \mu_0$, $Z_n \dot{\sim} \text{Normal}(0,1)$ by the Central Limit Theorem; hence,

$$\begin{aligned}
\mathbf{p} &= P_{\mu_0}\left(|\bar{X}_n - \mu_0| \geq |\bar{x}_n - \mu_0|\right) \\
&= 1 - P_{\mu_0}\left(-|\bar{x}_n - \mu_0| < \bar{X}_n - \mu_0 < |\bar{x}_n - \mu_0|\right) \\
&= 1 - P_{\mu_0}\left(-\frac{|\bar{x}_n - \mu_0|}{\sigma/\sqrt{n}} < \frac{\bar{X}_n - \mu_0}{\sigma/\sqrt{n}} < \frac{|\bar{x}_n - \mu_0|}{\sigma/\sqrt{n}}\right) \\
&= 1 - P_{\mu_0}\left(-|z| < Z_n < |z|\right) \\
&\approx 1 - \left[\Phi(|z|) - \Phi(-|z|)\right] \\
&= 2\Phi(-|z|),
\end{aligned}$$

which can be computed by the following R command:

```
> 2*pnorm(-abs(z))
```

An illustration of the normal probability of interest is sketched in Figure 9.1.

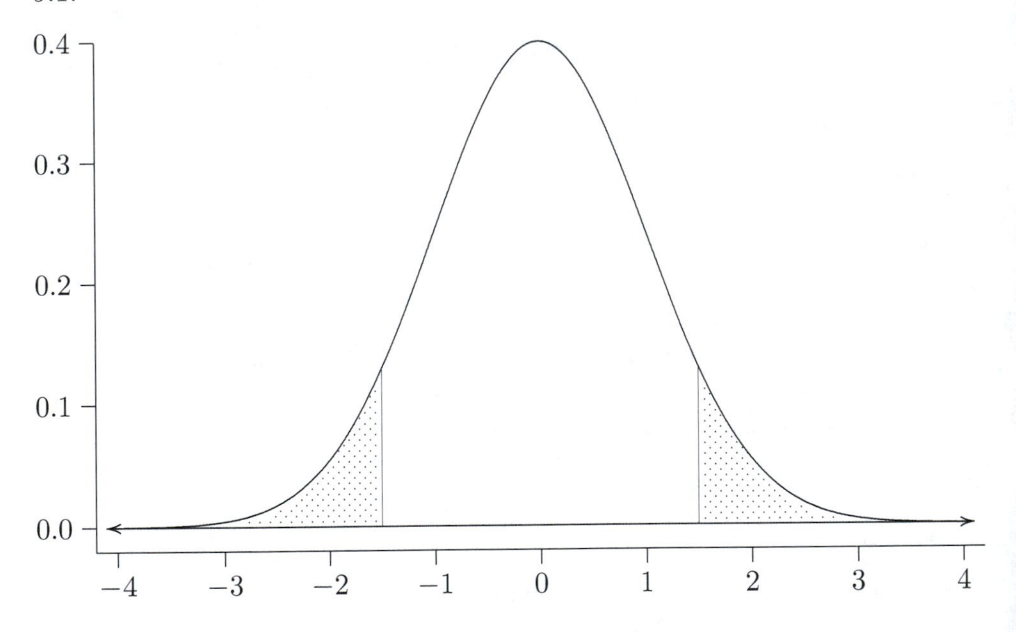

Figure 9.1: $P(|Z| \geq |z| = 1.5)$

An important example of Case 1 occurs when $X_i \sim \text{Bernoulli}(\mu)$. In this case, $\sigma^2 = \text{Var}\, X_i = \mu(1-\mu)$; hence, under the null hypothesis that $\mu = \mu_0$, $\sigma^2 = \mu_0(1-\mu_0)$ and

$$z = \frac{\bar{x}_n - \mu_0}{\sqrt{\mu_0(1-\mu_0)/n}}.$$

Example 9.1 *To test $H_0 : \mu = 0.5$ versus $H_1 : \mu \neq 0.5$ at significance level $\alpha = 0.05$, we perform $n = 2500$ trials and observe 1200 successes. Should H_0 be rejected?*

The observed proportion of successes is $\bar{x}_n = 1200/2500 = 0.48$, so the value of the test statistic is

$$z = \frac{0.48 - 0.50}{\sqrt{0.5(1-0.5)/2500}} = \frac{-0.02}{0.5/50} = -2$$

and the significance probability is

$$\mathbf{p} \approx 2\Phi(-2) \doteq 0.0456 < 0.05 = \alpha.$$

Because $\mathbf{p} \leq \alpha$, we reject H_0.

Case 2: The population variance is unknown.

Because σ^2 is unknown, we must estimate it from the sample. We will use the estimator introduced in Section 9.2,

$$S_n^2 = \frac{1}{n-1} \sum_{i=1}^{n} \left(X_i - \bar{X}_n\right)^2,$$

and define

$$T_n = \frac{\bar{X}_n - \mu_0}{S_n/\sqrt{n}}.$$

Because S_n^2 is a consistent estimator of σ^2, i.e., $S_n^2 \xrightarrow{P} \sigma^2$, it follows from Theorem 8.3 that

$$\lim_{n\to\infty} P\left(T_n \leq z\right) = \Phi(z).$$

Just as we could use a normal approximation to compute probabilities involving Z_n, so can we use a normal approximation to compute probabilities involving T_n. The fact that we must estimate σ^2 slightly degrades the quality of the approximation; however, because n is large, we should observe an accurate estimate of σ^2 and the approximation should not suffer much. Accordingly, we proceed as in Case 1, using

$$t = \frac{\bar{x}_n - \mu_0}{s_n/\sqrt{n}}$$

instead of z.

Example 9.2 *To test $H_0 : \mu = 20$ versus $H_1 : \mu \neq 20$ at significance level $\alpha = 0.05$, we collect $n = 400$ observations, observing $\bar{x}_n = 21.82935$ and $s_n = 24.70037$. Should H_0 be rejected?*

The value of the test statistic is

$$t = \frac{21.82935 - 20}{24.70037/20} = 1.481234$$

and the significance probability is

$$\mathbf{p} \approx 2\Phi(-1.481234) \doteq 0.1385 > 0.05 = \alpha.$$

Because $\mathbf{p} > \alpha$, we decline to reject H_0.

9.4.1 One-Sided Hypotheses

In Section 9.3 we suggested that, if Robynne is not interested in whether or not penny-spinning is fair but rather in whether or not it favors her brother, then appropriate hypotheses would be $p < 0.5$ (penny-spinning favors Arlen) and $p \geq 0.5$ (penny-spinning does not favor Arlen). These are examples of one-sided (as opposed to two-sided) hypotheses.

More generally, we will consider two canonical cases:

$$\begin{array}{lcl} H_0 : \mu \leq \mu_0 & \text{versus} & H_1 : \mu > \mu_0 \\ H_0 : \mu \geq \mu_0 & \text{versus} & H_1 : \mu < \mu_0 \end{array}$$

Notice that the possibility of equality, $\mu = \mu_0$, belongs to the null hypothesis in both cases. This is a technical necessity that arises because we compute significance probabilities using the μ in H_0 that is nearest H_1. For such a μ to exist, the boundary between H_0 and H_1 must belong to H_0. We will return to this necessity later in this section.

Instead of memorizing different formulas for different situations, we will endeavor to understand which values of our test statistic tend to undermine the null hypothesis in question. Such reasoning can be used on a case-by-case basis to determine the relevant significance probability. In so doing, sketching crude pictures can be quite helpful!

Consider testing each of the following:

$$\begin{array}{llll} \text{(a)} & H_0: \mu = \mu_0 & \text{versus} & H_1: \mu \neq \mu_0 \\ \text{(b)} & H_0: \mu \leq \mu_0 & \text{versus} & H_1: \mu > \mu_0 \\ \text{(c)} & H_0: \mu \geq \mu_0 & \text{versus} & H_1: \mu < \mu_0 \end{array}$$

Qualitatively, we will be inclined to reject the null hypothesis if and only if:

(a) We observe $\bar{x}_n \ll \mu_0$ or $\bar{x}_n \gg \mu_0$, i.e., if we observe $|\bar{x}_n - \mu_0| \gg 0$.

This is equivalent to observing $|t| \gg 0$, so the significance probability is

$$\mathbf{p}_a = P_{\mu_0}\left(|T_n| \geq |t|\right).$$

(b) We observe $\bar{x}_n \gg \mu_0$, i.e., if we observe $\bar{x}_n - \mu_0 \gg 0$.

This is equivalent to observing $t \gg 0$, so the significance probability is

$$\mathbf{p}_b = P_{\mu_0}\left(T_n \geq t\right).$$

(c) We observe $\bar{x}_n \ll \mu_0$, i.e., if we observe $\bar{x}_n - \mu_0 \ll 0$.

This is equivalent to observing $t \ll 0$, so the significance probability is

$$\mathbf{p}_c = P_{\mu_0}\left(T_n \leq t\right).$$

Example 9.2 (continued) Applying the above reasoning, we obtain the significance probabilities sketched in Figure 9.2. Notice that $\mathbf{p}_b = \mathbf{p}_a/2$ and that $\mathbf{p}_b + \mathbf{p}_c = 1$. The probability $\mathbf{p}_b$ is fairly small, about 7%. This makes sense: we observed $\bar{x}_n \doteq 21.8 > 20 = \mu_0$, so the sample does contain *some* evidence that $\mu > 20$. However, the statistical test reveals that the strength of this evidence is not sufficiently compelling to reject $H_0: \mu \leq 20$.

In contrast, the probability of $\mathbf{p}_c$ is quite large, about 93%. This also makes sense, because the sample contains *no* evidence that $\mu < 20$. In such instances, performing a statistical test confirms only that which is transparent from comparing the sample and hypothesized means.

9.4.2 Formulating Suitable Hypotheses

Examples 9.1 and 9.2 illustrated the mechanics of hypothesis testing. Once understood, the above techniques for calculating significance probabilities are fairly straightforward and can be applied routinely to a wide variety of problems. In contrast, determining suitable hypotheses to be tested requires one to carefully consider each situation presented. These determinations

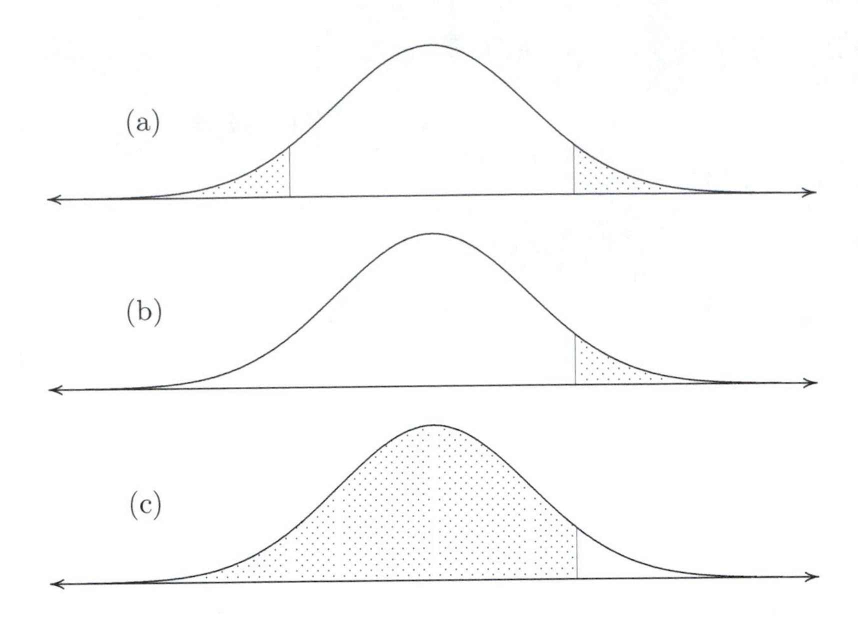

Figure 9.2: Significance probabilities for Example 9.2. Each significance probability is the area of the corresponding shaded region.

cannot be reduced to formulas. To make them requires good judgment, which can only be acquired through practice.

We now consider some examples that illustrate some important issues that arise when formulating hypotheses. In each case, there are certain key questions that must be answered: *Why was the experiment performed? Who needs to be convinced of what? Is one type of error perceived as more important than the other?*

Example 9.3 *A group of concerned parents wants speed humps installed in front of a local elementary school, but the city traffic office is reluctant to allocate funds for this purpose. Both parties agree that humps should be installed if the average speed of all motorists who pass the school while it is in session exceeds the posted speed limit of* 15 *miles per hour (mph). Let* μ *denote the average speed of the motorists in question. A random sample of* $n = 150$ *of these motorists was observed to have a sample mean of* $\bar{x} = 15.3$ *mph with a sample standard deviation of* $s = 2.5$ *mph.*

(a) State null and alternative hypotheses that are appropriate from the parents' perspective.

(b) *State null and alternative hypotheses that are appropriate from the city traffic office's perspective.*

(c) *Compute the value of an appropriate test statistic.*

(d) *Adopting the parents' perspective and assuming that they are willing to risk a 1% chance of committing a Type I error, what action should be taken? Why?*

(e) *Adopting the city traffic office's perspective and assuming that they are willing to risk a 10% chance of committing a Type I error, what action should be taken? Why?*

Solution

(a) The parents would prefer to err on the side of protecting their children, so they would rather build unnecessary speed humps than forgo necessary speed humps. Hence, they would like to see the hypotheses formulated so that forgoing necessary speed humps is a Type I error. Because speed humps will be built if it is concluded that $\mu > 15$ and will not be built if it is concluded that $\mu < 15$, the parents would prefer a null hypothesis of $H_0 : \mu \geq 15$ and an alternative hypothesis of $H_1 : \mu < 15$.

Equivalently, if we suppose that the purpose of the experiment is to provide evidence to the parents, then it is clear that the parents need to be persuaded that speed humps are unnecessary. The null hypothesis to which they will default in the absence of compelling evidence is $H_0 : \mu \geq 15$. They will require compelling evidence to the contrary, $H_1 : \mu < 15$.

(b) The city traffic office would prefer to err on the side of conserving their budget for important public works, so they would rather forgo necessary speed humps than build unnecessary speed humps. Hence, they would like to see the hypotheses formulated so that building unnecessary speed humps is a Type I error. Because speed humps will be built if it is concluded that $\mu > 15$ and will not be built if it is concluded that $\mu < 15$, the city traffic office would prefer a null hypothesis of $H_0 : \mu \leq 15$ and an alternative hypothesis of $H_1 : \mu > 15$.

Equivalently, if we suppose that the purpose of the experiment is to provide evidence to the city traffic, then it is clear that the office needs to be persuaded that speed humps are necessary. The null hypothesis to which it will default in the absence of compelling evidence is $H_0 : \mu \leq 15$. It will require compelling evidence to the contrary, $H_1 : \mu > 15$.

(c) Because the population variance is unknown, the appropriate test statistic is

$$t = \frac{\bar{x} - \mu_0}{s/\sqrt{n}} = \frac{15.3 - 15}{2.5/\sqrt{150}} \doteq 1.47.$$

(d) We would reject the null hypothesis in (a) if $\bar{x}$ is sufficiently smaller than $\mu_0 = 15$. Because $\bar{x} = 15.3 > 15$, there is no evidence against $H_0 : \mu \geq 15$. The null hypothesis is retained and speed humps are installed.

(e) We would reject the null hypothesis in (b) if $\bar{x}$ is sufficiently larger than $\mu_0 = 15$, i.e., for sufficiently large positive values of t. Hence, the significance probability is

$$\mathbf{p} = P\left(T_n \geq t\right) \approx P(Z \geq 1.47) = 1 - \Phi(1.47) \doteq 0.071 < 0.10 = \alpha.$$

Because $\mathbf{p} \leq \alpha$, the traffic office should reject $H_0 : \mu \leq 15$ and install speed humps.

Example 9.4 *Imagine a variant of the Lanarkshire milk experiment described in Section 1.2. Suppose that it is known that 10-year-old Scottish schoolchildren gain an average of* 0.5 *pounds per month. To study the effect of daily milk supplements, a random sample of* $n = 1000$ *such children is drawn. Each child receives a daily supplement of 3/4 cups pasteurized milk. The study continues for four months and the weight gained by each student during the study period is recorded. Formulate suitable null and alternative hypotheses for testing the effect of daily milk supplements.*

Solution Let $X_1, \ldots, X_n$ denote the weight gains and let $\mu = EX_i$. Then milk supplements are effective if $\mu > 2$ and ineffective if $\mu < 2$. One of these possibilities will be declared the null hypothesis, the other will be declared the alternative hypothesis. The possibility $\mu = 2$ will be incorporated into the null hypothesis.

The alternative hypothesis should be the one for which compelling evidence is desired. Who needs to be convinced of what? The parents and teachers already believe that daily milk supplements are beneficial and would have to be convinced otherwise. But this is not the purpose of the study!

The study is performed for the purpose of obtaining objective scientific evidence that supports prevailing popular wisdom. It is performed to convince government bureaucrats that spending money on daily milk supplements for schoolchildren will actually have a beneficial effect. The parents and teachers hope that the study will provide compelling evidence of this effect. Thus, the appropriate alternative hypothesis is $H_1 : \mu > 2$ and the appropriate null hypothesis is $H_0 : \mu \leq 2$.

9.4.3 Statistical Significance and Material Significance

The significance probability is the probability that a coincidence at least as extraordinary as the phenomenon observed can be produced by chance. The smaller the significance probability, the more confidently we reject the null hypothesis. However, it is one thing to be convinced that the null hypothesis is incorrect—it is something else to assert that the true state of nature is very different from the state(s) specified by the null hypothesis.

Example 9.5 A government agency requires prospective advertisers to provide statistical evidence that documents their claims. In order to claim that a gasoline additive increases mileage, an advertiser must fund an independent study in which n vehicles are tested to see how far they can drive, first without and then with the additive. Let X_i denote the increase in miles per gallon (mpg with the additive minus mpg without the additive) observed for vehicle i and let $\mu = EX_i$. The null hypothesis $H_0 : \mu \leq 1$ is tested against the alternative hypothesis $H_1 : \mu > 1$ and advertising is authorized if H_0 is rejected at a significance level of $\alpha = 0.05$.

Consider the experiences of two prospective advertisers:

1. A large corporation manufactures an additive that increases mileage by an average of $\mu = 1.01$ miles per gallon. The corporation funds a large study of $n = 900$ vehicles in which $\bar{x} = 1.01$ and $s = 0.1$ are observed. This results in a test statistic of

$$t = \frac{\bar{x} - \mu_0}{s/\sqrt{n}} = \frac{1.01 - 1.00}{0.1/\sqrt{900}} = 3$$

and a significance probability of

$$\mathbf{p} = P(T_n \geq t) \approx P(Z \geq 3) = 1 - \Phi(3) \doteq 0.00135 < 0.05 = \alpha.$$

The null hypothesis is decisively rejected and advertising is authorized

2. An amateur automotive mechanic invents an additive that increases mileage by an average of $\mu = 1.21$ miles per gallon. The mechanic funds a small study of $n = 9$ vehicles in which $\bar{x} = 1.21$ and $s = 0.4$ are observed. This results in a test statistic of

$$t = \frac{\bar{x} - \mu_0}{s/\sqrt{n}} = \frac{1.21 - 1.00}{0.4/\sqrt{9}} = 1.575$$

and (assuming that the normal approximation remains valid) a significance probability of

$$\mathbf{p} = P\left(T_n \geq t\right) \approx P(Z \geq 1.575) = 1 - \Phi(1.575) \doteq 0.05763 > 0.05 = \alpha.$$

The null hypothesis is not rejected and advertising is not authorized.

These experiences are highly illuminating. Although the corporation's mean increase of $\mu = 1.01$ mpg is much closer to the null hypothesis than the mechanic's mean increase of $\mu = 1.21$ mpg, the corporation's study resulted in a much smaller significance probability. This occurred because of the smaller standard deviation and larger sample size in the corporation's study. As a result, the government could be more confident that the corporation's product had a mean increase of more than 1.0 mpg than they could be that the mechanic's product had a mean increase of more than 1.0 mpg.

The preceding example illustrates that a small significance probability does not imply a large physical effect and that a large physical effect does not imply a small significance probability. To avoid confusing these two concepts, statisticians distinguish between statistical significance and *material significance* (importance). To properly interpret the results of hypothesis testing, it is essential that one remember:

Statistical significance is not the same as material significance.

9.5 Set Estimation

Hypothesis testing is concerned with situations that demand a binary decision, e.g., whether or not to install speed humps in front of an elementary school. The relevance of hypothesis testing in situations that do not demand a binary decision is somewhat less clear. For example, many statisticians feel that the scientific community overuses hypothesis testing and that other types of statistical inference are often more appropriate. As we have discussed, a typical application of hypothesis testing in science partitions the states of nature into two sets, one that corresponds to a theory and one that corresponds to chance. Usually the theory encompasses a great many possible states of nature and the mere conclusion that the theory is true only begs the question of which states of nature are actually plausible. Furthermore, it is a rather fanciful conceit to imagine that a single scientific article should attempt to decide whether a theory is or is not true. A more sensible enterprise for the authors to undertake is simply to set forth the evidence that they have discovered and allow evidence to accumulate until the scientific community reaches a consensus. One way to accomplish this is for each article to identify what its authors consider a set of plausible values for the population quantity in question.

To construct a set of plausible values of μ, we imagine testing $H_0 : \mu = \mu_0$ versus $H_1 : \mu \neq \mu_0$ for *every* $\mu_0 \in (-\infty, \infty)$ and eliminating those μ_0 for which $H_0 : \mu = \mu_0$ is rejected. To see where this leads, let us examine our decision criterion in the case that σ is known: we reject $H_0 : \mu = \mu_0$ if and only if

$$\mathbf{p} = P_{\mu_0}\left(\left|\bar{X}_n - \mu_0\right| \geq \left|\bar{x}_n - \mu_0\right|\right) \approx 2\Phi\left(-\left|z\right|\right) \leq \alpha, \tag{9.4}$$

where $z = (\bar{x}_n - \mu_0)/(\sigma/\sqrt{n})$. Using the symmetry of the normal distribution, we can rewrite condition (9.4) as

$$\alpha/2 \geq \Phi\left(-\left|z\right|\right) = P\left(Z < -\left|z\right|\right) = P\left(Z > \left|z\right|\right),$$

which in turn is equivalent to the condition

$$\Phi\left(\left|z\right|\right) = P\left(Z < \left|z\right|\right) = 1 - P\left(Z > \left|z\right|\right) \geq 1 - \alpha/2, \tag{9.5}$$

where $Z \sim \text{Normal}(0, 1)$.

Now let q denote the $1 - \alpha/2$ quantile of Normal$(0, 1)$, so that

$$\Phi(q) = 1 - \alpha/2.$$

Then condition (9.5) obtains if and only if $|z| \geq q$. We express this by saying that q is the *critical value* of the test statistic $|Z_n|$, where $Z_n = (\bar{X}_n - \mu_0)/(\sigma/\sqrt{n})$. For example, suppose that $\alpha = 0.05$, so that $1 - \alpha/2 = 0.975$. Then the critical value is computed in R as follows:

```
> qnorm(.975)
[1] 1.959964
```

Given a significance level α and the corresponding q, we have determined that q is the critical value of $|Z_n|$ for testing $H_0 : \mu = \mu_0$ versus $H_1 : \mu \neq \mu_0$

at significance level α. Thus, we reject $H_0 : \mu = \mu_0$ if and only if (iff)

$$\begin{aligned} & \left| \frac{\bar{x}_n - \mu_0}{\sigma/\sqrt{n}} \right| = |z| \geq q \\ \text{iff} \quad & |\bar{x}_n - \mu_0| \geq q\sigma/\sqrt{n} \\ \text{iff} \quad & \mu_0 \notin \left(\bar{x}_n - q\sigma/\sqrt{n},\ \bar{x}_n + q\sigma/\sqrt{n}\right). \end{aligned}$$

Thus, the desired set of plausible values is the interval

$$\left(\bar{x}_n - q\frac{\sigma}{\sqrt{n}},\ \bar{x}_n + q\frac{\sigma}{\sqrt{n}}\right). \tag{9.6}$$

If σ is unknown, then the argument is identical except that we estimate σ^2 as

$$s_n^2 = \frac{1}{n-1}\sum_{i=1}^{n}(x_i - \bar{x}_n)^2,$$

obtaining as the set of plausible values the interval

$$\left(\bar{x}_n - q\frac{s_n}{\sqrt{n}},\ \bar{x}_n + q\frac{s_n}{\sqrt{n}}\right). \tag{9.7}$$

Example 9.2 (continued) *A random sample of $n = 400$ observations is drawn from a population with unknown mean μ and unknown variance σ^2, resulting in $\bar{x}_n = 21.82935$ and $s_n = 24.70037$. Using a significance level of $\alpha = 0.05$, determine a set of plausible values of μ.*

First, because $\alpha = 0.05$ is the significance level, $q = 1.959964$ is the critical value. From (9.7), an interval of plausible values is

$$21.82935 \pm 1.959964 \cdot 24.70037/\sqrt{400} = (19.40876, 24.24994).$$

Notice that $20 \in (19.40876, 24.24994)$, meaning that (as we discovered in Section 9.4) we would not reject $H_0 : \mu = 20$ at significance level $\alpha = 0.05$.

Now consider the random interval I, defined in Case 1 (population variance known) by

$$I = \left(\bar{X}_n - q\frac{\sigma}{\sqrt{n}},\ \bar{X}_n - q\frac{\sigma}{\sqrt{n}}\right)$$

and in Case 2 (population variance unknown) by

$$I = \left(\bar{X}_n - q\frac{S_n}{\sqrt{n}},\ \bar{X}_n - q\frac{S_n}{\sqrt{n}}\right).$$

The probability that this random interval covers the real number μ_0 is

$$P_\mu(I \supset \mu_0) = 1 - P_\mu(\mu_0 \notin I) = 1 - P_\mu(\text{reject } H_0 : \mu = \mu_0).$$

If $\mu = \mu_0$, then the probability of coverage is

$$1 - P_{\mu_0}(\text{reject } H_0 : \mu = \mu_0) = 1 - P_{\mu_0}(\text{Type I error}) \geq 1 - \alpha.$$

Thus, the probability that I covers the true value of the population mean is at least $1-\alpha$, which we express by saying that I is a $(1-\alpha)$-*level confidence interval* for μ. The level of confidence, $1-\alpha$, is also called the *confidence coefficient*.

We emphasize that the confidence interval I is random and the population mean μ is fixed, albeit unknown. Each time that the experiment in question is performed, a random sample is observed and an interval is constructed from it. As the sample varies, so does the interval. Any one such interval, constructed from a single sample, either does or does not contain the population mean. However, if this procedure is repeated a great many times, then the proportion of such intervals that contain μ will be at least $1-\alpha$. Observing one sample and constructing one interval from it amounts to randomly selecting one of the many intervals that might or might not contain μ. Because most (at least $1-\alpha$) of the intervals do, we can be "confident" that the interval that was actually constructed does contain the unknown population mean.

9.5.1 Sample Size

Confidence intervals are often used to determine sample sizes for future experiments. Typically, the researcher specifies a desired confidence level, $1-\alpha$, and a desired interval length, L. After determining the appropriate critical value, q, one equates L with $2q\sigma/\sqrt{n}$ and solves for n, obtaining

$$n = (2q\sigma/L)^2. \tag{9.8}$$

Of course, this formula presupposes knowledge of the population variance. In practice, it is usually necessary to replace σ with an estimate—which may be easier said than done if the experiment has not yet been performed. This is one reason to perform a pilot study: to obtain a preliminary estimate of the population variance and use it to design a better study.

Several useful relations can be deduced from equation (9.8):

1. Higher levels of confidence $(1-\alpha)$ correspond to larger critical values (q), which result in larger sample sizes (n).

2. Smaller interval lengths (L) result in larger sample sizes (n).

3. Larger variances (σ^2) result in larger sample sizes (n).

In summary, if a researcher desires high confidence that the true mean of a highly variable population is covered by a small interval, then s/he should plan on collecting a great deal of data!

Example 9.5 (continued) *A rival corporation purchases the rights to the amateur mechanic's additive. How large a study is required to determine this additive's mean increase in mileage to within* 0.05 *mpg with a confidence coefficient of* $1 - \alpha = 0.99$*?*

The desired interval length is $L = 2 \cdot 0.05 = 0.1$ and the critical value that corresponds to $\alpha = 0.01$ is computed in R as follows:

```
> qnorm(1-.01/2)
[1] 2.575829
```

From the mechanic's small pilot study, we estimate σ to be $s = 0.4$. Then

$$n = (2 \cdot 2.575829 \cdot 0.4/0.1)^2 \doteq 424.6,$$

so the desired study will require $n = 425$ vehicles.

9.5.2 One-Sided Confidence Intervals

The set of μ_0 for which we would retain the null hypothesis $H_0 : \mu = \mu_0$ when tested against the two-sided alternative hypothesis $H_1 : \mu \neq \mu_0$ is a traditional, 2-sided confidence interval. In situations where 1-sided alternatives are appropriate, we can construct corresponding 1-sided confidence intervals by determining the set of μ_0 for which the appropriate null hypothesis would be retained.

Example 9.5 (continued) The government test has a significance level of $\alpha = 0.05$. It rejects the null hypothesis $H_0 : \mu \leq \mu_0$ if and only if (iff)

$$\begin{aligned} & \mathbf{p} = P(Z \geq t) \leq 0.05 \\ \text{iff} \quad & P(Z < t) \geq 0.95 \\ \text{iff} \quad & t \geq \texttt{qnorm(0.95)} \doteq 1.645. \end{aligned}$$

Equivalently, the null hypothesis $H_0 : \mu \leq \mu_0$ is retained if and only if

$$\begin{aligned} & t = \frac{\bar{x} - \mu_0}{s/\sqrt{n}} < 1.645 \\ \text{iff} \quad & \bar{x} < \mu_0 + 1.645 \cdot \frac{s}{\sqrt{n}} \\ \text{iff} \quad & \mu_0 > \bar{x} - 1.645 \cdot \frac{s}{\sqrt{n}}. \end{aligned}$$

1. In the case of the large corporation, the null hypothesis $H_0 : \mu \leq \mu_0$ is retained if and only if

$$\mu_0 > 1.01 - 1.645 \cdot \frac{0.1}{\sqrt{900}} \doteq 1.0045,$$

so the 1-sided confidence interval with confidence coefficient $1 - \alpha = 0.95$ is $(1.0045, \infty)$.

2. In the case of the amateur mechanic, the null hypothesis $H_0 : \mu \leq \mu_0$ is retained if and only if

$$\mu_0 > 1.21 - 1.645 \cdot \frac{0.4}{\sqrt{9}} \doteq 0.9967,$$

so the 1-sided confidence interval with confidence coefficient $1 - \alpha = 0.95$ is $(0.9967, \infty)$.

9.6 Exercises

1. According to *The Justice Project*, "John Spirko was sentenced to death on the testimony of a witness who was '70 percent certain' of his identification." Formulate this case as a problem in hypothesis testing. What can be deduced about the significance level used to convict Spirko? Does this choice of significance level strike you as suitable for a capital murder trial?

2. Blaise Pascal, the French theologian and mathematician, argued that we cannot know whether or not God exists, but that we must behave as though we do. He submitted that the consequences of wrongly behaving as though God does not exist are greater than the consequences of wrongly behaving as though God does exist, concluding that it is better to err on the side of caution and act as though God exists.

This argument is known as Pascal's Wager. Formulate Pascal's Wager as a hypothesis testing problem. What are the Type I and Type II errors? On whom did Pascal place the burden of proof, believers or nonbelievers?

3. Dorothy owns a lovely glass dreidl. Curious as to whether or not it is fairly balanced, she spins her dreidl ten times, observing five gimels and five hehs. Surprised by these results, Dorothy decides to compute the probability that a fair dreidl would produce such aberrant results. Which of the probabilities specified in Exercise 3.7.6 is the most appropriate choice of a significance probability for this investigation? Why?

4. The U.S. Food and Drug Administration requires evaporated milk to contain "not less than 23 percent by weight of total milk solids." A company that sells evaporated milk is sued by a group of consumers who are concerned that the company's product does not meet FDA standards. The two parties agree to binding arbitration. If the consumers win, the company will pay damages and enhance its product; if the company wins, then the consumers will issue a public apology.

   To resolve the dispute, the arbiter commissions a neutral study in which the percent by weight of total milk solids will be measured in a random sample of $n = 225$ packages produced by the company. Both parties agree to a standard of proof ($\alpha = 0.05$), but they disagree on which party should bear the burden of proof.

   (a) State appropriate null and alternative hypotheses from the perspective of the consumers.

   (b) State appropriate null and alternative hypotheses from the perspective of the company.

   (c) Suppose that the random sample reveals a sample mean of $\bar{x} = 22.8$ percent with a sample standard deviation of $s = 3$ percent. Compute t, the value of the test statistic.

   (d) From the consumers' perspective, what action should be taken? Why?

   (e) From the company's perspective, what action should be taken? Why?

5. It is thought that human influenza viruses originate in birds. It is quite possible that, several years ago, a human influenza pandemic

was averted by slaughtering 1.5 million chickens brought to market in Hong Kong. Because it is impossible to test each chicken individually, such decisions are based on samples. Suppose that a boy has already died of a bird flu virus apparently contracted from a chicken. Several diseased chickens have already been identified. The health officials would prefer to err on the side of caution and destroy all chickens that might be infected; the farmers do not want this to happen unless it is absolutely necessary. Suppose that both the farmers and the health officals agree that all chickens should be destroyed if more than 2 percent of the population is diseased. A random sample of $n = 1000$ chickens reveals 40 diseased chickens.

(a) Let $X_i = 1$ if chicken i is diseased and $X_i = 0$ if it is not. Assume that $X_1, \ldots, X_n \sim P$. To what family of probability distributions does P belong? What population parameter indexes this family? Use this parameter to state formulas for $\mu = EX_i$ and $\sigma^2 = \operatorname{Var} X_i$.

(b) State appropriate null and alternative hypotheses from the perspective of the health officials.

(c) State appropriate null and alternative hypotheses from the perspective of the farmers.

(d) Use the value of μ_0 in the above hypotheses to compute the value of σ^2 under H_0. Then compute z, the value of the test statistic.

(e) Adopting the health officials' perspective, and assuming that they are willing to risk a 0.1% chance of committing a Type I error, what action should be taken? Why?

(f) Adopting the farmers' perspective, and assuming that they are willing to risk a 10% chance of committing a Type I error, what action should be taken? Why?

6. A company that manufactures light bulbs has advertised that its 75-watt bulbs burn an average of 800 hours before failing. In reaction to the company's advertising campaign, several dissatisfied customers have complained to a consumer watchdog organization that they believe the company's claim to be exaggerated. The consumer organization must decide whether or not to allocate some of its financial resources to countering the company's advertising campaign. So that it can make an informed decision, it begins by purchasing and testing 100 of the disputed light bulbs. In this experiment, the 100 light bulbs

burned an average of $\bar{x} = 745.1$ hours before failing, with a sample standard deviation of $s = 238.0$ hours. Formulate null and alternative hypotheses that are appropriate for this situation. Calculate a significance probability. Do these results warrant rejecting the null hypothesis at a significance level of $\alpha = 0.05$?

7. To study the effects of Alzheimer's disease (AD) on cognition, a scientist administers two batteries of neuropsychological tasks to 60 mildly demented AD patients. One battery is administered in the morning, the other in the afternoon. Each battery includes a task in which discourse is elicited by showing the patient a picture and asking the patient to describe it. The quality of the discourse is measured by counting the number of "information units" conveyed by the patient. The scientist wonders if asking a patient to describe Picture A in the morning is equivalent to asking the same patient to describe Picture B in the afternoon, after having described Picture A several hours earlier. To investigate, she computes the number of information units for Picture A minus the number of information units for Picture B for each patient. She finds an average difference of $\bar{x} = -0.1833$, with a sample standard deviation of $s = 5.18633$. Formulate null and alternative hypotheses that are appropriate for this situation. Calculate a significance probability. Do these results warrant rejecting the null hypothesis at a significance level of $\alpha = 0.05$?

8. Each student in a large statistics class of 600 students is asked to toss a fair coin 100 times, count the resulting number of `Heads`, and construct a 0.95-level confidence interval for the probability of `Heads`. Assume that each student uses a fair coin and constructs the confidence interval correctly. *True or False: We would expect approximately* 570 *of the confidence intervals to contain the number* 0.5. Explain.

9. Mt. Wrightson, the fifth highest summit in Arizona and the highest in Pima County, has a reputed elevation of 9453 feet. To amuse its members, the Southern Arizona Hiking Club (SAHC) decides to construct its own confidence interval for μ, the true elevation of Mt. Wrightson's summit. SAHC acquires an altimeter whose measurements will have an expected value of μ with a standard deviation of 6 feet. How many measurements should SAHC plan to take if it wants to construct a 0.99-level confidence interval for μ that has a length of 2 feet?

10. Professor Johnson is interested in the probability that a certain type

of randomly generated matrix has a positive determinant. His student attempts to calculate the probability exactly, but runs into difficulty because the problem requires her to evaluate an integral in 9 dimensions. Professor Johnson therefore decides to obtain an approximate probability by simulation, i.e., by randomly generating some matrices and observing the proportion that have positive determinants. His preliminary investigation reveals that the probability is roughly 0.05. At this point, Professor Park decides to undertake a more comprehensive simulation experiment that will, with 0.95-level confidence, correctly determine the probability of interest to within ± 0.00001. How many random matrices should he generate to achieve the desired accuracy?

11. Refer to Exercise 8.4.5. A third Math 351 student wants to use the function `urn.model` to construct a 0.95-level confidence interval for $p = P(170.5 < Y < 199.5)$. If he desires an interval of length L, then how many times should he plan to evaluate `urn.model`?

    Hint: How else might the student estimate p?

12. In September 2003, Lena spun a penny 89 times and observed 2 `Heads`. Let p denote the true probability that one spin of her penny will result in `Heads`.

    (a) The significance probability for testing $H_0 : p \geq 0.3$ versus $H_1 : p < 0.3$ is $\mathbf{p} = P(Y \leq 2)$, where $Y \sim \text{Binomial}(89; 0.3)$.

    i. Compute $\mathbf{p}$ as in Section 9.1, using the binomial distribution and `pbinom`.

    ii. Approximate $\mathbf{p}$ as in Section 9.4, using the normal distribution and `pnorm`. How good is this approximation?

    (b) Construct a 1-sided confidence interval for p by determining for which values of p_0 the null hypothesis $H_0 : p \geq p_0$ would be retained at a significance level of (approximately) $\alpha = 0.05$.

Chapter 10

1-Sample Location Problems

The basic ideas associated with statistical inference were introduced in Chapter 9. We developed these ideas in the context of drawing inferences about a single population mean, and we assumed that the sample was large enough to justify appeals to the Central Limit Theorem for normal approximations. The population mean is a natural measure of centrality, but it is not the only one. Furthermore, even if we are interested in the population mean, our sample may be too small to justify the use of a large-sample normal approximation. The purpose of the next several chapters is to explore more thoroughly how statisticians draw inferences about measures of centrality.

Measures of centrality are sometimes called location parameters. The title of this chapter indicates an interest in a location parameter of *one* population. More specifically, we assume that $X_1, \ldots, X_n \sim P$ are independently and identically distributed, we observe a random sample $\vec{x} = \{x_1, \ldots, x_n\}$, and we attempt to draw an inference about a location parameter of P. Because it is not always easy to identify the relevant population in a particular experiment, we begin with some examples. Our analysis of these examples is clarified by posing the following four questions:

1. What are the experimental units, i.e., what are the objects that are being measured?

2. From what population (or populations) were the experimental units drawn?

3. What measurements were taken on each experimental unit?

4. What random variables are relevant to the specified inference?

For the sake of specificity, we assume that the location parameter of interest in the following examples is the population median, $q_2(P)$.

Example 10.1 A recycled printing paper is supposed to have a basis weight of 24 pounds for 500 basic sheets[1] and a caliper (thickness) of 0.0048 inches per sheet. To determine if the paper is being correctly manufactured, a sample of sheets is drawn and the caliper of each sheet is measured with a micrometer. For this experiment:

1. An experimental unit is a sheet of paper. Notice that we are distinguishing between experimental units, the objects being measured (sheets of paper), and units of measurement (inches).

2. There is one population, viz., all sheets of paper that might be produced by the designated manufacturing process.

3. One measurement (caliper) is taken on each experimental unit.

4. Let X_i denote the caliper of sheet i. Then $X_1, \ldots, X_n \sim P$ and we are interested in drawing inferences about $q_2(P)$, the population median caliper. For example, we might test $H_0 : q_2(P) = 0.0048$ against $H_1 : q_2(P) \neq 0.0048$.

Example 10.2 A drug is supposed to lower blood pressure.[2] To determine if it does, a number of hypertensive patients are administered the drug for two months. Each person's mean arterial pressure is measured before and after the two month period. For this experiment:

1. An experimental unit is a patient.

2. There is one population of hypertensive patients. (It may be difficult to discern the precise population that was actually sampled. All hypertensive patients? All Hispanic male hypertensive patients who live in Houston, TX? All Hispanic male hypertensive patients who live in Houston, TX, and who are sufficiently well-disposed to the medical

[1] A basic sheet of bond paper is 17 inches by 22 inches, so 500 basic sheets is equivalent to 2000 letter-size sheets.

[2] Blood pressure is the force exerted by circulating blood in the body's large arteries. Arterial blood pressure is measured by a sphygmomanometer, usually in units of millimeters of mercury (mm Hg). Systolic arterial pressure is the maximum pressure attained during a cardiac cycle, diastolic arterial pressure is the minimum, and mean arterial pressure is the average.

establishment to participate in the study? In published journal articles, scientists are often rather vague about just what population was actually sampled.)

3. Two measurements (mean arterial pressure before and after treatment) are taken on each experimental unit. Let B_i and A_i denote the mean arterial pressures of patient i before and after treatment.

4. Let $X_i = B_i - A_i$, the decrease in mean arterial pressure for patient i. Then $X_1, \ldots, X_n \sim P$ and we are interested in drawing inferences about $q_2(P)$, the population median decrease. For example, we might test $H_0 : q_2(P) \leq 0$ against $H_1 : q_2(P) > 0$.

Example 10.3 Nancy Solomon and Thomas Hixon investigated the effect of Parkinson's disease (PD) on speech breathing.[3] Nancy recruited 14 PD patients to participate in the study. She also recruited 14 normal control (NC) subjects. Each NC subject was carefully matched to one PD patient with respect to sex, age, height, and weight. The lung volume of each study participant was measured. For this experiment:

1. An experimental unit was a matched PD-NC pair.

2. The population comprises all possible PD-NC pairs that satisfy the study criteria.

3. Two measurements (PD and NC lung volume) were taken on each experimental unit. Let D_i and C_i denote the PD and NC lung volumes of pair i.

4. Let $X_i = \log(D_i/C_i) = \log D_i - \log C_i$, the logarithm of the PD proportion of NC lung volume. (This is not the only way of comparing D_i and C_i, but it worked well in this investigation. As explained in Section 7.6, ratios can be difficult to analyze and logarithms convert ratios to differences. Furthermore, lung volume data tend to be skewed to the right. As in Exercise 7.7.4, logarithmic transformations of such data often have a symmetrizing effect.) Then $X_1, \ldots, X_n \sim P$ and we are interested in drawing inferences about $q_2(P)$. For example, to test the theory that PD restricts lung volume, we might test $H_0 : q_2(P) \geq 0$ against $H_1 : q_2(P) < 0$.

[3] N. P. Solomon and T. J. Hixon. Speech breathing in Parkinson's disease. *Journal of Speech and Hearing Research*, 36:294–310, April 1993.

This chapter is divided into sections according to distributional assumptions about the X_i and (consequently) the location parameter of interest.

10.1 If the data are assumed to be normally distributed, then we will be interested in inferences about the population's mean. Because normal distributions are symmetric, the population mean is also the population median.

10.2 If the data are only assumed to be continuously distributed (not necessarily normally, or even symmetrically distributed), then we will be interested in inferences about the population median.

10.3 If the data are assumed to be continuously and symmetrically distributed (but not necessarily normally distributed), then we will be interested in inferences about the population's center of symmetry.

Each section is subdivided into subsections, according to the type of inference (point estimation, hypothesis testing, set estimation) at issue.

10.1 The Normal 1-Sample Location Problem

In this section we assume that $P = \text{Normal}(\mu, \sigma^2)$. As necessary, we will distinguish between cases in which σ is known and cases in which σ is unknown.

10.1.1 Point Estimation

Because normal distributions are symmetric, the location parameter μ is the center of symmetry and therefore both the population mean and the population median. Hence, there are (at least) two natural estimators of μ, the sample mean $\bar{X}_n$ and the sample median $q_2(\hat{P}_n)$. Both are consistent, unbiased estimators of μ. We will compare them by considering their *asymptotic relative efficiency* (ARE). A rigorous definition of ARE is beyond the scope of this book, but the concept is easily interpreted.

If the true distribution is $P = \text{Normal}(\mu, \sigma^2)$, then the ARE of the sample median to the sample mean for estimating μ is

$$e(P) = \frac{2}{\pi} \doteq 0.64.$$

This statement has the following interpretation: for large samples, using the sample median to estimate a normal population mean is equivalent to randomly discarding approximately 36% of the observations and calculating the sample mean of the remaining 64%. Thus, the sample mean is substantially more efficient than is the sample median at extracting location information from a normal sample.

In fact, if $P = \text{Normal}(\mu, \sigma^2)$, then the ARE of *any* estimator of μ to the sample mean is ≤ 1. This is sometimes expressed by saying that the sample mean is *asymptotically efficient* for estimating a normal mean. The sample mean also enjoys a number of other optimal properties in this case. The sample mean is unquestionably the preferred estimator for the normal 1-sample location problem.

10.1.2 Hypothesis Testing

If σ is known, then the possible distributions of X_i are

$$\left\{ \text{Normal}(\mu, \sigma^2) : -\infty < \mu < \infty \right\}.$$

If σ is unknown, then the possible distributions of X_i are

$$\left\{ \text{Normal}(\mu, \sigma^2) : -\infty < \mu < \infty, \sigma > 0 \right\}.$$

We partition the possible distributions into two subsets, the null and alternative hypotheses. For example, if σ is known then we might specify

$$H_0 = \left\{ \text{Normal}(0, \sigma^2) \right\} \quad \text{and} \quad H_1 = \left\{ \text{Normal}(\mu, \sigma^2) : \mu \neq 0 \right\},$$

which we would typically abbreviate as $H_0 : \mu = 0$ and $H_1 : \mu \neq 0$. Analogously, if σ is unknown then we might specify

$$H_0 = \left\{ \text{Normal}(0, \sigma^2) : \sigma > 0 \right\}$$

and

$$H_1 = \left\{ \text{Normal}(\mu, \sigma^2) : \mu \neq 0, \ \sigma > 0 \right\},$$

which we would also abbreviate as $H_0 : \mu = 0$ and $H_1 : \mu \neq 0$.

More generally, for any real number μ_0 we might specify

$$H_0 = \left\{ \text{Normal}(\mu_0, \sigma^2) \right\} \quad \text{and} \quad H_1 = \left\{ \text{Normal}(\mu, \sigma^2) : \mu \neq \mu_0 \right\}$$

if σ is known, or

$$H_0 = \left\{ \text{Normal}(\mu_0, \sigma^2) : \sigma > 0 \right\}$$

and

$$H_1 = \left\{ \text{Normal}(\mu, \sigma^2) : \mu \neq \mu_0, \ \sigma > 0 \right\}$$

if σ in unknown. In both cases, we would typically abbreviate these hypotheses as $H_0 : \mu = \mu_0$ and $H_1 : \mu \neq \mu_0$.

The preceding examples involve two-sided alternative hypotheses. Of course, as in Section 9.4, we might also specify one-sided hypotheses. However, the material in the present section is so similar to the material in Section 9.4 that we will discuss only two-sided hypotheses.

The intuition that underlies testing $H_0 : \mu = \mu_0$ versus $H_1 : \mu \neq \mu_0$ was discussed in Section 9.4:

- If H_0 is true, then we would expect the sample mean to be close to the population mean μ_0.
- Hence, if $\bar{X}_n = \bar{x}_n$ is observed far from μ_n, then we are inclined to reject H_0.

To make this reasoning precise, we reject H_0 if and only if the significance probability

$$\mathbf{p} = P_{\mu_0}\left(\left|\bar{X}_n - \mu_0\right| \geq \left|\bar{x}_n - \mu_0\right|\right) \leq \alpha. \tag{10.1}$$

The first equation in (10.1) is a formula for a significance probability. Notice that this formula is identical to equation (9.2). The one difference between the material in Section 9.4 and the present material lies in how one computes $\mathbf{p}$. For emphasis, we recall the following:

1. The hypothesized mean μ_0 is a fixed number specified by the null hypothesis.
2. The estimated mean, $\bar{x}_n$, is a fixed number computed from the sample. Therefore, so is $|\bar{x}_n - \mu_0|$, the difference between the estimated mean and the hypothesized mean.
3. The estimator, $\bar{X}_n$, is a random variable.
4. The subscript in P_{μ_0} reminds us to compute the probability under $H_0 : \mu = \mu_0$.
5. The significance level α is a fixed number specified by the researcher, preferably before the experiment was performed.

To apply (10.1), we must compute $\mathbf{p}$. In Section 9.4, we overcame that technical difficulty by appealing to the Central Limit Theorem. This allowed us to approximate $\mathbf{p}$ even when we did not know the distribution of the X_i, but only for reasonably large sample sizes. However, if we know that $X_1, \ldots, X_n$ are normally distributed, then it turns out that we can calculate $\mathbf{p}$ exactly, even when n is small.

Case 1: The Population Variance is Known

Under the null hypothesis that $\mu = \mu_0$, $X_1, \ldots, X_n \sim \text{Normal}(\mu_0, \sigma^2)$ and

$$\bar{X}_n \sim \text{Normal}\left(\mu_0, \frac{\sigma^2}{n}\right).$$

This is the exact distribution of $\bar{X}_n$, not an asymptotic approximation. We convert $\bar{X}_n$ to standard units, obtaining

$$Z = \frac{\bar{X}_n - \mu_0}{\sigma/\sqrt{n}} \sim \text{Normal}(0, 1). \tag{10.2}$$

The observed value of Z is

$$z = \frac{\bar{x}_n - \mu_0}{\sigma/\sqrt{n}}.$$

The significance probability is

$$\begin{aligned}
\mathbf{p} &= P_{\mu_0}\left(\left|\bar{X}_n - \mu_0\right| \geq |\bar{x}_n - \mu_0|\right) \\
&= P_{\mu_0}\left(\left|\frac{\bar{X}_n - \mu_0}{\sigma/\sqrt{n}}\right| \geq \left|\frac{\bar{x}_n - \mu_0}{\sigma/\sqrt{n}}\right|\right) \\
&= P\left(|Z| \geq |z|\right) \\
&= 2P\left(Z \geq |z|\right).
\end{aligned}$$

In this case, the test that rejects H_0 if and only if $\mathbf{p} \leq \alpha$ is sometimes called the 1-*sample* z-*test*. The random variable Z is the *test statistic*.

Before considering the case of an unknown population variance, we remark that it is possible to derive point estimators from hypothesis tests. For testing $H_0 : \mu = \mu_0$ versus $H_1 : \mu \neq \mu_0$, the test statistics are

$$Z(\mu_0) = \frac{\bar{X}_n - \mu_0}{\sigma/\sqrt{n}}.$$

If we observe $\bar{X}_n = \bar{x}_n$, then what value of μ_0 minimizes $|z(\mu_0)|$? Clearly, the answer is $\mu_0 = \bar{x}_n$. Thus, our preferred point estimate of μ is the μ_0 for which it is most difficult to reject $H_0 : \mu = \mu_0$. This type of reasoning will be extremely useful for analyzing situations in which we know how to test but don't know how to estimate.

Case 2: The Population Variance is Unknown

Statement (10.2) remains true if σ is unknown, but it is no longer possible to compute z. Therefore, we require a different test statistic for this case. A natural approach is to modify Z by replacing the unknown σ with an estimator of it. Toward that end, we introduce the test statistic

$$T_n = \frac{\bar{X}_n - \mu_0}{S_n/\sqrt{n}},$$

where S_n^2 is the unbiased estimator of the population variance defined by equation (9.1). Because T_n and Z are different random variables, they have different probability distributions and our first order of business is to determine the distribution of T_n.

We begin by stating a useful fact.

Theorem 10.1 *If $X_1, \ldots, X_n \sim Normal(\mu, \sigma^2)$, then*

$$\frac{(n-1)S_n^2}{\sigma^2} = \sum_{i=1}^{n} \left(X_i - \bar{X}_n\right)^2 / \sigma^2 \sim \chi^2(n-1).$$

The χ^2 (chi-squared) distribution was described in Section 5.5 and Theorem 10.1 is closely related to Theorem 5.3.

Next we write

$$\begin{aligned} T_n &= \frac{\bar{X}_n - \mu_0}{S_n/\sqrt{n}} = \frac{\bar{X}_n - \mu_0}{\sigma/\sqrt{n}} \cdot \frac{\sigma/\sqrt{n}}{S_n/\sqrt{n}} \\ &= Z \cdot \frac{\sigma}{S_n} = Z/\sqrt{S_n^2/\sigma^2} \\ &= Z/\sqrt{\left[(n-1)S_n^2/\sigma^2\right]/(n-1)}. \end{aligned}$$

Using Theorem 10.1, we see that T_n can be written in the form

$$T_n = \frac{Z}{\sqrt{Y/\nu}},$$

where $Z \sim \text{Normal}(0,1)$ and $Y \sim \chi^2(\nu)$. If Z and Y are independent random variables, then it follows from Definition 5.7 that $T_n \sim t(n-1)$.

Both Z and $Y = (n-1)S_n^2/\sigma^2$ depend on $X_1, \ldots, X_n$, so one would be inclined to think that Z and Y are dependent. This is usually the case, but it turns out that they are independent if $X_1, \ldots, X_n \sim \text{Normal}(\mu, \sigma^2)$. This is another remarkable property of normal distributions, usually stated as follows:

Theorem 10.2 *If $X_1, \ldots, X_n \sim Normal(\mu, \sigma^2)$, then $\bar{X}_n$ and S_n^2 are independent random variables.*

The result that interests us can then be summarized as follows:

Corollary 10.1 *If $X_1, \ldots, X_n \sim Normal(\mu_0, \sigma^2)$, then*

$$T_n = \frac{\bar{X}_n - \mu_0}{S_n/\sqrt{n}} \sim t(n-1).$$

Now let

$$t_n = \frac{\bar{x}_n - \mu_0}{s_n/\sqrt{n}},$$

the observed value of the test statistic T_n. The significance probability is

$$\mathbf{p} = P_{\mu_0}\left(|T_n| \geq |t_n|\right) = 2P_{\mu_0}\left(T_n \geq |t_n|\right).$$

In this case, the test that rejects H_0 if and only if $\mathbf{p} \leq \alpha$ is called *Student's 1-sample t-test*. Because it is rarely the case that the population variance is known when the population mean is not, Student's 1-sample t-test is used much more frequently than the 1-sample z-test. We will use the `R` function `pt` to compute significance probabilities for Student's 1-sample t-test, as illustrated in the following examples.

Example 10.4 Suppose that, to test $H_0 : \mu = 0$ versus $H_1 : \mu \neq 0$ (a 2-sided alternative), we draw a sample of size $n = 25$ and observe $\bar{x} = 1$ and $s = 3$. Then $t = (1-0)/(3/\sqrt{25}) = 5/3$ and the 2-tailed significance probability is computed using both tails of the $t(24)$ distribution, i.e., $\mathbf{p} = 2 * \texttt{pt}(-5/3, \texttt{df} = 24) \doteq 0.1086$.

Example 10.5 Suppose that, to test $H_0 : \mu \leq 0$ versus $H_1 : \mu > 0$ (a 1-sided alternative), we draw a sample of size $n = 25$ and observe $\bar{x} = 2$ and $s = 5$. Then $t = (2-0)/(5/\sqrt{25}) = 2$ and the 1-tailed significance probability is computed using one tail of the $t(24)$ distribution, i.e., $\mathbf{p} = 1 - \texttt{pt}(2, \texttt{df} = 24) \doteq 0.0285$.

10.1.3 Set Estimation

As in Section 9.5, we will derive confidence intervals from tests. We imagine testing $H_0 : \mu = \mu_0$ versus $H_1 : \mu \neq \mu_0$ for every $\mu_0 \in (-\infty, \infty)$. The μ_0 for which $H_0 : \mu = \mu_0$ is rejected are implausible values of μ; the μ_0 for which $H_0 : \mu = \mu_0$ is not rejected constitute the confidence interval. To accomplish this, we will have to derive the critical values of our tests. A significance level of α will result in a confidence coefficient of $1 - \alpha$.

Case 1: The Population Variance is Known

If σ is known, then we reject $H_0 : \mu = \mu_0$ if and only if

$$\mathbf{p} = P_{\mu_0}\left(\left|\bar{X}_n - \mu_0\right| \geq \left|\bar{x}_n - \mu_0\right|\right) = 2\Phi\left(-\left|z_n\right|\right) \leq \alpha,$$

where $z_n = (\bar{x}_n - \mu_0)/(\sigma/\sqrt{n})$. By the symmetry of the normal distribution, this condition is equivalent to the condition

$$1 - \Phi\left(-\left|z_n\right|\right) = P\left(Z > -\left|z_n\right|\right) = P\left(Z < \left|z_n\right|\right) = \Phi\left(\left|z_n\right|\right) \geq 1 - \alpha/2,$$

where $Z \sim \text{Normal}(0,1)$, and therefore to the condition $|z_n| \geq q_z$, where q_z denotes the $1 - \alpha/2$ quantile of $\text{Normal}(0,1)$. The quantile q_z is the critical value of the two-sided 1-sample z-test. Thus, given a significance level α and a corresponding critical value q_z, we reject $H_0 : \mu = \mu_0$ if and only if (iff)

$$\begin{aligned} & \left|\frac{\bar{x}_n - \mu_0}{\sigma/\sqrt{n}}\right| = |z_n| \geq q_z \\ \text{iff} \quad & |\bar{x}_n - \mu_0| \geq q_z\sigma/\sqrt{n} \\ \text{iff} \quad & \mu_0 \notin \left(\bar{x}_n - q_z\sigma/\sqrt{n},\ \bar{x}_n + q_z\sigma/\sqrt{n}\right) \end{aligned}$$

and we conclude that the desired set of plausible values is the interval

$$\left(\bar{x}_n - q_z\frac{\sigma}{\sqrt{n}},\ \bar{x}_n + q_z\frac{\sigma}{\sqrt{n}}\right).$$

Notice that both the preceding derivation and the resulting confidence interval are identical to the derivation and confidence interval in Section 9.5. The only difference is that, because we are now assuming that $X_1, \ldots, X_n \sim \text{Normal}(\mu, \sigma^2)$ instead of relying on the Central Limit Theorem, no approximation is required.

Example 10.6 Suppose that we desire 90% confidence about μ and $\sigma = 3$ is known. Then $\alpha = 0.10$ and $q_z \doteq 1.645$. Suppose that we draw $n = 25$ observations and observe $\bar{x}_n = 1$. Then

$$1 \pm 1.645\frac{3}{\sqrt{25}} = 1 \pm 0.987 = (0.013, 1.987)$$

is a 0.90-level confidence interval for μ.

Case 2: The Population Variance is Unknown

If σ is unknown, then it must be estimated from the sample. The reasoning in this case is the same, except that we rely on Student's 1-sample t-test.

As before, we use S_n^2 to estimate σ^2. The critical value of the 2-sided 1-sample t-test is q_t, the $1 - \alpha/2$ quantile of a t distribution with $n - 1$ degrees of freedom, and the confidence interval is

$$\left(\bar{x}_n - q_t \frac{s_n}{\sqrt{n}}, \; \bar{x}_n + q_t \frac{s_n}{\sqrt{n}} \right).$$

Example 10.7 Suppose that we desire 90% confidence about μ and σ is unknown. Suppose that we draw $n = 25$ observations and observe $\bar{x}_n = 1$ and $s = 3$. Then $q_t = \texttt{qt}(.95, \texttt{df} = 24) \doteq 1.711$ and

$$1 \pm 1.711 \times 3/\sqrt{25} = 1 \pm 1.027 = (-0.027, 2.027)$$

is a 90% confidence interval for μ. Notice that the confidence interval is larger when we use $s = 3$ instead of $\sigma = 3$.

10.2 The General 1-Sample Location Problem

In Section 10.1 we assumed that $X_1, \ldots, X_n \sim P$ and $P = \text{Normal}(\mu, \sigma^2)$. In this section, we again assume that $X_1, \ldots, X_n \sim P$, but now we assume only that the X_i are continuous random variables.

Because P is not assumed to be symmetric, we must decide which location parameter to study. The population median, $q_2(P)$, enjoys several advantages. Unlike the population mean, the population median always exists and is not sensitive to the influence of outliers. Furthermore, it turns out that one can develop fairly elementary ways to study medians, even when little is known about the probability distribution P. For simplicity, we now denote the population median by θ.

10.2.1 Hypothesis Testing

It is convenient to begin our study of the general 1-sample location problem with a discussion of hypothesis testing. As in Section 10.1, we initially consider testing a 2-sided alternative, $H_0 : \theta = \theta_0$ versus $H_1 : \theta \neq \theta_0$. We will explicate a procedure known as the *sign test*.

The intuition that underlies the sign test is elementary. If the population median is $\theta = \theta_0$, then when we sample P we should observe roughly half the x_i above θ_0 and half the x_i below θ_0. Hence, if we observe proportions of x_i above/below θ_0 that are very different from one half, then we are inclined to reject the possibility that $\theta = \theta_0$.

More formally, let $p_+ = P_{H_0}(X_i > \theta_0)$ and $p_- = P_{H_0}(X_i < \theta_0)$. Because the X_i are continuous, $P_{H_0}(X_i = \theta_0) = 0$ and therefore $p_+ = p_- = 0.5$. Hence, under H_0, observing whether $X_i > \theta_0$ or $X_i < \theta_0$ is equivalent to tossing a fair coin, i.e., to observing a Bernoulli trial with success probability $p = 0.5$. The sign test is the following procedure:

1. Let $\vec{x} = \{x_1, \ldots, x_n\}$ denote the observed sample. If the X_i are continuous random variables, then $P(X_i = \theta_0) = 0$ and it should be that each $x_i \neq \theta_0$. In practice, of course, it may happen that we do observe one or more $x_i = \theta_0$. For the moment, we assume that $\vec{x}$ contains no such values.

2. Let
$$Y = \#\{X_i > \theta_0\} = \#\{X_i - \theta_0 > 0\}$$
be the test statistic. Under $H_0 : \theta = \theta_0$, $Y \sim \text{Binomial}(n; p = 0.5)$. The observed value of the test statistic is
$$y = \#\{x_i > \theta_0\} = \#\{x_i - \theta_0 > 0\}.$$

3. Notice that $EY = n/2$. The significance probability is
$$\mathbf{p} = P_{\theta_0}\left(\left|Y - \frac{n}{2}\right| \geq \left|y - \frac{n}{2}\right|\right).$$
The sign test rejects $H_0 : \theta = \theta_0$ if and only if $\mathbf{p} \leq \alpha$.

4. To compute $\mathbf{p}$, we first note that
$$\left|Y - \frac{n}{2}\right| \geq \left|y - \frac{n}{2}\right|$$
is equivalent to the event
   (a) $\{Y \leq y \text{ or } Y \geq n - y\}$ if $y \leq n/2$;
   (b) $\{Y \geq y \text{ or } Y \leq n - y\}$ if $y \geq n/2$.

   To accommodate both cases, let $c = \min(y, n - y)$. Then
$$\mathbf{p} = P_{\theta_0}(Y \leq c) + P_{\theta_0}(Y \geq n - c) = 2P_{\theta_0}(Y \leq c) = \texttt{2*pbinom(c,n,.5)}.$$

Example 10.8(a) Suppose that we want to test $H_0 : \theta = 100$ versus $H_1 : \theta \neq 100$ at significance level $\alpha = 0.05$, having observed the sample

$$\vec{x} = \{98.73, 97.17, 100.17, 101.26, 94.47, 96.39, 99.67, 97.77, 97.46, 97.41\}.$$

Here $n = 10$, $y = \#\{x_i > 100\} = 2$, and $c = \min(2, 10 - 2) = 2$, so

$$\mathbf{p} = \texttt{2*pbinom(2,10,.5)} = 0.109375 > 0.05$$

and we decline to reject H_0.

Example 10.8(b) Now suppose that we want to test $H_0 : \theta \leq 97$ versus $H_1 : \theta > 97$ at significance level $\alpha = 0.05$, using the same data. Here $n = 10$, $y = \#\{x_i > 97\} = 8$, and $c = \min(8, 10 - 8) = 2$. Because large values of Y are evidence against $H_0 : \theta \leq 97$,

$$\begin{aligned} \mathbf{p} &= P_{\theta_0}(Y \geq y) = P_{\theta_0}(Y \geq 8) = 1 - P_{\theta_0}(Y \leq 7) \\ &= \texttt{1-pbinom(7,10,.5)} = 0.0546875 > 0.05 \end{aligned}$$

and we decline to reject H_0.

Thus far we have assumed that the sample contains no values for which $x_i = \theta_0$. In practice, we may well observe such values. For example, if the measurements in Example 10.8(a) were made less precisely, then we might have observed the following sample:

$$\vec{x} = \{99, 97, 100, 101, 94, 96, 100, 98, 97, 97\}. \tag{10.3}$$

If we want to test $H_0 : \theta = 100$ versus $H_1 : \theta \neq 100$, then we have two values that equal θ_0 and the sign test requires modification.

We assume that $\#\{x_i = \theta_0\}$ is fairly small; otherwise, the assumption that the X_i are continuous is questionable. We consider two possible ways to proceed:

1. Perhaps the most satisfying solution is to compute all of the significance probabilities that correspond to different ways of counting the $x_i = \theta_0$ as larger or smaller than θ_0. If there are k observations $x_i = \theta_0$, then this will produce 2^k significance probabilities, which we might average to obtain a single $\mathbf{p}$.

2. Alternatively, let $\mathbf{p}_0$ denote the significance probability obtained by counting in the way that is most favorable to H_0 (least favorable to

H_1). This is the largest of the possible significance probabilities, so if $\mathbf{p}_0 \le \alpha$ then we reject H_0. Similarly, let $\mathbf{p}_1$ denote the significance probability obtained by counting in the way that is least favorable to H_0 (most favorable to H_1). This is the smallest of the possible significance probabilities, so if $\mathbf{p}_1 > \alpha$ then we decline to reject H_0. If $\mathbf{p}_0 > \alpha \ge \mathbf{p}_1$, then we simply declare the results to be equivocal.

Example 10.8(c) Suppose that we want to test $H_0 : \theta = 100$ versus $H_1 : \theta \neq 100$ at significance level $\alpha = 0.05$, having observed the sample (10.3). Here $n = 10$ and $y = \#\{x_i > 100\}$ depends on how we count the observations $x_3 = x_7 = 100$. There are $2^2 = 4$ possibilities:

possibility	$y = \#\{x_i > 100\}$	$c = \min(y, 10 - y)$	$\mathbf{p}$
$y_3 < 100, y_7 < 100$	1	1	0.021484
$y_3 < 100, y_7 > 100$	2	2	0.109375
$y_3 > 100, y_7 < 100$	2	2	0.109375
$y_3 > 100, y_7 > 100$	3	3	0.343750

Noting that $\mathbf{p}_0 \doteq 0.344 > 0.05 > 0.021 \doteq \mathbf{p}_1$, we might declare the results to be equivocal. However, noting that only 1 of the 4 possibilities lead us to reject H_0 (and that the average $\mathbf{p} \doteq 0.146$), we might conclude—somewhat more decisively—that there is insufficient evidence to reject H_0. The distinction between these two interpretations is largely rhetorical, as the fundamental logic of hypothesis testing requires that we decline to reject H_0 unless there is compelling evidence against it.

10.2.2 Point Estimation

Next we consider the problem of estimating the population median. A natural estimate is the plug-in estimate, the sample median. Another approach begins by posing the following question: For what value of θ_0 is the sign test least inclined to reject $H_0 : \theta = \theta_0$ in favor of $H_1 : \theta \neq \theta_0$? The answer to this question is also a natural estimate of the population median.

In fact, the plug-in and sign-test approaches lead to the same estimation procedure. To understand why, we focus on the case that n is even, in which case $n/2$ is a possible value of $Y = \#\{X_i > \theta_0\}$. If $|y - n/2| = 0$, then

$$\mathbf{p} = P\left(\left| Y - \frac{n}{2} \right| \ge 0 \right) = 1.$$

We see that the sign test produces the maximal significance probability of $\mathbf{p} = 1$ when $y = n/2$, i.e., when θ_0 is chosen so that precisely half the observations exceed θ_0. This means that the sign test is least likely to reject $H_0 : \theta = \theta_0$ when θ_0 is the sample median. (A similar argument leads to the same conclusion when n is odd.)

Thus, using the sign test to test hypotheses about population medians corresponds to using the sample median to estimate population medians, just as using Student's t-test to test hypotheses about population means corresponds to using the sample mean to estimate population means. One consequence of this remark is that, when the population mean and median are identical, the "Pitman efficiency" of the sign test to Student's t-test equals the asymptotic relative efficiency of the sample median to the sample mean. For example, using the sign test on normal data is asymptotically equivalent to randomly discarding 36% of the observations, then using Student's t-test on the remaining 64%.

10.2.3 Set Estimation

Finally, we consider the problem of constructing a $(1-\alpha)$-level confidence interval for the population median. Again we rely on the sign test, determining for which θ_0 the level-α sign test of $H_0 : \theta = \theta_0$ versus $H_1 : \theta \neq \theta_0$ does not reject H_0.

The sign test rejects $H_0 : \theta = \theta_0$ if and only if

$$y(\theta_0) = \#\{x_i > \theta_0\}$$

is either too large or too small. Equivalently, the sign test declines to reject H_0 if and only if θ_0 is such that the numbers of observations above and below θ_0 are roughly equal.

To determine the critical value for the desired sign test, we suppose that $Y \sim \text{Binomial}(n; 0.5)$. We would like to find k such that $\alpha = 2P(Y \leq k)$, or $\alpha/2 = \texttt{pbinom}(k, n, 0.5)$. In practice, we won't be able to solve this equation exactly. Binomial distributions are discrete; hence, `pbinom(k,n,0.5)` only attains certain values. We will use the `qbinom` function plus trial-and-error to find k such that $\texttt{pbinom}(k, n, 0.5) \approx \alpha/2$, then modify our choice of α accordingly.

Having determined a suitable (α, k), the sign test rejects $H_0 : \theta = \theta_0$ at level α if and only if either $y(\theta_0) \leq k$ or $y(\theta_0) \geq n-k$. We would like to translate these inequalities into an interval of plausible values of θ_0. To do so, it is helpful to sort the values observed in the sample.

Definition 10.1 *The* order statistics *of* $\vec{x} = \{x_1, \ldots, x_n\}$ *are any permutation of the* x_i *such that*

$$x_{(1)} \leq x_{(2)} \leq \cdots \leq x_{(n-1)} \leq x_{(n)}.$$

If $\vec{x}$ contains n distinct values, then there is a unique set of order statistics and the above inequalities are strict; otherwise, we say that $\vec{x}$ contains ties.

Thus, $x_{(1)}$ is the smallest value in $\vec{x}$ and $x_{(n)}$ is the largest. If $n = 2m + 1$ (n is odd), then the sample median is $x_{(m+1)}$; if $n = 2m$ (n is even), then the sample median is $[x_{(m)} + x_{(m+1)}]/2$.

For simplicity we assume that $\vec{x}$ contains no ties. If $\theta_0 < x_{(k+1)}$, then at least $n - k$ observations exceed θ_0 and the sign test rejects $H_0 : \theta = \theta_0$. Similarly, if $\theta_0 > x_{(n-k)}$, then no more than k observations exceed θ_0 and the sign test rejects $H_0 : \theta = \theta_0$. We conclude that the sign test does not reject $H_0 : \theta = \theta_0$ if and only if θ_0 lies in the $(1-\alpha)$-level confidence interval

$$\left(x_{(k+1)}, x_{(n-k)}\right).$$

Example 10.8(d) Using the 10 observations from Example 10.8(a), we endeavor to construct a 0.90-level confidence interval for the population median. We begin by determining a suitable choice of (α, k). If $1-\alpha = 0.90$, then $\alpha/2 = 0.05$. The R command `qbinom(.05,10,.5)` returns $k = 2$. Next we experiment:

k	pbinom$(k, 10, 0.5)$
2	0.0546875
1	0.01074219

We choose $k = 2$, resulting in a confidence level of

$$1 - \alpha = 1 - 2 \cdot 0.0546875 = 0.890625 \doteq 0.89,$$

nearly equal to the requested level of 0.90. Now, upon sorting the data (the `sort` function in R may be useful), we quickly discern that the desired confidence interval is

$$\left(x_{(3)}, x_{(8)}\right) = (97.17, 99.67).$$

10.3 The Symmetric 1-Sample Location Problem

Again we assume that $X_1, \ldots, X_n \sim P$. In Section 10.1 we made a strong assumption, that $P = \text{Normal}(\mu, \sigma^2)$. The procedures that we derived under the assumption of normality, e.g., the sample mean estimator and Student's 1-sample t-test, are efficient when P is normal, but may not perform well if P is not normal. For example, the sample mean is sensitive to the effect of outliers.

In Section 10.2 we made a weak assumption, that the X_i are continuous random variables with pdf f. The procedures that we derived, e.g., the sample median estimator and the sign test, perform well in a variety of circumstances, but are somewhat inefficient when P is normal.

The present section describes one possible compromise between the normal methods of Section 10.1 and the general methods of Section 10.2. To the assumption that the X_i are continuous random variables, we add the assumption that their pdf, f, is symmetric. The assumption of symmetry permits comparison of procedures for means and procedures for medians, as the population mean and the population median necessarily equal the population center of symmetry, which we denote by θ. We follow the same reasoning that we deployed in Section 10.2, first describing a test of $H_0 : \theta = \theta_0$ versus $H_1 : \theta \neq \theta_0$, then using it to derive point and set estimators of θ.

10.3.1 Hypothesis Testing

Let $D_i = X_i - \theta_0$, the amount by which X_i exceeds the hypothesized center of symmetry. The numerator of Student's 1-sample t statistic is the sample mean of the D_i, whereas the sign test statistic is the number of observations for which $D_i > 0$. We now describe a third test based on $D_1, \ldots, D_n$. The *Wilcoxon signed rank test* extracts less information from $D_1, \ldots, D_n$ than does Student's 1-sample t-test, but more information than does the sign test.

Because the X_i are continuous random variables, each $P(D_i = 0)$ and each $P(|D_i| = |D_j|) = 0$ for $i \neq j$. Therefore, with probability one, there is a unique ordering of the absolute differences:

$$|D_{i_1}| < |D_{i_2}| < \cdots < |D_{i_n}|.$$

The Wilcoxon signed rank test is based on this ordering. (In practice we may encounter samples with $d_i = 0$ and/or $|d_i| = |d_j|$, in which case we will consider multiple orderings.)

Let R_i denote the rank of $|D_i|$, i.e., $R_{i_1} = 1$, $R_{i_2} = 2$, etc. Consider two possible test statistics,

$$T_+ = \sum_{D_{i_k} > 0} k = \sum_{D_i > 0} R_i,$$

the sum of the "positive ranks," and

$$T_- = \sum_{D_{i_k} < 0} k = \sum_{D_i < 0} R_i,$$

the sum of the "negative ranks." Because

$$T_+ + T_- = \sum_{k=1}^{n} k = n(n+1)/2,$$

T_- is determined by T_+ (and vice versa), so it suffices to restrict attention to T_+.

If θ_0 is the population center of symmetry, then each D_i is equally likely to be positive or negative and

$$ET_+ = \sum_{i=1}^{n} iP\left(D_i > 0\right) = \sum_{i=1}^{n} i/2 = n(n+1)/4.$$

The Wilcoxon signed rank test rejects $H_0 : \theta = \theta_0$ if and only if we observe T_+ sufficiently different from ET_+.

Let $\vec{x} = \{x_1, \ldots, x_n\}$ denote the observed sample and let $d_i = x_i - \theta_0$ denote the observed differences. For now, we assume that $d_i \neq 0$ and $|d_i| \neq |d_j|$. Let t_+ denote the observed value of T_+. Then the Wilcoxon signed rank test rejects $H_0 : \theta = \theta_0$ if and only if

$$\mathbf{p} = P_{H_0}\left(|T_+ - n(n+1)/4| \geq |t_+ - n(n+1)/4|\right) \leq \alpha.$$

The challenge lies in computing $\mathbf{p}$. To illustrate the nature of this challenge, we consider a simple example.

Example 10.9 We test $H_0 : \theta = 10$ versus $H_1 : \theta \neq 10$ with a significance level of $\alpha = 0.15$. We draw $n = 4$ observations, obtaining a sample of $\{12.4, 11.0, 11.3, 11.7\}$. The corresponding differences are $\{2.4, 1.0, 1.3, 1.7\}$, each of which is positive, and the sum of the positive ranks is $t_+ = 4+1+2+3 = 10$. Because $0 \leq T_+ \leq 10$ and $ET_+ = n(n+1)/4 = 5$, the significance probability is

$$\mathbf{p} = P_{H_0}\left(|T_+ - 5| \geq |10 - 5|\right) = P_{H_0}\left(T_+ = 10\right) + P_{H_0}\left(T_+ = 0\right).$$

To compute $\mathbf{p}$, we require the pmf of the discrete random variable T_+ under the null hypothesis that $\theta_0 = 10$ is the population center of symmetry.

Under $H_0 : \theta = 10$, each D_i is equally likely to be positive or negative; hence, each of the $2^4 = 16$ sign patterns in Table 10.1 are equally likely. The pmf of T_+ is as follows:

k	0	1	2	3	4	5	6	7	8	9	10
$16P(T_+ = k)$	1	1	1	2	2	2	2	2	1	1	1

R_1	R_2	R_3	R_4	T_+
+	+	+	+	10
+	+	+	-	6
+	+	-	+	7
+	+	-	-	3
+	-	+	+	8
+	-	+	-	4
+	-	-	+	5
+	-	-	-	1
-	+	+	+	9
-	+	+	-	5
-	+	-	+	6
-	+	-	-	2
-	-	+	+	7
-	-	+	-	3
-	-	-	+	4
-	-	-	-	0

Table 10.1: Behavior of T_+ under $H_0 : \theta = \theta_0$ with $n = 4$ observations.

Thus,

$$\mathbf{p} = P_{H_0}(T_+ = 10) + P_{H_0}(T_+ = 0) = \frac{1}{16} + \frac{1}{16} = 0.125.$$

Because $\mathbf{p} \leq \alpha$, we reject $H_0 : \theta = 10$.

Although conceptually straightforward, it is evidently cumbersome to compute significance probabilities for the Wilcoxon signed rank test unless n is quite small. For $n = 20$, there are $2^{20} = 1,048,576$ possible sign patterns! Rather than compute exact significance probabilities, we consider two ways of approximating $\mathbf{p}$:

1. Simulation.

   Using R, it is easy to generate random sign patterns and compute the value of T_+ for each pattern. The observed proportion of sign patterns for which

   $$|T_+ - n(n+1)/4| \geq |t_+ - n(n+1)/4|$$

   estimates the true significance probability. (The more sign patterns that we generate, the more accurate our estimate of $\mathbf{p}$.) I wrote an R function, `W1.p.sim`, that implements this procedure. This function is described in Appendix R and can be obtained from the web page for this book.

2. Normal Approximation.

   It turns out that, for n sufficiently large, the discrete distribution of T_+ under $H_0 : \theta = \theta_0$ can be approximated by a normal distribution.

   Theorem 10.3 *Suppose that $X_1, \ldots, X_n$ are symmetric continuous random variables with center of symmetry θ_0. Let T_+ denote the test statistic for the Wilcoxon signed rank test of $H_0 : \theta = \theta_0$ versus $H_1 : \theta \neq \theta_0$. Under H_0, $ET_+ = n(n+1)/4$, $Var T_+ = n(n+1)(2n+1)/24$, and*

   $$P_{H_0}(T_+ \leq c) \to P\left(Z \leq \frac{c - ET_+}{\sqrt{Var T_+}}\right)$$

   as $n \to \infty$, where $Z \sim Normal(0,1)$.

   I wrote an `R` function, `W1.p.norm`, that uses Theorem 10.3 to compute approximate significance probabilities. This function is described in Appendix R and can be obtained from the web page for this book.

Example 10.9 (continued) Five replications of the simulation procedure `W1.p.sim(4,10)` resulted in estimated significance probabilities of 0.117, 0.114, 0.128, 0.124, and 0.125. The normal approximation of 0.100, obtained from `W1.p.norm(4,10)`, is surprisingly good.

Example 10.10 Suppose that $n = 20$ observations produce $t_+ = 50$. Two replications of `W1.p.sim(20,50,10000)` resulted in estimated significance probabilities of 0.0395 and 0.0366. The normal approximation, obtained from `W1.p.norm(20,50)`, is 0.0412.

Finally, we consider the case of samples with $d_i = 0$ and/or $|d_i| = |d_j|$. We will estimate the average significance probability that results from all of the plausible rankings of the $|d_i|$. To do so, we simply modify the simulation procedure described above by subjecting each x_i to small random perturbations. The $|d_i|$ are recalculated on the perturbed samples, each of which results in a unique ordering. By choosing the magnitude of the perturbations sufficiently small, we can guarantee that each complete ordering generated from a perturbed sample is consistent with the partial ordering derived from the original sample. I wrote an R function, `W1.p.ties`, that implements this procedure. This function is described in Appendix R and can be obtained from the web page for this book.

Example 10.11 To test $H_0 : \theta = 5$ versus $H_1 : \theta \neq 5$ at $\alpha = 0.05$, the following x_i were observed:

1.5	9.7	3.9	7.6	8.0	7.3	5.0	9.7	2.3	2.3
6.6	9.4	8.6	7.7	8.4	2.7	9.1	5.3	3.1	9.4

The corresponding $d_i = x_i - 5$ are the following:

−3.5	4.7	−1.1	2.6	3.0	2.3	0.0	4.7	−2.7	−2.7
1.6	4.4	3.6	2.7	3.4	−2.3	4.1	0.3	−1.9	4.4

Ranked by $|d_i|$, we observe the following signed ranks:

$\text{sign}(d_i)$	?		+		−		+		−	
d_i	0.0	<	0.3	<	1.1	<	1.6	<	1.9	<
r_i	1		2		3		4		5	
$\text{sign}(d_i)$	−/+		+		−/−/+		+		+	
d_i	2.3, 2.3	<	2.6	<	2.7, 2.7, 2.7	<	3.0	<	3.4	<
r_i	6/7		8		9/10/11		12		13	
$\text{sign}(d_i)$	−		+		+		+/+		+/+	
d_i	3.5	<	3.6	<	4.1	<	4.4, 4.4	<	4.7, 4.7	
r_i	14		15		16		17/18		19/20	

The value of t_+ depends on

1. whether $r_7 = 1$ is counted as a positive or a negative rank;
2. whether $d_6 = +2.3$ is ranked 6 or 7; and
3. whether $d_{14} = +2.7$ is ranked 9, 10, or 11.

Thus, t_+ might be as small as

$$\begin{aligned} t_+ &= 2 + 4 + 6 + 8 + 9 + 12 + 13 + 15 + 16 + 17 + 18 + 19 + 20 \\ &= 20 \cdot 21/2 - (1 + 3 + 5 + 7 + 10 + 11 + 14) = 159 \end{aligned}$$

or as large as

$$\begin{aligned} t_+ &= 1 + 2 + 4 + 7 + 8 + 11 + 12 + 13 + 15 + 16 + 17 + 18 + 19 + 20 \\ &= 20 \cdot 21/2 - (3 + 5 + 6 + 9 + 10 + 14) = 163. \end{aligned}$$

We might use `W1.p.sim` to estimate, or `W1.p.norm` to approximate, the significance probabilities associated with $t_+ = 159$ and $t_+ = 163$. Alternatively, we might use `W1.p.ties` to estimate the average significance probability associated with all of the possible ways of assigning signed ranks to these data. Five replications of `W1.p.ties(x,5)` resulted in estimated $\mathbf{p}$-values of 0.035, 0.024, 0.038, 0.033, and 0.020. As each $\mathbf{p} \leq \alpha$, it is reasonable to reject $H_0 : \theta = 5$.

10.3.2 Point Estimation

In Section 10.2.2 we showed that the sample median is the value of θ_0 that the sign test is least inclined to reject as the population median. Now we derive an estimator, $\hat{\theta}$, of the population center of symmetry, θ. We do so by determining the value of θ_0 for which the Wilcoxon signed rank test is least inclined to reject $H_0 : \theta = \theta_0$ in favor of $H_1 : \theta \neq \theta_0$.

Our derivation relies on a clever representation of t_+. First, given θ_0, we order the observations:

$$x_{(1)} < \cdots < x_{(k)} < \theta_0 < x_{(k+1)} < \cdots < x_{(n)}$$

Let $d_j = x_{(j)} - \theta_0$ and let r_j denote the rank of $|d_j|$, so that

$$t_+ = r_{k+1} + \cdots + r_n.$$

For $j = k+1, \ldots, n$,

$$|d_j| = \left| x_{(j)} - \theta_0 \right| = x_{(j)} - \theta_0 = d_j.$$

Hence, if $i \leq j$, then

$$d_i = x_{(i)} - \theta_0 \leq x_{(j)} - \theta_0 = d_j = |d_j|$$

and

$$\begin{aligned} |d_i| \leq |d_j| \quad &\text{iff} \quad -\left(x_{(i)} - \theta_0\right) \leq x_{(j)} - \theta_0 \\ &\text{iff} \quad \left(x_{(i)} - \theta_0\right) + \left(x_{(j)} - \theta_0\right) \geq 0. \end{aligned}$$

It follows that

$$\begin{aligned} r_j &= \#\{i : i \leq j, |d_i| \leq |d_j|\} \\ &= \#\left\{i : i \leq j, \left(x_{(i)} - \theta_0\right) + \left(x_{(j)} - \theta_0\right) \geq 0\right\}. \end{aligned}$$

Finally, notice that, if $i \le j \le k$, then

$$\left(x_{(i)} - \theta_0\right) + \left(x_{(j)} - \theta_0\right) < 0.$$

Hence, we can write the Wilcoxon test statistic, the sum of the positive ranks, as

$$\begin{aligned} t_+ = \sum_{j=k+1}^{n} r_j &= \sum_{j=k+1}^{n} \#\left\{i : i \le j, \left(x_{(i)} - \theta_0\right) + \left(x_{(j)} - \theta_0\right) \ge 0\right\} \\ &= \sum_{1}^{n} \#\left\{i : i \le j, \left(x_{(i)} - \theta_0\right) + \left(x_{(j)} - \theta_0\right) \ge 0\right\} \\ &= \#\left\{(i,j) : i \le j, \left(x_{(i)} - \theta_0\right) + \left(x_{(j)} - \theta_0\right) \ge 0\right\} \\ &= \#\left\{(i,j) : i \le j, (x_i - \theta_0) + (x_j - \theta_0) \ge 0\right\}. \end{aligned}$$

We know that $H_0 : \theta = \theta_0$ is most difficult to reject when θ_0 is such that $t_+ = ET_+ = n(n+1)/4$. There are $n(n+1)/2$ (i,j) pairs, so the Wilcoxon signed rank test will be least inclined to reject θ_0 when $(x_i - \theta_0) + (x_j - \theta_0)$ is positive for half the pairs and negative for the other half. This condition is equivalent to the condition that $2\theta_0$ is the median of the pairwise sums, $x_i + x_j$, or that θ_0 is the median of the pairwise averages, $(x_i + x_j)/2$.

The pairwise averages, $(x_i + x_j)/2$ for $1 \le i \le j \le n$, are called the *Walsh averages.* The median of the Walsh averages is the estimator, $\hat{\theta}$, of the population center of symmetry, θ, that corresponds to the Wilcoxon signed rank test. This estimator is an excellent compromise between the sample mean and the sample median. For any symmetric distribution with finite variance, the asymptotic relative efficiency (ARE) of $\hat{\theta}$ to $\bar{X}$ is guaranteed to be at least 0.864. For normal distributions, the ARE is $3/\pi \doteq 0.955$, so using $\hat{\theta}$ instead of $\bar{X}$ entails only a slight loss of efficiency. For some symmetric distributions, the ARE is considerably greater than 1.

Although cumbersome to calculate by hand, the median of the Walsh averages is easily computed in `R`. I wrote a function, `W1.walsh`, that does so. This function is described in Appendix R and can be obtained from the web page for this book.

Example 10.11 (continued) To estimate the center of symmetry, we compute the median of the Walsh averages.

```
> W1.walsh(x)
[1] 6.3
```

10.3.3 Set Estimation

In Section 10.2.3 we constructed confidence intervals for the population median by including θ_0 if and only if the sign test failed to reject θ_0 as a plausible population median. We follow the same reasoning here. Let θ denote the population center of symmetry. We use the Wilcoxon signed rank test to test $H_0 : \theta = \theta_0$ versus $H_1 : \theta \neq \theta_0$ at significance level α. The set of θ_0 that are not rejected are the plausible populations centers of symmetry, a $(1 - \alpha)$-level confidence interval for θ.

As in the preceding section, we represent the Wilcoxon test statistic, T_+, as the number of Walsh averages that exceed θ_0. Because the Wilcoxon signed rank test rejects $H_0 : \theta = \theta_0$ if and only if T_+ is either sufficiently large or sufficiently small, we deem θ_0 a plausible value of θ if and only if there are sufficient numbers of Walsh averages below and above θ_0. Thus, the desired confidence interval for θ must consist of those θ_0 for which at least k Walsh averages are $\leq \theta_0$ and at least k Walsh averages are $\geq \theta_0$. The quantity k is determined by the level of confidence that is desired. As with the confidence intervals for population medians that we derived from the sign test, not all confidence levels are possible.

The fact that we must approximate the distribution of the discrete random variable T_+ under $H_0 : \theta = \theta_0$ complicates our efforts to construct confidence intervals. We proceed in three steps:

1. Use the normal approximation to guess a reasonable value of k, i.e., a value for which $P(T_+ \leq k - 1) \approx \alpha/2$. Recall that

$$\begin{aligned} P\left(T_+ \leq k-1\right) &= P\left(T_+ \leq k - 0.5\right) \\ &= P\left(\frac{T_+ - ET_+}{\sqrt{\operatorname{Var} T_+}} \leq \frac{k - 0.5 - ET_+}{\sqrt{\operatorname{Var} T_+}}\right) \\ &\approx P\left(Z \leq \frac{k - 0.5 - ET_+}{\sqrt{\operatorname{Var} T_+}}\right), \end{aligned}$$

where $Z \sim \text{Normal}(0,1)$; hence, given $\alpha \in (0,1)$, a reasonable value of k is obtained by solving

$$P\left(Z \leq \frac{k - 0.5 - ET_+}{\sqrt{\operatorname{Var} T_+}}\right) = \frac{\alpha}{2},$$

resulting in

$$k = 0.5 + ET_+ - q_z\sqrt{\operatorname{Var} T_+} = 0.5 + \frac{n(n+1)}{4} - q_z\sqrt{\frac{n(n+1)(2n+1)}{24}},$$

where $q_z = \texttt{qnorm}(1 - \alpha/2)$.

2. Use simulation to estimate the confidence coefficients, $1 - 2P(T_+ \leq k - 1)$, associated with several reasonable choices of k.

3. Finalize the choice of k (and thereby the confidence coefficient) and construct the corresponding confidence interval. The lower endpoint of the interval is the kth Walsh average; the upper endpoint is the $[n(n+1)/2 + 1 - k]$th Walsh average.

I implemented these steps in the R function `W1.ci`, described in Appendix R and available from the web page for this book. This function returns a 5×4 matrix. The first column contains possible choices of k, the second and third columns contain the lower and upper endpoints of the corresponding confidence interval, and the fourth column contains the estimated confidence coefficients.

Example 10.11 (continued) To construct a confidence interval with confidence coefficient $1 - \alpha \approx 0.90$ for θ, the population center of symmetry, I obtained the following results:

```
> W1.ci(x,.1,10000)
      k Lower Upper Coverage
[1,] 59  5.25  8.00   0.9195
[2,] 60  5.30  7.95   0.9158
[3,] 61  5.30  7.85   0.9083
[4,] 62  5.35  7.85   0.8913
[5,] 63  5.35  7.85   0.8851
```

The estimated confidence coefficient for $k = 61$ (which happens to be the value of k produced by the normal approximation) is nearest 0.90, so the desired confidence interval is $(5.30, 7.85)$.

10.4 Case Study: Deficit Unawareness in Alzheimer's Disease

Many clinical descriptions of Alzheimer's disease (AD) include the phenomenon of anosognosia, an unawareness of cognitive deficit. One way to obtain experimental evidence of this phenomenon is to perform a *predicted performance experiment*, the hallmark feature of which is that each AD patient is asked to predict his performance on a cognitive task. The hope is that one can infer deficit unawareness from discrepancies between actual and predicted performance, but one must first control for other possible explanations of such discrepancies. In this section we describe a predicted performance experiment introduced by McGlynn and Kaszniak (1991), adopt a measure of deficit unawareness proposed by Trosset and Kaszniak (1996), and analyze data reported by Kaszniak, DiTraglia and Trosset (1993).[4]

The most basic predicted performance experiment simply examines the discrepancy between a patient's predicted and actual performance on a cognitive task. In such an experiment, the exerimental unit is an individual patient. Two measurements are made on each patient: the patient's prediction of his performance, denoted *ppp*, and the patient's actual score, denoted *pscor*. Following McGlynn and Kaszniak (1991), we measure the discrepancy between these two quantities by forming their ratio, $ppp/pscor$. As described in Section 7.6, we then transform the ratios to an additive scale, e.g., by computing $\log_2(ppp/pscor)$.

A fatal flaw with the basic experiment described above is that one cannot attribute overpredicted performance, $\log_2(ppp/pscor) > 0$, to deficit unawareness. Overpredicted performance *might* mean that AD patients are not aware of impaired cognitive function, but it might also mean that they tend to underestimate the difficulty of the task. To distinguish these possibilites, one must construct a more elaborate experiment.

Suppose that we ask each subject to predict his own performance (*ppp*) and the performance of his spousal caregiver (*ppc*), then measure the actual performance of both patient (*pscor*) and caregiver (*cscor*). In this experiment, the experimental unit is a patient-caregiver pair and four measurements are made on each pair. One can compare the extent to which the patient overpredicted his own performance and the extent to which the patient overpredicted his caregiver's performance, e.g., by computing

$$\log_2\left[(ppp/pscor) \div (ppc/cscor)\right].$$

Yet even knowing that the patient overpredicted his own performance while correctly predicting his caregiver's performance is insufficient to infer deficit unawareness. Suppose that $ppp/pscor = 2$ and $ppc/cscor = 1$. These results

[4]S. M. McGlynn and A. W. Kaszniak (1991). When metacognition fails: impaired awareness of deficit in Alzheimer's disease. *Journal of Cognitive Neuroscience*, 3:183–189.

M. W. Trosset and A. W. Kaszniak (1996). Measures of deficit unawareness for predicted performance experiments. *Journal of the International Neuropsychological Society*, 2:315–322.

A. W. Kaszniak, G. DiTraglia and M. W. Trosset (1993). Self-awareness of cognitive deficit in patients with probably Alzheimer's disease. *Journal of Clinical and Experimental Neuropsychology*, 15:32. Abstract.

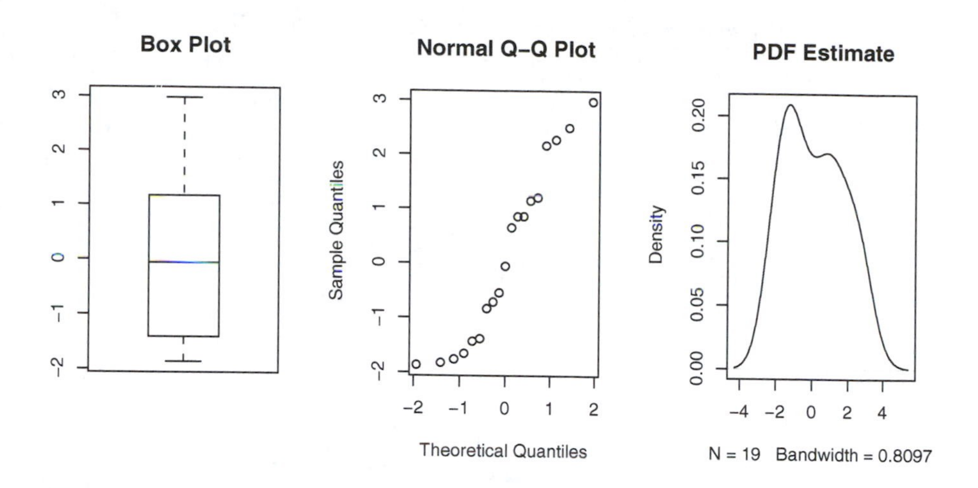

Figure 10.1: Three representations of $n = 19$ values of $\log_2(\text{CPA})$.

might obtain because AD patients are not aware of impaired cognitive function, but they might also obtain because of a general human tendency to overestimate one's own ability. Accordingly, McGlynn and Kaszniak (1991) proposed an even more elaborate predicted performance experiment.

In the McGlynn-Kaszniak experiment, the experimental unit is a patient-caregiver pair. The patient predicts his own performance (*ppp*) and his caregiver's performance (*ppc*). The caregiver predicts her own performance (*cpc*) and the patient's performance (*cpp*). Finally, the actual performance of both patient (*pscor*) and caregiver (*cscor*) is measured. Thus, six measurements are made on each experiment unit.

It is not at all obvious how to combine the six measurements in the McGlynn-Kaszniak experiment to obtain a single measure of deficit unawareness. Trosset and Kaszniak (1996) discussed several flawed possibilities and proposed the Comparative Prediction Accuracy (CPA) measure,

$$\text{CPA} = \frac{(ppp/pscor) \div (ppc/cscor)}{(cpc/cscor) \div (cpp/pscor)}.$$

They envisioned analyzing $\log(\text{CPA})$ by performing Student's 1-sample t-test. For ease of interpretation, we replace $\log(\text{CPA})$ with $\log_2(\text{CPA})$.

Table 10.2 contains data from a predicted performance experiment reported by Kaszniak, DiTraglia and Trosset (1993). This particular experiment involved $n = 19$ AD patient-caregiver pairs and used a generative

i	PMMSE	*ppp*	*pscor*	*ppc*	*cscor*	*cpc*	*cpp*	CPA	$\log_2$(CPA)
1	26	10	7	10	18	15	5	2.2041	1.1402
2	21	6	21	6	11	10	10	0.2744	−1.8658
3	25	8	10	16	19	6	6	1.8050	0.8520
4	19	5	8	8	17	20	2	0.2822	−1.8251
5	15	5	8	10	17	10	8	1.8062	0.8530
6	19	2	5	20	20	10	6	0.9600	−0.0589
7	18	5	6	35	20	25	5	0.3175	−1.6554
8	23	10	17	15	16	10	5	0.2953	−1.7599
9	18	8	16	18	16	30	25	0.3704	−1.4330
10	20	5	8	20	14	10	5	0.3828	−1.3853
11	13	3	2	30	26	15	7	7.8867	2.9794
12	22	6	7	6	21	8	5	5.6250	2.4919
13	27	12	11	15	26	12	12	4.4694	2.1601
14	10	20	5	30	19	10	5	4.8133	2.2670
15	19	8	13	10	17	10	5	0.6840	−0.5479
16	19	6	13	7	12	12	10	0.6086	−0.7164
17	23	10	13	10	23	8	4	1.5651	0.6462
18	21	6	15	8	12	6	7	0.5600	−0.8365
19	8	12	4	24	14	8	3	2.2969	1.1997

Table 10.2: Data from a predicted performance experiment on $n = 19$ AD patient-caregiver pairs. This experiment used a generative naming task. The quantity PMMSE is the patient's score on the Mini-Mental State Examination, a widely used measure of AD severity (lower scores indicate greater impairment). The quantities *ppp*, *pscor*, *ppc*, *cscor*, *cpc* and *cpp* are the six measures from which CPA (Comparative Prediction Accuracy) is computed.

naming task. The final column contains the values of $\log_2(\text{CPA})$, the sample $\vec{x}$ that we will analyze using techniques for 1-sample location problems.

A box plot, a normal probability plot, and a kernel density estimate constructed from $\vec{x}$ are displayed in Figure 10.1. As is so often the case in practice, none of the methods that we have described is entirely satisfying. The tails in the normal probability plot suggest systematic departures from normality, but even the weaker assumption of symmetry appears suspect. On the other hand, the departure from symmetry is not so great that one can be very enthusiastic about forgoing procedures based on the Wilcoxon signed rank test for less powerful procedures based on the sign test. In such situations, one should do several things. First, one should generate several simulated data sets to obtain a sense of how sampling variation may be affecting one's interpretation of the diagnostic displays in Figure 10.1. Second, one should perform different analyses under different assumptions. If different procedures yield comparable conclusions, then one can only be encouraged that the conclusions are valid. These practices are explored in Exercise Set B in Section 10.5.

It is important to appreciate that reasonable minds may disagree about the best ways to analyze actual data. In practice, one cannot know the true distribution from which the sample was drawn—one can only make reasonable assumptions about it. Even professional statisticians may disagree about what assumptions are most reasonable. For the present $\vec{x}$, my personal preference is to use procedures based on an assumption of symmetry, i.e., procedures derived from the Wilcoxon signed rank test.

Assuming symmetry, let θ denote the population center of symmetry. Because positive values of $\log_2(\text{CPA})$ provide nominal evidence of deficit unawareness, it is natural to test $H_0 : \theta \leq 0$ versus $H_1 : \theta > 0$. The expected sum of the positive ranks is $ET_+ = n(n+1)/4 = 95$. The sum of the positive ranks is $t_+ = 103$, computed by the following R commands:

```
> pos <- which(x>0)
> d <- abs(x)
> r <- rank(d)
> tplus <- sum(r[pos])
```

Is $t_+ = 103$ enough larger than $ET_+ = 95$ to reject $H_0 : \theta \leq 0$ in favor of $H_1 : \theta > 0$? The significance probability for testing $H_0 : \theta = 0$ against the two-sided alternative $H_1 : \theta \neq 0$ is approximately 0.77, obtained as follows:

```
> W1.p.sim(19,103,100000)
[1] 0.76859
```

The significance probability for testing $H_0 : \theta \leq 0$ against the one-sided alternative $H_1 : \theta > 0$ is half the two-sided signficance probability, too great to reject $H_0 : \theta \leq 0$ at the conventional significance level of $\alpha = 0.05$.

Finally, to construct a two-sided confidence interval for θ with confidence coefficient $1 - \alpha \approx 0.90$, we obtain the following results:

```
> W1.ci(x,.05,100000)
      k      Lower     Upper Coverage
[1,] 53 -0.5568219 0.7970628  0.91157
[2,] 54 -0.5478818 0.7753275  0.90406
```

```
[3,] 55 -0.5068864 0.7732281  0.89421
[4,] 56 -0.5063870 0.7496210  0.88785
[5,] 57 -0.5045537 0.7491217  0.87826
```

The estimated confidence coefficient for $k = 54$ slightly exceeds 0.90, so a reasonable confidence interval for θ is approximately $(-0.548, 0.775)$. Considering that these numbers are (base 2) logarithms, this is quite a range of values. The corresponding range of plausible CPA values is approximately $(0.684, 1.712)$.

10.5 Exercises

Problem Set A

1. Assume that $n = 400$ observations are independently drawn from a normal distribution with unknown population mean μ and unknown population variance σ^2. The resulting sample, $\vec{x}$, is used to test $H_0 : \mu \leq 0$ versus $H_1 : \mu > 0$ at significance level $\alpha = 0.05$.

   (a) What test should be used in this situation? If we observe $\vec{x}$ that results in $\bar{x} = 3.194887$ and $s^2 = 104.0118$, then what is the value of the test statistic?

   (b) If we observe $\vec{x}$ that results in a test statistic value of 1.253067, then which of the following R expressions best approximates the significance probability?

      i. `2*pnorm(1.253067)`
      ii. `2*pnorm(-1.253067)`
      iii. `1-pnorm(1.253067)`
      iv. `1-pt(1.253067,df=399)`
      v. `pt(1.253067,df=399)`

   (c) True or False: if we observe $\vec{x}$ that results in a significance probability of $\mathbf{p} = 0.03044555$, then we should reject the null hypothesis.

2. A device counts the number of ions that arrive in a given time interval, unless too many arrive. An experiment that relies on this device produces the following counts, where `Big` means that the count exceeded 255.

251	238	249	Big	243	248	229	Big	235	244
254	251	252	244	230	222	224	246	Big	239

Use these data to construct a confidence interval for the population median number of ions with a confidence coefficient of approximately 0.95.

Problem Set B The following exercises elaborate on the case study explicated in Section 10.4. The sample, $\vec{x}$, is the final column in Table 10.2.

1. Refer to Figure 10.1. The tails in the normal probability plot do not appear to be consistent with an assumption of normality, but the lack of linearity might be due to sampling variation. To investigate whether or not this is the case, please do the following:

   (a) Use `rnorm` to generate four samples from a normal distribution, each with $n = 19$ observations.

   (b) Construct a normal probability plot for each simulated sample. Compare these plots to the normal probability plot of $\vec{x}$ in Figure 10.1.

   (c) Compute the ratio of the sample interquartile range to the sample standard deviation for $\vec{x}$ and for each simulated sample.

   (d) Reviewing the available evidence, are you comfortable assuming that $\vec{x}$ was drawn from a normal distribution? Do four simulated samples provide enough information to answer the preceding question?

2. Assuming that $X_1, \ldots, X_{19} \sim \text{Normal}(\mu, \sigma^2)$, use Student's 1-sample test to test $H_0 : \mu \leq 0$ versus $H_1 : \mu > 0$ with a significance level of $\alpha = 0.05$. Construct a two-sided confidence interval for μ with a confidence coefficient of $1 - \alpha = 0.90$. Compare the results of these analyses to the results obtained in Section 10.4.

3. Assuming only that $X_1, \ldots, X_{19}$ are continuous random variables with population median θ, use the sign test to test $H_0 : \theta \leq 0$ versus $H_1 : \theta > 0$ with a significance level of $\alpha \approx 0.05$. Construct a two-sided confidence interval for θ with a confidence coefficient of $1 - \alpha \approx 0.90$. Compare the results of these analyses to the results obtained in Section 10.4.

Problem Set C Table 10.3 displays a famous data set studied by Charles Darwin.[5] These data appear as Data Set 3 in *A Handbook of Small Data*

---

[5] C. Darwin (1876) *The Effect of Cross- and Self-Fertilization in the Vegetable Kingdom*, Second Edition. John Murray, London.

Sets, accompanied by the following description:

> Pairs of seedlings of the same age, one produced by cross-fertilization and the other by self-fertilization, were grown together so that the members of each pair were reared under nearly identical conditions. The aim was to demonstrate the greater vigour of the cross-fertilized plants. The data are the final heights [in inches] of each plant after a fixed period of time. Darwin consulted [Francis] Galton about the analysis of these data, and they were discussed further in [Ronald] Fisher's *Design of Experiments.*

Pair	Fertilized	
	Cross	Self
1	23.5	17.4
2	12.0	20.4
3	21.0	20.0
4	22.0	20.0
5	19.1	18.4
6	21.5	18.6
7	22.1	18.6
8	20.4	15.3
9	18.3	16.5
10	21.6	18.0
11	23.3	16.3
12	21.0	18.0
13	22.1	12.8
14	23.0	15.5
15	12.0	18.0

Table 10.3: Darwin's data on the heights (in inches) of cross- and self-fertilized seedlings.

1. Show that this problem can be formulated as a 1-sample location problem. To do so, you should:

   (a) Identify the experimental units and the measurement(s) taken on each unit.

(b) Define appropriate random variables $X_1, \ldots, X_n \sim P$. Remember that the statistical procedures that we will employ assume that these random variables are independent and identically distributed.

(c) Let θ denote the location parameter (measure of centrality) of interest. Depending on which statistical procedure we decide to use, either $\theta = EX_i = \mu$ or $\theta = q_2(X_i)$. State appropriate null and alternative hypotheses about θ.

2. Does it seem reasonable to assume that the sample $\vec{x} = (x_1, \ldots, x_n)$, the observed values of $X_1, \ldots, X_n$, were drawn from:

   (a) a normal distribution? Why or why not?

   (b) a symmetric distribution? Why or why not?

3. Assume that $X_1, \ldots, X_n$ are normally distributed and let $\theta = EX_i = \mu$.

   (a) Test the null hypothesis derived above using Student's 1-sample t-test. What is the significance probability? If we adopt a significance level of $\alpha = 0.05$, should we reject the null hypothesis?

   (b) Construct a (2-sided) confidence interval for θ with a confidence coefficient of approximately 0.90.

4. Now we drop the assumption of normality. Assume that $X_1, \ldots, X_n$ are symmetric (but not necessarily normal), continuous random variables and let θ denote the center of symmetry.

   (a) Test the null hypothesis derived above using the Wilcoxon signed rank test. What is the significance probability? If we adopt a significance level of $\alpha = 0.05$, should we reject the null hypothesis?

   (b) Estimate θ by computing the median of the Walsh averages.

   (c) Construct a (2-sided) confidence interval for θ with a confidence coefficient of approximately 0.90.

5. Finally we drop the assumption of symmetry, assuming only that $X_1, \ldots, X_n$ are continuous random variables, and let $\theta = q_2(X_i)$.

   (a) Test the null hypothesis derived above using the sign test. What is the significance probability? If we adopt a significance level of $\alpha = 0.05$, should we reject the null hypothesis?

(b) Estimate θ by computing the sample median.

(c) Construct a (2-sided) confidence interval for θ with a confidence coefficient of approximately 0.90.

Problem Set D The ancient Greeks greatly admired rectangles with a height-to-width ratio of

$$1 : \frac{1+\sqrt{5}}{2} = 0.618034.$$

They called this number the "golden ratio" and used it repeatedly in their art and architecture, e.g., in building the Parthenon. Furthermore, golden rectangles are often found in the art of later western cultures.

A cultural anthropologist wondered if the Shoshoni, a native American civilization, also used golden rectangles.[6] The following measurements, which appear as Data Set 150 in *A Handbook of Small Data Sets*, are height-to-width ratios of beaded rectangles used by the Shoshoni in decorating various leather goods.

0.693	0.662	0.690	0.606	0.570
0.749	0.672	0.628	0.609	0.844
0.654	0.615	0.668	0.601	0.576
0.670	0.606	0.611	0.553	0.933

We will analyze the Shoshoni rectangles as a 1-sample location problem.

1. There are two natural scales that we might use in analyzing these data. One possibility is to analyze the ratios themselves; the other is to analyze the (natural) logarithms of the ratios. For which of these possibilities would an assumption of normality seem more plausible? Please justify your answer.

2. Choose the possibility (ratios or logarithms of ratios) for which an assumption of normality seems more plausible. Formulate suitable null and alternative hypotheses for testing the possibility that the Shoshoni were using golden rectangles. Using Student's 1-sample t-test, compute a significance probability for testing these hypotheses. Would you reject or accept the null hypothesis using a significance level of 0.05?

[6] *Lowie's Selected Papers in Anthropology*, edited by C. Dubois, University of California Press, Berkeley, 1970.

3. Suppose that we are unwilling to assume that either the ratios or the log-ratios were drawn from a normal distribution. Use the sign test to construct a 0.90-level confidence interval for the population median of the ratios.

Problem Set E Developed in 1975, the drug captopril is used to treat hypertension. Table 10.4 displays data on the effect of captopril on blood pressure. Researchers measured the supine systolic and diastolic blood pressures of $n = 15$ patients with moderate essential hypertension, immediately before and two hours after administering caprotil.[7]

Patient	Systolic		Diastolic	
	before	after	before	after
1	210	201	130	125
2	169	165	122	121
3	187	166	124	121
4	160	157	104	106
5	167	147	112	101
6	176	145	101	85
7	185	168	121	98
8	206	180	124	105
9	173	147	115	103
10	146	136	102	98
11	174	151	98	90
12	201	168	119	98
13	198	179	106	110
14	148	129	107	103
15	154	131	100	82

Table 10.4: Blood pressures of $n = 15$ patients immediately before and two hours after receiving captopril.

We consider the question of whether or not captopril affects systolic and diastolic blood pressure differently.

[7]G. A. MacGregor, N. D. Markandu, J. E. Roulston, and J. C. Jones (1979). Essential hypertension: effect of an oral inhibitor of angiotension-converting enzyme. *British Medical Journal*, 2:1100–1109. These data appear as Data Set 72 in *A Handbook of Small Data Sets*.

1. Let SB and SA denote before and after systolic blood pressure; let DB and DA denote before and after diastolic blood pressure. There are several random variables that might be of interest:

$$X_i = (SB_i - SA_i) - (DB_i - DA_i) \tag{10.4}$$

$$X_i = \frac{SB_i - SA_i}{SB_i} - \frac{DB_i - DA_i}{DB_i} \tag{10.5}$$

$$X_i = \frac{SB_i - SA_i}{SB_i} \div \frac{DB_i - DA_i}{DB_i} \tag{10.6}$$

$$X_i = \log\left(\frac{SB_i - SA_i}{SB_i} \div \frac{DB_i - DA_i}{DB_i}\right) \tag{10.7}$$

   Suggest rationales for considering each of these possibilities.

2. Which (if any) of the above random variables appear to be normally distributed? Which appear to be symmetrically distributed? Identify the variable that you will use in subsequent analyses.

3. Choose a suitable measure of centrality, $\theta = EX_i = \mu$ or $\theta = q_2(X_i)$, for subseqient analysis. What value of θ corresponds to the possibility that captopril affects systolic and diastolic blood pressure equally?

4. Test the null hypothesis that captopril affects systolic and diastolic blood pressure equally. Compute a significance probability. Should the the null hypothesis be rejected at a significance level of $\alpha = 0.05$?

5. Construct a confidence interval for θ that has a confidence coefficient of (approximately) $1 - \alpha = 0.90$.

Chapter 11

2-Sample Location Problems

Thus far, in Chapters 9 and 10, we have studied inferences about a single population. In contrast, the present chapter is concerned with comparing *two* populations with respect to a measure of centrality, either the population mean or the population median. We assume the following:

1. $X_1, \ldots, X_{n_1} \sim P_1$ and $Y_1, \ldots, Y_{n_2} \sim P_2$ are continuous random variables. The X_i and the Y_j are mutually independent. In particular, there is no natural pairing of X_1 with Y_1, X_2 with Y_2, etc.

2. P_1 has location parameter θ_1 and P_2 has location parameter θ_2. We assume that comparisons of θ_1 and θ_2 are meaningful. For example, we might compare population means, $\theta_1 = \mu_1 = EX_i$ and $\theta_2 = \mu_2 = EY_j$, or population medians, $\theta_1 = q_2(X_i)$ and $\theta_2 = q_2(Y_j)$, but we would not compare the mean of one population and the median of another population. The *shift parameter*, $\Delta = \theta_1 - \theta_2$, measures the difference in population location.

3. We observe random samples $\vec{x} = \{x_1, \ldots, x_{n_1}\}$ and $\vec{y} = \{y_1, \ldots, y_{n_2}\}$, from which we attempt to draw inferences about Δ. Notice that we do *not* assume that $n_1 = n_2$.

The same four questions that we posed at the beginning of Chapter 10 can be asked here. What distinguishes 2-sample problems from 1-sample problems is the number of populations from which the experimental units were drawn. The prototypical case of a 2-sample problem is the case of a treatment population and a control population. We begin by considering two examples.

Example 11.1 A researcher investigated the effect of Alzheimer's disease (AD) on ability to perform a confrontation naming task. She recruited 60 mildly demented AD patients and 60 normal elderly control subjects. The control subjects resembled the AD patients in that the two groups had comparable mean ages, years of education, and (estimated) IQ scores; however, the control subjects were not individually matched to the AD patients. Each person was administered the Boston Naming Test (BNT), on which higher scores represent better performance. For this experiment:

1. An experimental unit is a person.

2. The experimental units belong to one of two populations: AD patients or normal elderly persons.

3. One measurement (score on BNT) is taken on each experimental unit.

4. Let X_i denote the BNT score for AD patient i. Let Y_j denote the BNT score for control subject j. Then $X_1, \ldots, X_{n_1} \sim P_1$, $Y_1, \ldots, Y_{n_2} \sim P_2$, and we are interested in drawing inferences about $\Delta = \theta_1 - \theta_2$. Notice that $\Delta < 0$ if and only if $\theta_1 < \theta_2$. Thus, to document that AD compromises confrontation naming ability, we might test $H_0 : \Delta \geq 0$ against $H_1 : \Delta < 0$.

Example 11.2 A drug is supposed to lower blood pressure. To determine if it does, $n_1 + n_2$ hypertensive patients are recruited to participate in a *double-blind* study. The patients are randomly assigned to a treatment group of n_1 patients and a control group of n_2 patients. Each patient in the treatment group receives the drug for two months; each patient in the control group receives a *placebo* for the same period. Each patient's blood pressure is measured before and after the two month period, and neither the patient nor the technician know to which group the patient was assigned. For this experiment:

1. An experimental unit is a patient.

2. The experimental units belong to one of two populations: hypertensive patients who receive the drug and hypertensive patients who receive the placebo. Notice that there are two populations despite the fact that all $n_1 + n_2$ patients were initially recruited from a single population. *Different treatment protocols create different populations.*

3. Two measurements (blood pressure before and after treatment) are taken on each experimental unit.

4. Let B_{1i} and A_{1i} denote the before and after blood pressures of patient i in the treatment group. Similarly, let B_{2j} and A_{2j} denote the before and after blood pressures of patient j in the control group. Let $X_i = B_{1i} - A_{1i}$, the decrease in blood pressure for patient i in the treatment group, and let $Y_j = B_{2j} - A_{2j}$, the decrease in blood pressure for patient j in the control group. Then $X_1, \ldots, X_{n_1} \sim P_1$, $Y_1, \ldots, Y_{n_2} \sim P_2$, and we are interested in drawing inferences about $\Delta = \theta_1 - \theta_2$. Notice that $\Delta > 0$ if and only if $\theta_1 > \theta_2$, i.e., if the decrease in blood pressure is greater for the treatment group than for the control group. Thus, a drug company required to produce compelling evidence of the drug's efficacy might test $H_0 : \Delta \leq 0$ against $H_1 : \Delta > 0$.

This chapter describes two important approaches to 2-sample location problems:

11.1 If we assume that the data are normally distributed, then we will be interested in inferences about the difference in population means. We will distinguish three cases, corresponding to what is known about the population variances.

11.2 If we assume only that the data are continuously distributed, then we will be interested in inferences about the difference in population medians. We will assume a *shift model*; i.e., we will assume that P_1 and P_2 differ only with respect to location.

The first approach assumes that both populations are normal, but does not assume that their pdfs have the same shape. The second approach assumes that both pdfs have the same shape, but does not assume normality. The case of nonnormal populations with different shapes is more challenging, beyond the scope of this book. If the populations are symmetric, then one need not assume either normality or a shift model.[1]

11.1 The Normal 2-Sample Location Problem

In this section we assume that

$$P_1 = \text{Normal}\left(\mu_1, \sigma_1^2\right) \quad \text{and} \quad P_2 = \text{Normal}\left(\mu_2, \sigma_2^2\right).$$

[1] M. A. Fligner and G. E. Policello (1981). Robust rank procedures for the Behrens-Fisher problem. *Journal of the American Statistical Association*, 76:162–168.

In describing inferential methods for $\Delta = \mu_1 - \mu_2$, we emphasize connections with material in Chapter 9 and Section 10.1. For example, the natural estimator of a single normal population mean μ is the plug-in estimator $\hat{\mu}$, the sample mean, an unbiased, consistent, asymptotically efficient estimator of μ. In precise analogy, the natural estimator of $\Delta = \mu_1 - \mu_2$, the difference in populations means, is $\hat{\Delta} = \hat{\mu}_1 - \hat{\mu}_2 = \bar{X} - \bar{Y}$, the difference in sample means. Because

$$E\hat{\Delta} = E\bar{X} - E\bar{Y} = \mu_1 - \mu_2 = \Delta,$$

$\hat{\Delta}$ is an unbiased estimator of Δ. It is also consistent and asymptotically efficient.

In Chapter 9 and Section 10.1, hypothesis testing and set estimation for a single population mean were based on knowing the distribution of the standardized natural estimator, a random variable of the form

$$\frac{\begin{array}{c}\text{sample}\\\text{mean}\end{array} - \begin{array}{c}\text{hypothesized}\\\text{mean}\end{array}}{\begin{array}{c}\text{standard deviation}\\\text{of sample mean}\end{array}}.$$

The denominator of this random variable, often called the *standard error*, was either known or estimated, depending on our knowledge of the population variance σ^2. For σ^2 known, we learned that

$$Z = \frac{\bar{X} - \mu_0}{\sqrt{\sigma^2/n}} \quad \left\{ \begin{array}{lll} \sim & \text{Normal}(0,1) & \text{if} \quad X_1, \ldots, X_n \sim \text{Normal}\left(\mu_0, \sigma^2\right) \\ \approx & \text{Normal}(0,1) & \text{if} \quad n \text{ large} \end{array} \right\}.$$

For σ^2 unknown and estimated by S^2, we learned that

$$T = \frac{\bar{X} - \mu_0}{\sqrt{S^2/n}} \quad \left\{ \begin{array}{lll} \sim & t(n-1) & \text{if} \quad X_1, \ldots, X_n \sim \text{Normal}\left(\mu_0, \sigma^2\right) \\ \approx & \text{Normal}(0,1) & \text{if} \quad n \text{ large} \end{array} \right\}.$$

These facts allowed us to construct confidence intervals for and test hypotheses about the population mean. The confidence intervals were of the form

$$\left(\begin{array}{c}\text{sample}\\\text{mean}\end{array} \right) \pm q \cdot \left(\begin{array}{c}\text{standard}\\\text{error}\end{array} \right),$$

where the critical value q is the appropriate quantile of the distribution of Z or T. The tests also were based on Z or T, and the significance probabilities were computed using the corresponding distribution.

The logic for drawing inferences about two populations means is identical to the logic for drawing inferences about one population mean—we simply replace "mean" with "difference in means" and base inferences about Δ on the distribution of

$$\frac{\text{sample difference} - \text{hypothesized difference}}{\text{standard deviation of sample difference}} = \frac{\hat{\Delta} - \Delta_0}{\text{standard error}}.$$

Because $X_i \sim \text{Normal}(\mu_1, \sigma_1^2)$ and $Y_j \sim \text{Normal}(\mu_2, \sigma_2^2)$,

$$\bar{X} \sim \text{Normal}\left(\mu_1, \frac{\sigma_1^2}{n_1}\right) \quad \text{and} \quad \bar{Y} \sim \text{Normal}\left(\mu_2, \frac{\sigma_2^2}{n_2}\right).$$

Because $\bar{X}$ and $\bar{Y}$ are independent, it follows from Theorem 5.2 that

$$\hat{\Delta} = \bar{X} - \bar{Y} \sim \text{Normal}\left(\Delta = \mu_1 - \mu_2, \frac{\sigma_1^2}{n_1} + \frac{\sigma_2^2}{n_2}\right).$$

We now distinguish three cases:

1. Both σ_i are known (and possibly unequal). The inferential theory for this case is easy; unfortunately, population variances are rarely known.

2. The σ_i are unknown, but necessarily equal ($\sigma_1 = \sigma_2 = \sigma$). This case should strike the reader as somewhat implausible. If the population variances are not known, then under what circumstances might we reasonably assume that they are equal? Although such circumstances do exist, the primary importance of this case is that the corresponding theory is elementary. Nevertheless, it is important to study this case because the methods derived from the assumption of an unknown common variance are widely used—and abused.

3. The σ_i are unknown and possibly unequal. This is clearly the case of greatest practical importance, but the corresponding theory is somewhat unsatisfying. The problem of drawing inferences when the population variances are unknown and possibly unequal is sufficiently notorious that it has a name: the *Behrens-Fisher problem*.

11.1.1 Known Variances

If $\Delta = \Delta_0$, then

$$Z = \frac{\hat{\Delta} - \Delta_0}{\sqrt{\frac{\sigma_1^2}{n_1} + \frac{\sigma_2^2}{n_2}}} \sim \text{Normal}(0,1).$$

Given $\alpha \in (0,1)$, let q_z denote the $1-\alpha/2$ quantile of Normal(0, 1). We construct a $(1-\alpha)$-level confidence interval for Δ by writing

$$\begin{aligned}
1-\alpha &= P\left(|Z| < q_z\right) \\
&= P\left(|\hat{\Delta} - \Delta| < q_z \sqrt{\frac{\sigma_1^2}{n_1} + \frac{\sigma_2^2}{n_2}}\right) \\
&= P\left(\hat{\Delta} - q_z \sqrt{\frac{\sigma_1^2}{n_1} + \frac{\sigma_2^2}{n_2}} < \Delta < \hat{\Delta} - q_z \sqrt{\frac{\sigma_1^2}{n_1} + \frac{\sigma_2^2}{n_2}}\right).
\end{aligned}$$

The desired confidence interval is

$$\hat{\Delta} \pm q_z \sqrt{\frac{\sigma_1^2}{n_1} + \frac{\sigma_2^2}{n_2}}.$$

Example 11.3 For the first population, suppose that we know that the population standard deviation is $\sigma_1 = 5$ and that we observe a sample of size $n_1 = 60$ with sample mean $\bar{x} = 7.6$. For the second population, suppose that we know that the population standard deviation is $\sigma_2 = 2.5$ and that we observe a sample of size $n_2 = 15$ with sample mean $\bar{y} = 5.2$. To construct a 0.95-level confidence interval for Δ, we first compute

$$q_z = \texttt{qnorm}(.975) = 1.959964 \doteq 1.96,$$

then

$$(7.6 - 5.2) \pm 1.96\sqrt{\frac{5^2}{60} + \frac{2.5^2}{15}} \doteq 2.4 \pm 1.79 = (0.61, 4.21).$$

Example 11.4 For the first population, suppose that we know that the population variance is $\sigma_1^2 = 8$ and that we observe a sample of size $n_1 = 10$ with sample mean $\bar{x} = 9.7$. For the second population, suppose that we know that the population variance is $\sigma_2^2 = 96$ and that we observe a sample of size $n_2 = 5$ with sample mean $\bar{y} = 2.6$. To construct a 0.95-level confidence interval for Δ, we first compute

$$q_z = \texttt{qnorm}(.975) = 1.959964 \doteq 1.96,$$

then

$$(9.7 - 2.6) \pm 1.96\sqrt{\frac{8}{10} + \frac{96}{5}} \doteq 7.1 \pm 8.765 = (-1.665, 15.865).$$

To test $H_0 : \Delta = \Delta_0$ versus $H_1 : \Delta \neq \Delta_0$, we exploit the fact that $Z \sim \text{Normal}(0, 1)$ under H_0. Let z denote the observed value of Z. Then a natural level-α test is the test that rejects H_0 if and only if

$$\mathbf{p} = P_{\Delta_0}(|Z| \geq |z|) \leq \alpha,$$

which is equivalent to rejecting H_0 if and only if $|z| \geq q_z$. This test is sometimes called the 2-sample z-test.

Example 11.3 (continued) To test $H_0 : \Delta = 0$ versus $H_1 : \Delta \neq 0$, we compute

$$z = \frac{(7.6 - 5.2) - 0}{\sqrt{5^2/60 + 2.5^2/15}} \doteq 2.629.$$

Because $|2.629| > 1.96$, we reject H_0 at significance level $\alpha = 0.05$. The significance probability is

$$\mathbf{p} = P_{\Delta_0}(|Z| \geq |2.629|) = 2 * \texttt{pnorm}(-2.629) \doteq 0.008562.$$

Example 11.4 (continued) To test $H_0 : \Delta = 0$ versus $H_1 : \Delta \neq 0$, we compute

$$z = \frac{(9.7 - 2.6) - 0}{\sqrt{8/10 + 96/5}} \doteq 1.5876.$$

Because $|1.5876| < 1.96$, we decline to reject H_0 at significance level $\alpha = 0.05$. The significance probability is

$$\mathbf{p} = P_{\Delta_0}(|Z| \geq |1.5876|) = 2 * \texttt{pnorm}(-1.5876) \doteq 0.1124.$$

11.1.2 Unknown Common Variance

Now we assume that $\sigma_1 = \sigma_2 = \sigma$, but that the common variance σ^2 is unknown. Because σ^2 is unknown, we must estimate it. Let

$$S_1^2 = \frac{1}{n_1 - 1} \sum_{i=1}^{n_1} (X_i - \bar{X})^2$$

denote the sample variance for the X_i and let

$$S_2^2 = \frac{1}{n_2 - 1} \sum_{j=1}^{n_2} (Y_j - \bar{Y})^2$$

denote the sample variance for the Y_j. If we sampled only the first population, then we would use S_1^2 to estimate the first population variance, σ_1^2. Likewise, if we sampled only the second population, then we would use S_2^2 to estimate the second population variance, σ_2^2. Neither is appropriate in the present situation, as S_1^2 does not use the second sample and S_2^2 does not use the first sample. Therefore, we create a weighted average of the separate sample variances,

$$\begin{aligned} S_P^2 &= \frac{(n_1-1)S_1^2 + (n_2-1)S_2^2}{(n_1-1)+(n_2-1)} \\ &= \frac{1}{n_1+n_2-2}\left[\sum_{i=1}^{n_1}(X_i-\bar{X})^2 + \sum_{j=1}^{n_2}(Y_j-\bar{Y})^2\right], \end{aligned}$$

the *pooled sample variance*. Then

$$ES_P^2 = \frac{(n_1-1)ES_1^2 + (n_2-1)ES_2^2}{(n_1-1)+(n_2-1)} = \frac{(n_1-1)\sigma^2 + (n_2-1)\sigma^2}{(n_1-1)+(n_2-1)} = \sigma^2,$$

so the pooled sample variance is an unbiased estimator of a common population variance. It is also consistent and asymptotically efficient for estimating a common normal variance.

Instead of

$$Z = \frac{\hat{\Delta} - \Delta_0}{\sqrt{\frac{\sigma_1^2}{n_1} + \frac{\sigma_2^2}{n_2}}} = \frac{\hat{\Delta} - \Delta_0}{\sqrt{\left(\frac{1}{n_1} + \frac{1}{n_2}\right)\sigma^2}},$$

we now rely on

$$T = \frac{\hat{\Delta} - \Delta_0}{\sqrt{\left(\frac{1}{n_1} + \frac{1}{n_2}\right) S_P^2}}.$$

The following result allows us to construct confidence intervals and test hypotheses about the shift parameter $\Delta = \mu_1 - \mu_2$.

Theorem 11.1 *If $\Delta = \Delta_0$, then $T \sim t(n_1 + n_2 - 2)$.*

Given $\alpha \in (0, 1)$, let q_t denote the $1 - \alpha/2$ quantile of $t(n_1 + n_2 - 2)$. Exploiting Theorem 11.1, a $(1 - \alpha)$-level confidence interval for Δ is

$$\hat{\Delta} \pm q_t \sqrt{\left(\frac{1}{n_1} + \frac{1}{n_2}\right) S_P^2}.$$

Example 11.3 (continued) Now suppose that, instead of knowing population standard deviations $\sigma_1 = 5$ and $\sigma_2 = 2.5$, we observe sample standard deviations $s_1 = 5$ and $s_2 = 2.5$. The ratio of sample variances, $s_1^2/s_2^2 = 4 \neq 1$, strongly suggests that the population variances are unequal. We proceed under the assumption that $\sigma_1 = \sigma_2$ for the purpose of illustration. The pooled sample variance is

$$S_P^2 = \frac{59 \cdot 5^2 + 14 \cdot 2.5^2}{59 + 14} = 21.40411.$$

To construct a 0.95-level confidence interval for Δ, we first compute

$$q_t = \texttt{qt}(.975, 73) = 1.992997 \doteq 1.993,$$

then

$$(7.6 - 5.2) \pm 1.993\sqrt{\left(\frac{1}{60} + \frac{1}{15}\right) \cdot 21.40411} \doteq 2.4 \pm 2.66 = (-0.26, 5.06).$$

Example 11.4 (continued) Now suppose that, instead of knowing population variances $\sigma_1^2 = 8$ and $\sigma_2^2 = 96$, we observe sample variances $s_1^2 = 8$ and $s_2^2 = 96$. Again, the ratio of sample variances, $s_2^2/s_1^2 = 12 \neq 1$, strongly suggests that the population variances are unequal. We proceed under the assumption that $\sigma_1 = \sigma_2$ for the purpose of illustration. The pooled sample variance is

$$S_P^2 = \frac{9 \cdot 8 + 4 \cdot 96}{9 + 4} = 35.07692.$$

To construct a 0.95-level confidence interval for Δ, we first compute

$$q_t = \texttt{qt}(.975, 13) = 2.160369 \doteq 2.16,$$

then

$$(9.7 - 2.6) \pm 2.16\sqrt{\left(\frac{1}{10} + \frac{1}{5}\right) \cdot 35.07692} \doteq 7.1 \pm 7.01 = (0.09, 14.11).$$

To test $H_0 : \Delta = \Delta_0$ versus $H_1 : \Delta \neq \Delta_0$, we exploit the fact that $T \sim t(n_1 + n_2 - 2)$ under H_0. Let t denote the observed value of T. Then a natural level-α test is the test that rejects H_0 if and only if

$$\mathbf{p} = P_{\Delta_0}(|T| \geq |t|) \leq \alpha,$$

which is equivalent to rejecting H_0 if and only if $|t| \geq q_t$. This test is called *Student's 2-sample t-test*.

Example 11.3 (continued) To test $H_0 : \Delta = 0$ versus $H_1 : \Delta \neq 0$, we compute

$$t = \frac{(7.6 - 5.2) - 0}{\sqrt{(1/60 + 1/15) \cdot 21.40411}} \doteq 1.797.$$

Because $|1.797| < 1.993$, we decline to reject H_0 at significance level $\alpha = 0.05$. The significance probability is

$$\mathbf{p} = P_{\Delta_0}(|T| \geq |1.797|) = 2 * \texttt{pt}(-1.797, 73) \doteq 0.0764684.$$

Example 11.4 (continued) To test $H_0 : \Delta = 0$ versus $H_1 : \Delta \neq 0$, we compute

$$t = \frac{(9.7 - 2.6) - 0}{\sqrt{(1/10 + 1/5) \cdot 35.07692}} \doteq 2.19.$$

Because $|2.19| > 2.16$, we reject H_0 at significance level $\alpha = 0.05$. The significance probability is

$$\mathbf{p} = P_{\Delta_0}(|T| \geq |2.19|) = 2 * \texttt{pt}(-2.19, 13) \doteq 0.04747.$$

11.1.3 Unknown Variances

Now we drop the assumption that $\sigma_1 = \sigma_2$. We must then estimate each population variance separately, σ_1^2 with S_1^2 and σ_2^2 with S_2^2. Instead of

$$Z = \frac{\hat{\Delta} - \Delta_0}{\sqrt{\frac{\sigma_1^2}{n_1} + \frac{\sigma_2^2}{n_2}}}$$

we now rely on

$$T_W = \frac{\hat{\Delta} - \Delta_0}{\sqrt{\frac{S_1^2}{n_1} + \frac{S_2^2}{n_2}}}.$$

Unfortunately, there is no analogue of Theorem 11.1—the exact distribution of T_W is not known.

The exact distribution of T_W appears to be intractable, but B. L. Welch[2] argued that $T_W \approx t(\nu)$, with

$$\nu = \frac{\left(\frac{\sigma_1^2}{n_1} + \frac{\sigma_2^2}{n_2}\right)^2}{\frac{(\sigma_1^2/n_1)^2}{n_1 - 1} + \frac{(\sigma_2^2/n_2)^2}{n_2 - 1}}.$$

[2] B. L. Welch (1937). The significance of the difference between two means when the population variances are unequal. *Biometrika*, 29:350–362.

B. L. Welch (1947). The generalization of "Student's" problem when several different population variances are involved. *Biometrila*, 34:28–35.

Because σ_1^2 and σ_2^2 are unknown, we estimate ν by

$$\hat{\nu} = \frac{\left(\frac{S_1^2}{n_1} + \frac{S_2^2}{n_2}\right)^2}{\frac{(S_1^2/n_1)^2}{n_1-1} + \frac{(S_2^2/n_2)^2}{n_2-1}}.$$

Simulation studies have revealed that the approximation $T_W \approx t(\hat{\nu})$ works well in practice.

Given $\alpha \in (0,1)$, let q_t denote the $1-\alpha/2$ quantile of $t(\hat{\nu})$. Using Welch's approximation, an approximate $(1-\alpha)$-level confidence interval for Δ is

$$\hat{\Delta} \pm q_t \sqrt{\frac{S_1^2}{n_1} + \frac{S_2^2}{n_2}}.$$

Example 11.3 (continued) Now we estimate the unknown population variances separately, σ_1^2 by $s_1^2 = 5^2$ and σ_2^2 by $s_2^2 = 2.5^2$. Welch's approximation involves

$$\hat{\nu} = \frac{\left(\frac{5^2}{60} + \frac{2.5^2}{15}\right)^2}{\frac{(5^2/60)^2}{60-1} + \frac{(2.5^2/15)^2}{15-1}} = 45.26027 \doteq 45.26$$

degrees of freedom. To construct a 0.95-level confidence interval for Δ, we first compute

$$q_t = \texttt{qt}(.975, 45.26) \doteq 2.014,$$

then

$$(7.6 - 5.2) \pm 2.014\sqrt{5^2/60 + 2.5^2/15} \doteq 2.4 \pm 1.84 = (0.56, 4.24).$$

Example 11.4 (continued) Now we estimate the unknown population variances separately, σ_1^2 by $s_1^2 = 8$ and σ_2^2 by $s_2^2 = 96$. Welch's approximation involves

$$\hat{\nu} = \frac{\left(\frac{8}{10} + \frac{96}{5}\right)^2}{\frac{(8/10)^2}{10-1} + \frac{(96/5)^2}{5-1}} = 4.336931 \doteq 4.337$$

degrees of freedom. To construct a 0.95-level confidence interval for Δ, we first compute

$$q_t = \texttt{qt}(.975, 4.337) = 2.6934,$$

then

$$(9.7 - 2.6) \pm 2.6934\sqrt{8/10 + 96/5} \doteq 7.1 \pm 13.413 = (-6.313, 20.513).$$

To test $H_0 : \Delta = \Delta_0$ versus $H_1 : \Delta \neq \Delta_0$, we exploit the approximation $T_W \approx t(\hat{\nu})$ under H_0. Let t_W denote the observed value of T_W. Then a natural approximate level-α test is the test that rejects H_0 if and only if

$$\mathbf{p} = P_{\Delta_0}(|T_W| \geq |t_W|) \leq \alpha,$$

which is equivalent to rejecting H_0 if and only if $|t_W| \geq q_t$. This test is sometimes called *Welch's approximate t-test.*

Example 11.3 (continued) To test $H_0 : \Delta = 0$ versus $H_1 : \Delta \neq 0$, we compute

$$t_W = \frac{(7.6 - 5.2) - 0}{\sqrt{5^2/60 + 2.5^2/15}} \doteq 2.629.$$

Because $|2.629| > 2.014$, we reject H_0 at significance level $\alpha = 0.05$. The significance probability is

$$\mathbf{p} = P_{\Delta_0}(|T_W| \geq |2.629|) = 2 * \texttt{pt}(-2.629, 45.26) \doteq 0.011655.$$

Example 11.4 (continued) To test $H_0 : \Delta = 0$ versus $H_1 : \Delta \neq 0$, we compute

$$t_W = \frac{(9.7 - 2.6) - 0}{\sqrt{8/10 + 96/5}} \doteq 1.4257.$$

Because $|1.4257| < 2.6934$, we decline to reject H_0 at significance level $\alpha = 0.05$. The significance probability is

$$\mathbf{p} = P_{\Delta_0}(|T_W| \geq |1.4257|) = 2 * \texttt{pt}(-1.4257, 4.337) \doteq 0.2218.$$

Examples 11.3 and 11.4 were carefully constructed to reveal the sensitivity of Student's 2-sample t-test to the assumption of equal population variances. Welch's approximation is good enough that we can use it to benchmark Student's test when variances are unequal. In Example 11.3, Welch's approximate t-test produced a significance probability of $\mathbf{p} \doteq 0.012$, leading us to reject the null hypothesis at $\alpha = 0.05$. Student's 2-sample t-test produced a misleading significance probability of $\mathbf{p} \doteq 0.076$, leading us to commit a Type II error. In Example 11.4, Welch's approximate t-test produced a significance probability of $\mathbf{p} \doteq 0.222$, leading us to retain the null hypothesis at $\alpha = 0.05$. Student's 2-sample t-test produced a misleading significance probability of $\mathbf{p} \doteq 0.047$, leading us to commit a Type I error.

Evidently, Student's 2-sample t-test (and the corresponding procedure for constructing confidence intervals) should not be used unless one is convinced that the population variances are identical. The consequences of using Student's test when the population variances are unequal may be exacerbated when the sample sizes are unequal. In general:

- If $n_1 = n_2$, then $t = t_W$.
- If the population variances are (approximately) equal, then t and t_W tend to be (approximately) equal.
- If the larger sample is drawn from the population with the larger variance, then t will tend to be less than t_W. All else equal, this means that Student's test will tend to produce significance probabilities that are too large.
- If the larger sample is drawn from the population with the smaller variance, then t will tend to be greater than t_W. All else equal, this means that Student's test will tend to produce significance probabilities that are too small.
- If the population variances are (approximately) equal, then $\hat{\nu}$ will be (approximately) $n_1 + n_2 - 2$.
- It will *always* be the case that $\hat{\nu} \leq n_1+n_2-2$. All else equal, this means that Student's test will tend to produce significance probabilities that are too large.

From these observations we draw the following conclusions:

1. If the population variances are unequal, then Student's 2-sample t-test may produce misleading significance probabilities.
2. If the population variances are equal, then Welch's approximate t-test is approximately equivalent to Student's 2-sample t-test. Thus, if one uses Welch's test in the situation for which Student's test is appropriate, one is not likely to be led astray.

3. *Don't use Student's 2-sample t-test!* I remember how shocked I was when I first heard this advice as a first-year graduate student in a course devoted to the theory of hypothesis testing. The instructor, Erich Lehmann, one of the great statisticians of the 20th century and the author of a famous book on hypothesis testing,[3] told us: "If you get just one thing out of this course, I'd like it to be that you should *never* use Student's 2-sample t-test."

11.2 The Case of a General Shift Family

In the preceding section we assumed that $X_i \sim P_1 = \text{Normal}(\mu_1, \sigma_1^2)$ and $Y_j \sim \text{Normal}(\mu_2, \sigma_2^2)$. Notice that, if $\sigma_1^2 = \sigma_2^2$ and $\Delta = \mu_1 - \mu_2$, then

$$X_i - \Delta \sim \text{Normal}\left(\mu_1 - \Delta, \sigma_1^2\right) = \text{Normal}\left(\mu_2, \sigma_2^2\right) = P_2.$$

In this case, P_1 and P_2 differ only with respect to location. This observation leads us to introduce the concept of a shift family.

Definition 11.1 *A family of distributions,* $\mathcal{P} = \{P_\theta : \theta \in \Re\}$, *is a* shift family *if and only if*

$$X \sim P_\theta \quad \textit{entails} \quad X - \theta \sim P_0.$$

If P_0 has a pdf, then a shift family is a family in which all of the pdfs have a common shape, differing only with respect to location.

Example 11.5 Fix $\sigma^2 > 0$. The family $\{\text{Normal}(\theta, \sigma^2) : \theta \in \Re\}$ is a shift family because, if $X \sim \text{Normal}(\theta, \sigma^2)$, then $X - \theta \sim \text{Normal}(0, \sigma^2)$.

Section 11.1.2 concerned the case of a normal shift family. Section 11.1.3 retained the assumption of normality, but, by allowing unequal variances, relaxed the assumption of a shift family. The present section does the reverse, relaxing the assumption of normality but retaining the assumption of a shift family. The 2-sample location problem for a general shift family is the problem of drawing inferences about the shift parameter, Δ, in the following statistical model.

[3] E. L. Lehmann (1959). *Testing Statistical Hypotheses.* John Wiley & Sons, New York.

1. Let $\mathcal{P} = \{P_\theta : \theta \in \Re\}$ denote a shift family in which P_0 has a pdf. Without loss of generality, we assume that θ is the population median of P_θ.

2. Assume that $X_1, \ldots, X_{n_1} \sim P_{\theta_1}$ and $Y_1, \ldots, Y_{n_2} \sim P_{\theta_2}$ are mutually independent. Without loss of generality, we assume that $n_1 \leq n_2$.

3. We observe random samples $\vec{x} = \{x_1, \ldots, x_{n_1}\}$ and $\vec{y} = \{y_1, \ldots, y_{n_2}\}$, from which we attempt to draw inferences about $\Delta = \theta_1 - \theta_2$, the difference in population medians.

As in Sections 10.2 and 10.3, we will proceed by first developing a test for $H_0 : \Delta = \Delta_0$ versus $H_1 : \Delta \neq \Delta_0$, then using it to derive point and set estimators of Δ.

11.2.1 Hypothesis Testing

We begin by noting that it suffices to test $H_0 : \Delta = 0$ versus $H_1 : \Delta \neq 0$. Given such a test, we can test $H_0 : \Delta = \Delta_0$ versus $H_1 : \Delta \neq \Delta_0$ by replacing X_i with $X_i' = X_i - \Delta_0$, then testing $H_0 : \Delta = 0$ versus $H_1 : \Delta \neq 0$.

Because the X_i and Y_j are continuous random variables, they assume $N = n_1 + n_2$ distinct values with probability one. Assuming that the observed x_i and y_j are in fact distinct, the *Wilcoxon rank sum test* is the following procedure:

1. Rank the N observed values from smallest to largest.

2. Let T_x denote the sum of the ranks that correspond to the X_i and let T_y denote the sum of the ranks that correspond to the Y_j. Notice that

$$T_x + T_y = \sum_{k=1}^{N} k = N(N+1)/2$$

and that we can write

$$T_x = \sum_{k=1}^{N} k I_k,$$

where $I_1, \ldots, I_N$ are Bernoulli random variables defined by $I_k = 1$ if rank k corresponds to an X_i and $I_k = 0$ if rank k corresponds to a Y_j.

3. Under $H_0 : \Delta = 0$, the X_i and the Y_j have the same distribution. Hence, for each k, the probability that rank k corresponds to an X_i is n_1/N. Because each k is equally likely to contribute to T_x,

the distribution of T_x must be symmetric. Furthermore, noting that $P(I_k = 1) = n_1/N$ and $EI_k = n_1/N$, we see that

$$\begin{aligned} ET_x &= E\left(\sum_{k=1}^{N} kI_k\right) = \sum_{k=1}^{N} kET_k = \frac{n_1}{N}\sum_{k=1}^{N} k \\ &= \frac{n_1}{N}\frac{N(N+1)}{2} = \frac{n_1(N+1)}{2}. \end{aligned}$$

4. Let t_x denote the observed value of T_x. The Wilcoxon rank sum test rejects $H_0 : \Delta = 0$ at significance level α if and only if t_x differs sufficiently from ET_x, i.e., if and only if

$$\mathbf{p} = P_{H_0}\left(|T_x - ET_x| \geq |t_x - ET_x|\right) \leq \alpha.$$

As in Section 10.3.1, the challenge lies in computing $\mathbf{p}$. Again we consider a simple example that illustrates the nature of this challenge.

Example 11.6 We test $H_0 : \Delta = 0$ versus $H_1 : \Delta \neq 0$ with a significance level of $\alpha = 0.15$. We draw $n_1 = 2$ observations from one population and $n_2 = 4$ observations from the other, obtaining samples $\vec{x} = \{9.1, 8.3\}$ and $\vec{y} = \{11.9, 10.0, 10.5, 11.3\}$. The pooled ranks are $\{2, 1\}$ and $\{6, 3, 4, 5\}$, so $t_x = 2 + 1 = 3$. Because $T_x \geq 1 + 2 = 3$, $T_x \leq 5 + 6 = 11$, and $ET_x = n_1(N + 1)/2 = 7$, the significance probability is

$$\mathbf{p} = P_{H_0}\left(|T_+ - 7| \geq |3 - 7|\right) = P_{H_0}\left(T_+ = 3\right) + P_{H_0}\left(T_+ = 11\right).$$

To compute $\mathbf{p}$, we require the pmf of the discrete random variable T_x under the null hypothesis that $\Delta = 0$.

Under $H_0 : \Delta = 0$, each of the $\binom{6}{2}$ possible ways of drawing two ranks from $\{1, \ldots, 6\}$ is equally likely to produce the ranks that correspond to $\{x_1, x_2\}$. These possibilities are

	1	1	1	1	1	2	2	2	2	3	3	3	4	4	5
	2	3	4	5	6	3	4	5	6	4	5	6	5	6	6
t_x	3	4	5	6	7	5	6	7	8	7	8	9	9	10	11

and the pmf of T_x is as follows:

k	3	4	5	6	7	8	9	10	11
$15P(T_x = k)$	1	1	2	2	3	2	2	1	1

Thus,

$$\mathbf{p} = P_{H_0}(T_+ = 3) + P_{H_0}(T_+ = 11) = \frac{1}{15} + \frac{1}{15} \doteq 0.1333.$$

Because $\mathbf{p} \leq \alpha$, we reject $H_0 : \Delta = 0$.

As was the case for the Wilcoxon signed rank test, it is cumbersome to compute significance probabilities for the Wilcoxon rank sum test unless n_1 and n_2 are quite small. As in Section 10.3.1, we consider two ways of approximating $\mathbf{p}$:

1. Simulation.

   Using `R`, it is easy to draw subsets of n_1 ranks from $\{1, \ldots, N\}$ and compute the value of T_x for each subset. The observed proportion of subsets for which

   $$|T_x - n_1(N+1)/2| \geq |t_+ - n_1(N+1)/2|$$

   estimates the true significance probability. (The more subsets that we draw, the more accurate our estimate of $\mathbf{p}$.) I wrote an `R` function, `W2.p.sim`, that implements this procedure. This function is described in Appendix R and can be obtained from the web page for this book.

2. Normal Approximation.

   It turns out that, for n_1 and n_2 sufficiently large, the discrete distribution of T_x under $H_0 : \Delta = 0$ can be approximated by a normal distribution.

Theorem 11.2 *Suppose that* $X_1, \ldots, X_{n_1} \sim P_{\theta_1}$ *and* $Y_1, \ldots, Y_{n_2} \sim P_{\theta_2}$ *satisfy the assumptions of a shift model. Let* $\Delta = \theta_1 - \theta_2$ *and* $N = n_1 + n_2$. *Let* T_x *denote the test statistic for the Wilcoxon rank sum test of* $H_0 : \Delta = 0$ *versus* $H_1 : \Delta \neq 0$. *Under* H_0, $ET_x = n_1(N+1)/4$, $\mathit{Var}\, T_x = n_1 n_2 (N+1)^2/12$, *and*

$$P_{H_0}(T_x \leq c) \to P\left(Z \leq \frac{c - ET_x}{\sqrt{\mathit{Var}\, T_x}}\right)$$

as $n_1, n_2 \to \infty$, *where* $Z \sim \mathit{Normal}(0, 1)$.

I wrote an `R` function, `W2.p.norm`, that uses Theorem 11.2 to compute approximate significance probabilities. This function is described in Appendix R and can be obtained from the web page for this book

Example 11.6 (continued) Five replications of the simulation procedure `W2.p.sim(2,4,3)` resulted in approximate significance probabilities of 0.132, 0.126, 0.128, 0.136, and 0.135. The normal approximation of 0.105, obtained from `W2.p.norm(2,4,3)`, is surprisingly good.

Example 11.7 Suppose that $n_1 = 20$ and $n_2 = 25$ observations produce $t_x = 50$. Two replications of `W2.p.sim(20,25,400,10000)` resulted in approximate significance probabilities of 0.1797 and 0.1774. The normal approximation, obtained from `W2.p.norm(20,25,400)`, is 0.1741.

Finally, we consider the case of ties in the pooled sample. As in Section 10.3.1, we will estimate the average significance probability that results from all of the plausible rankings of the pooled sample. To do so, we simply modify the simulation procedure described above by subjecting each observation to small random perturbations. Each perturbed pooled sample results in a unique ordering. By choosing the magnitude of the perturbations sufficiently small, we can guarantee that each complete ordering generated from a perturbed sample is consistent with the partial ordering derived from the original sample. I wrote an `R` function, `W2.p.ties`, that implements this procedure. This function is described in Appendix R and can be obtained from the web page for this book.

Example 11.8 To test $H_0 : \Delta = 3$ versus $H_1 : \Delta \neq 3$ at $\alpha = 0.05$, the following x_i and y_j were observed:

$\vec{x}$	6.6	14.7	15.7	11.1	7.0	9.0	9.6	8.2	6.8	7.2
$\vec{y}$	4.2	3.6	2.3	2.4	13.4	1.3	2.0	2.9	8.8	3.8

Replacing x_i with $x_i' = x_i - 3$, the pooled sample has the following ranks:

$\vec{x}'$	3.6	11.7	12.7	8.1	4.0	6.0	6.6	5.2	3.8	4.2
r_k	6/7	18	19	16	10	14	15	13	8/9	11/12
$\vec{y}$	4.2	3.6	2.3	2.4	13.4	1.3	2.0	2.9	8.8	3.8
r_k	11/12	6/7	3	4	20	1	2	5	17	8/9

The test statistic, t_x, might be as small as

$$t_x = 6 + 18 + 19 + 16 + 10 + 14 + 15 + 13 + 8 + 11 = 130$$

or as large as

$$t_x = 7 + 18 + 19 + 16 + 10 + 14 + 15 + 13 + 9 + 12 = 133.$$

We might use `W2.p.sim` to estimate, or `W2.p.norm` to approximate, the significance probabilities associated with $t_x = 130$ and $t_+ = 133$. Alternatively, we might use `W2.p.ties` to estimate the average significance probability associated with all of the possible ways of ranking the pooled sample. The latter approach resulted in the following estimated $\mathbf{p}$:

```
W2.p.ties(x,y,3,100000)
[1] 0.05281
```

As $\mathbf{p} > \alpha$, the evidence against $H_0 : \Delta = 3$ is not sufficiently compelling to reject the null hypothesis.

11.2.2 Point Estimation

Following the reasoning that we deployed in Sections 10.2.2 and 10.3.2, we estimate Δ by determining the value of Δ_0 for which the Wilcoxon rank sum test is least inclined to reject $H_0 : \Delta = \Delta_0$ in favor of $H_1 : \Delta \neq \Delta_0$. The key to determining this value is representing the Wilcoxon rank sum test in a slightly different form.

The smallest possible value of T_x is $1 + \cdots + n_1 = n_1(n_1 + 1)/2$. This value of T_x is attained if and only if each Y_j exceeds each X_i, i.e.,

$$W_{yx} = \# \{(X_i, Y_j) : Y_j < X_i\} = 0.$$

Pursuing this insight, we write

$$x_{(1)} < \cdots < x_{(n_1)}$$

and let r_k denote the rank of $x_{(k)}$ in the pooled sample. Then

$$r_k = k + \# \left\{ Y_j : Y_j < X_{(k)} \right\}$$

and

$$\begin{aligned} T_x &= \sum_{k=1}^{n_1} r_k = \sum_{k=1}^{n_1} k + \sum_{k=1}^{n_1} \# \left\{ Y_j : Y_j < X_{(k)} \right\} \\ &= n_1 (n_1 + 1)/2 + \# \{(X_i, Y_j) : Y_j < X_i\} \\ &= n_1 (n_1 + 1)/2 + W_{yx}. \end{aligned}$$

Similarly, reversing the roles of x and y, we have

$$T_y = n_2 (n_2 + 1)/2 + W_{xy},$$

where

$$W_{xy} = \# \{(X_i, Y_j) : X_i < Y_j\}.$$

We can use W_{yx} or W_{xy} instead of T_x or T_y, in which form the Wilcoxon rank sum test is called the Mann-Whitney test.

Recall that we test $H_0 : \Delta = \Delta_0$ versus $H_1 : \Delta \neq \Delta_0$ by testing $H_0 : \Delta = 0$ versus $H_1 : \Delta \neq 0$ using $\vec{x} - \Delta_0$ and $\vec{y}$. We will be least inclined to reject the null hypothesis when

$$T_{x-\Delta_0} = ET_{x-\Delta_0} = n_1 (n_1 + n_2 + 1)/2 = n_1 (n_1 + 1)/2 + n_1 n_2/2$$

and (equivalently)

$$T_y = ET_y = n_2 (n_1 + n_2 + 1)/2 = n_2 (n_2 + 1)/2 + n_1 n_2/2.$$

These conditions are equivalent to requiring that

$$W_{y,x-\Delta_0} = n_1 n_2/2 = W_{x-\Delta_0,y},$$

i.e.,

$$\# \{(X_i, Y_j) : Y_j < X_i - \Delta_0\} = \# \{(X_i, Y_j) : X_i - \Delta_0 < Y_j\},$$

i.e.,

$$\# \{(X_i, Y_j) : X_i - Y_j > \Delta_0\} = \# \{(X_i, Y_j) : X_i - Y_j < \Delta_0\}.$$

This condition will be satisfied if and only if Δ_0 is the median of the pairwise differences, $X_i - Y_j$. The estimator $\hat{\Delta} = \text{median}(X_i - Y_j)$ is called the Hodges-Lehmann estimator. I wrote an `R` function, `W2.hl`, that computes Hodges-Lehmann estimates. This function is described in Appendix R and can be obtained from the web page for this book.

Example 11.8 (continued) To estimate the shift parameter, we compute the median of the pairwise differences:

```
> W2.hl(x,y)
[1] 5.2
```

11.2.3 Set Estimation

We construct a confidence interval for Δ by determining the set of Δ_0 for which the Wilcoxon rank sum test does not reject $H_0 : \Delta = \Delta_0$ in favor of $H_1 : \Delta \neq \Delta_0$. From Section 11.2.2, it is evident that this set consists of those Δ_0 for which at least k $X_i - Y_j$ are less than Δ_0 and at least k $X_i - Y_j$ are greater than Δ_0. The quantity k is determined by the level of confidence that we desire. As in Sections 10.2.3 and 10.3.3, not all confidence levels are possible.

As in Section 10.2.3, the fact that we must approximate the distribution of the discrete random variable T_x under $H_0 : \Delta = \Delta_0$ complicates our efforts to construct confidence intervals. Again, we proceed in three steps:

1. Use the normal approximation to guess a reasonable value of k, i.e., a value for which

$$\begin{aligned} P\left(W_{yx} \leq k-1\right) &= P\left(T_x \leq k-1+n_1\left(n_1+1\right)/2\right) \\ &= P\left(T_x \leq k-1+ET_x - n_1 n_2/2\right) \\ &\approx \alpha/2. \end{aligned}$$

Recall that

$$\begin{aligned} &P\left(T_x \leq k-1+ET_x - n_1 n_2/2\right) \\ &\quad= P\left(T_x \leq k-0.5+ET_x - n_1 n_2/2\right) \\ &\quad= P\left(\frac{T_x - ET_x}{\sqrt{\operatorname{Var} T_x}} \leq \frac{k-0.5-n_1 n_2/2}{\sqrt{\operatorname{Var} T_x}}\right) \\ &\quad\approx P\left(Z \leq \frac{k-0.5-n_1 n_2/2}{\sqrt{\operatorname{Var} T_x}}\right), \end{aligned}$$

where $Z \sim \text{Normal}(0,1)$; hence, given $\alpha \in (0,1)$, a reasonable value of k is obtained by solving

$$P\left(Z \leq \frac{k-0.5-n_1 n_2/2}{\sqrt{\operatorname{Var} T_+}}\right) = \frac{\alpha}{2},$$

resulting in

$$k = 0.5 + n_1 n_2/2 - q_z\sqrt{\operatorname{Var} T_+} = 0.5 + n_1 n_2/4 - q_z\sqrt{n_1 n_2(N+1)/12},$$

where $q_z = \texttt{qnorm}(1-\alpha/2)$.

2. Use simulation to estimate the confidence coefficients,

$$1 - 2P\left(W_{yx} \le k-1\right) = 1 - 2P\left(T_x \le k - 1 + n_1\left(n_1+1\right)/2\right),$$

associated with several reasonable choices of k.

3. Finalize the choice of k (and thereby the confidence coefficient) and construct the corresponding confidence interval. The lower endpoint of the interval is the kth $X_i - Y_j$; the upper endpoint is the $(n_1 n_2 + 1 - k)$th $X_i - Y_j$.

I implemented these steps in the `R` function `W2.ci`, described in Appendix R and available from the web page for this book. This function returns a 5×4 matrix. The first column contains possible choices of k, the second and third columns contain the lower and upper endpoints of the corresponding confidence interval, and the fourth column contains the estimated confidence coefficients.

Example 11.8 (continued) To construct a confidence interval with confidence coefficient $1 - \alpha \approx 0.90$ for Δ, the population shift parameter, we obtain the following results:

```
> W2.ci(x,y,.1,10000)
      k Lower Upper Coverage
[1,] 27   3.2   7.3   0.9234
[2,] 28   3.4   7.2   0.9131
[3,] 29   3.4   7.0   0.8973
[4,] 30   3.6   6.9   0.8805
[5,] 31   3.7   6.9   0.8592
```

The estimated confidence coefficient for $k = 29$ (which happens to be the value of k produced by the normal approximation) nearly equals 0.90, so the desired confidence interval is $(3.4, 7.0)$.

11.3 Case Study: Etruscan versus Italian Head Breadth

In a collection of essays on the origin of the Etruscan empire, N.A. Barnicott and D.R. Brothwell compared measurements on ancient and modern bones.[4]

[4] N.A. Barnicott and D.R. Brothwell (1959). The evaluation of metrical data in the comparison of ancient and modern bones. In *Medical Biology and Etruscan Origins*, edited by G.E.W. Wolstenholme and C.M. O'Connor, Little, Brown & Company, p. 136.

141	148	132	138	154	142	150	146	155	158	150	140
147	148	144	150	149	145	149	158	143	141	144	144
126	140	144	142	141	140	145	135	147	146	141	136
140	146	142	137	148	154	137	139	143	140	131	143
141	149	148	135	148	152	143	144	141	143	147	146
150	132	142	142	143	153	149	146	149	138	142	149
142	137	134	144	146	147	140	142	140	137	152	145
133	138	130	138	134	127	128	138	136	131	126	120
124	132	132	125	139	127	133	136	121	131	125	130
129	125	136	131	132	127	129	132	116	134	125	128
139	132	130	132	128	139	135	133	128	130	130	143
144	137	140	136	135	126	139	131	133	138	133	137
140	130	137	134	130	148	135	138	135	138		

Table 11.1: Maximum breadth (in millimeters) of 84 skulls of Etruscan males (top) and 70 skulls of modern Italian males (bottom).

Measurements of the maximum breadth of 84 Etruscan skulls and 70 modern Italian skulls were subsequently reproduced as Data Set 155 in *A Handbook of Small Data Sets* and are displayed in Table 11.1. We use these data to explore the difference (if any) between Etruscan and modern Italian males with respect to head breadth. In the discussion that follows, x will denote Etruscans and y will denote modern Italians.

We begin by asking if it is reasonable to assume that maximum skull breadth is normally distributed. Normal probability plots of our two samples are displayed in Figure 11.1. The linearity of these plots conveys the distinct impression of normality. Kernel density estimates constructed from the two samples are superimposed in Figure 11.2, created by the following R commands:

```
> plot(density(x),type="l",xlim=c(100,180),
+ xlab="Maximum Skull Breadth",
+ main="Kernel Density Estimates")
> lines(density(y),type="l")
```

Not only do the kernel density estimates reinforce our impression of normality, they also suggest that the two populations have comparable variances. (The ratio of sample variances is $s_1^2/s_2^2 = 1.07819$.) The difference

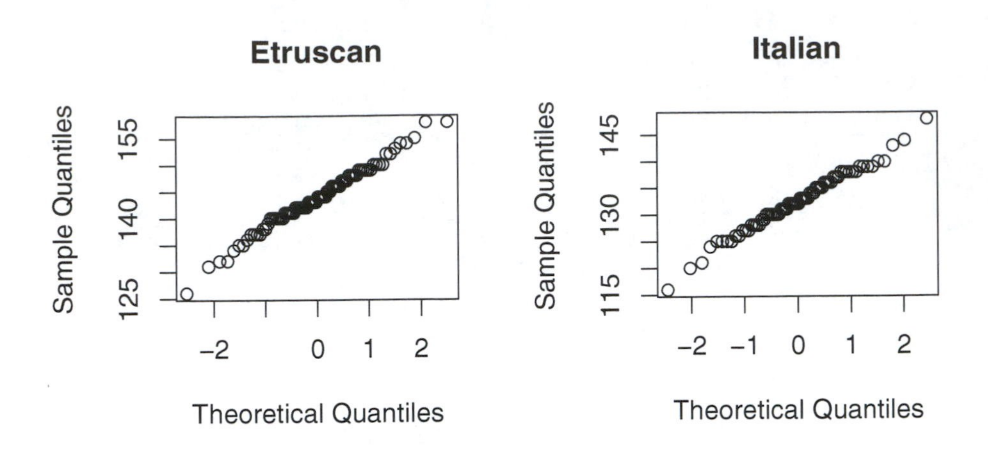

Figure 11.1: Normal probability plots of two samples of maximum skull breadth.

is maximum breadth between Etruscan and modern Italian skulls is nicely summarized by a shift parameter.

Now we construct a probability model. This is a 2-sample location problem in which an experimental unit is a skull. The skulls were drawn from two populations, Etruscan males and modern Italian males, and one measurement (maximum breadth) was made on each experimental unit. Let X_i denote the maximum breadth of Etruscan skull i and let Y_j denote the maximum breadth of Italian skull j. We assume that the X_i and Y_j are independent, with $X_i \sim \text{Normal}(\mu_1, \sigma_1^2)$ and $Y_j \sim \text{Normal}(\mu_2, \sigma_2^2)$. Notice that, although the sample variances are nearly equal, we do not assume that the population variances are identical. Instead, we will use Welch's approximation to construct an approximate 0.95-level confidence interval for $\Delta = \mu_1 - \mu_2$.

Because the confidence coefficient $1 - \alpha = 0.95$, $\alpha = 0.05$. The desired confidence interval is of the form

$$\hat{\Delta} \pm q\sqrt{\frac{s_1^2}{n_1} + \frac{s_2^2}{n_2}},$$

where q is the $1 - \alpha/2 = 0.975$ quantile of a t distribution with $\hat{\nu}$ degrees of freedom. We can easily compute these quantities in R. To compute $\hat{\Delta}$, the estimated shift parameter:

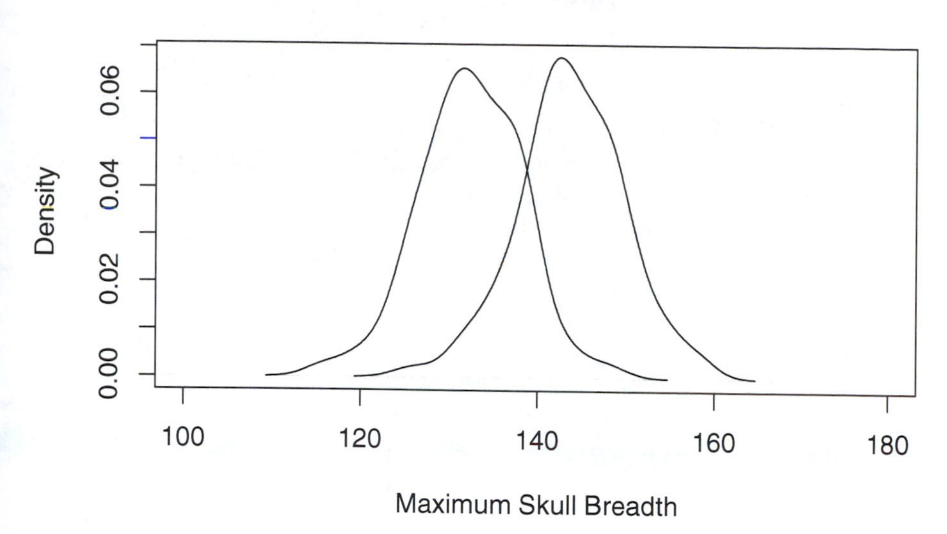

Figure 11.2: Kernel density estimates constructed from two samples of maximum skull breadth. The sample mean for the Etruscan skulls is $\bar{x} \doteq 143.8$; the sample mean for the modern Italian skulls is $\bar{y} \doteq 132.4$.

```
> Delta <- mean(x)-mean(y)
```

To compute the standard error:

```
> n1 <- length(x)
> n2 <- length(y)
> v1 <- var(x)/n1
> v2 <- var(y)/n2
> se <- sqrt(v1+v2)
```

To compute $\hat{\nu}$, the estimated degrees of freedom:

```
> nu <- (v1+v2)^2/(v1^2/(n1-1)+v2^2/(n2-1))
```

To compute q, the desired quantile:

```
> q <- qt(.975,df=nu)
```

Finally, to compute the lower and upper endpoints of the desired confidence interval:

```
> lower <- Delta-q*se
> upper <- Delta+q*se
```

These calculations result in a 0.95-level confidence interval for $\Delta = \mu_1 - \mu_2$ of $(9.459782, 13.20212)$, so that we can be fairly confident that the maximum breadth of Etruscan male skulls is, on average, roughly a centimeter greater than the maximum breadth of modern Italian male skulls.

11.4 Exercises

Problem Set A

1. We have been using various mathematical symbols in our study of 1- and 2-sample location problems. Each of the symbols listed below is used to represent a real number. State which of the following statements applies to each symbol:

   i. The real number represented by this symbol is an unknown population parameter.

   ii. The real number represented by this symbol is calculated from the observed data.

   iii. The real number represented by this symbol is specified by the experimenter.

   Here are the symbols:

$$\mu \quad \mu_0 \quad \bar{x} \quad s^2 \quad t \quad \alpha \quad \Delta \quad \Delta_0 \quad \mathbf{p} \quad \hat{\nu}$$

2. Assume that $X_1, \ldots, X_{10} \sim \text{Normal}(\mu_1, \sigma_1^2)$ and that $Y_1, \ldots, Y_{20} \sim \text{Normal}(\mu_2, \sigma_2^2)$. None of the population parameters are known. Let $\Delta = \mu_1 - \mu_2$. To test $H_0 : \Delta \geq 0$ versus $H_1 : \Delta < 0$ at significance level $\alpha = 0.05$, we observe samples $\vec{x}$ and $\vec{y}$.

   (a) What test should be used in this situation? If we observe $\vec{x}$ and $\vec{y}$ that result in $\bar{x} = -0.82$, $s_1 = 4.09$, $\bar{y} = 1.39$, and $s_2 = 1.22$, then what is the value of the test statistic?

   (b) If we observe $\vec{x}$ and $\vec{y}$ that result in $s_1 = 4.09$, $s_2 = 1.22$, and a test statistic value of 1.76, then which of the following R expressions best approximates the significance probability?

i. `2*pnorm(-1.76)`
ii. `pt(-1.76,df=28)`
iii. `pt(1.76,df=10)`
iv. `pt(-1.76,df=10)`
v. `2*pt(1.76,df=28)`

(c) True of False: if we observe $\vec{x}$ and $\vec{y}$ that result in a significance probability of $\mathbf{p} = 0.96$, then we should reject the null hypothesis.

3. In Section 11.3 we assumed that $P_1 = \text{Normal}(\mu_1, \sigma_1^2)$ and $P_2 = \text{Normal}(\mu_2, \sigma_2^2)$, where P_1 is the distribution of the maximum breadths of Etruscan male skulls and P_2 is the distribution of the maximum breadths of modern Italian male skulls. We then constructed a 0.95-level confidence interval for $\Delta = \mu_1 - \mu_2$, the difference in population means. In this exercise we replace the assumption of normality with the assumption that P_1 and P_2 belong to a shift family.

(a) Which of the following statements is correct? Explain.

i. Assuming that both P_1 and P_2 are normal is *stronger* than assuming that P_1 and P_2 belong to a shift family; i.e., normality is a special case of a shift family.

ii. Assuming that both P_1 and P_2 are normal is *weaker* than assuming that P_1 and P_2 belong to a shift family; i.e., a shift family is a special case of normality.

iii. Neither assumption is stronger than the other, i.e., it is possible for P_1 and P_2 to be normal without belonging to a shift family *and* it is possile for P_1 and P_2 to belong to a shift family without being normal.

(b) Assume that P_1 and P_2 belong to a shift family, with population medians θ_1 and θ_2. Let $\Delta = \theta_1 - \theta_2$. Construct a confidence interval for Δ that has a confidence coefficient of approximately 0.95.

Problem Set B Each of the following scenarios can be modelled as a 1- or 2-sample location problem. For 1-sample problems, let X_i denote the random variables of interest and let $\mu = EX_i$. For 2-sample problems, let X_i and Y_j denote the random variables of interest; let $\mu_1 = EX_i$, $\mu_2 = EY_j$, and $\Delta = \mu_1 - \mu_2$. For each scenario, you should answer/do the following:

(a) What is the experimental unit?

(b) From how many populations were the experimental units drawn? Identify the population(s). How many units were drawn from each population? Is this a 1- or a 2-sample problem?

(c) How many measurements were taken on each experimental unit? Identify them.

(d) Define the parameter(s) of interest for this problem. For 1-sample problems, this should be μ; for 2-sample problems, this should be Δ.

(e) State appropriate null and alternative hypotheses.

Here are the scenarios:

1. A mathematics/education concentrator theorizes that learning mathematics and statistics is sometimes impeded by the widespread use of odd symbols like α, χ, and ω. She reasons that, if her theory is correct, then students who belong to sororities and fraternities—who she presumes are more familiar with Greek letters—should have an easier time learning the mathematical subjects that use such symbols. To investigate, she obtains a list of all William & Mary students who are enrolled in Math 111 (calculus) and a list of all William & Mary students who belong to a sorority or fraternity. She uses this information to choose (at random) 20 calculus students who belong to a sorority or fraternity and 20 calculus students who do not. She persuades each of these students to take a calculus quiz, specially designed to use lots of Greek letters. How might she use the resulting data to test her theory? (Respond to (a)–(e) above.)

2. Umberto theorizes that living with a dog diminishes depression in the elderly, here defined as more than 70 years of age. To investigate his theory, he recruits 15 single elderly men who own dogs and 15 single elderly men who do not own any pets. The Hamilton instrument for measuring depressive tendency is administered to each subject. High scores indicate depression. How might Umberto use the resulting data to test his theory? (Respond to (a)–(e) above. Especially (d).)

3. Irmina, a professional massage/physical therapist and ski instructor, decides to moonlight as an areobics instructor. Her supervisor recommends that she begin each class with 10 minutes of static stretching, but Irmina believes that static stretching is detrimental to athletic

performance. She devises an experiment, for which she recruits 20 aerobics students, that consists of two protocols. In protocol S, a participant walks for 5 minutes, then does 10 minutes of static stretches of the hamstring, quadricep, and calf muscles, then rides a stationary bike for 30 minutes. Protocol D replaces static stretches with dynamic stretches. Each bike is equipped with a heart monitor and the ability to measure watts of power expended. To equalize level of exertion, each participant is asked to maintain a constant training heart rate calculated using the Karvonen formula[5] with an intensity of 0.80. The study participants perform protocol D one week and protocol S the following week. Irmina records the number of watts expended during each 30-minute ride. How might she use the resulting data to persuade her supervisor that dynamic stretching is superior to static stretching? (Respond to (a)–(e) above.)

4. The William & Mary women's tennis team uses championship balls in their matches and less expensive practice balls in their team practices. The players have formed a strong impression that the practice balls do not wear as well as the championship balls, i.e., that the practice balls lose their bounce more quickly than the championship balls. To investigate this perception, Nina and Delphine conceive the following experiment. Before one practice, the team opens new cans of championship balls and practice balls, which they then use for that day's practice. After practice, Nina and Delphine randomly select 10 of the used championship balls and 10 of the used practice balls. They drop each ball from a height of 2 meters and measure the height of its first bounce. How might Nina and Delphine test the team's impression that practice balls do not wear as well as championship balls? (Respond to (a)–(e) above.)

5. A political scientist theorizes that women tend to be more opposed to military intervention than do men. To investigate this theory, he devises an instrument on which a subject responds to several recent U.S. military interventions on a 5-point Likert scale (1="strongly support,"...,5="strongly oppose"). A subject's score on this instrument is the sum of his/her individual responses. The scientist randomly selects 50 married couples in which neither spouse has a registered party affiliation and administers the instrument to each of the 100 individu-

[5]Training Heart Rate = Resting Heart Rate + Intensity $\times$ (220 $-$ Age - Resting Heart Rate).

als so selected. How might he use his results to determine if his theory is correct? (Respond to (a)–(e) above.)

6. A shoe company claims that wearing its racing flats will typically improve one's time in a 10K road race by more than 30 seconds. A running magazine sponsors an event to test this claim. It arranges for 120 runners to enter two road races, held two weeks apart on the same course. For the second race, each of these runners is supplied with the new racing flat. How might the race results be used to determine the validity of the shoe company's claim? (Respond to (a)–(e) above.)

7. Susan theorizes that impregnating wood with an IGR (insect growth regulator) will reduce wood consumption by termites. To investigate this theory, she impregnates 60 wood blocks with a solvent containing the IGR and 60 wood blocks with just the solvent. Each block is weighed, then placed in a separate container with 100 ravenous termites. After two weeks, she removes the blocks and weighs them again to determine how much wood has been consumed. How might Susan use her results to determine if her theory is correct? (Respond to (a)–(e) above.)

8. To investigate the effect of swing dancing on cardiovascular fitness, an exercise physiologist recruits 20 couples enrolled in introductory swing dance classes. Each class meets once a week for ten weeks. Participants are encouraged to go out dancing on at least two additional occasions each week. In general, lower resting pulses are associated with greater cardiovascular fitness. Accordingly, each participant's resting pulse is measured at the beginning and at the end of the ten-week class. How might the resulting data be used to determine if swing dancing improves cardiovascular fitness? (Respond to (a)–(e) above.)

9. It is thought that Alzheimer's disease (AD) impairs short-term memory more than it impairs long-term memory. To test this theory, a psychologist studied 60 mildly demented AD patients and 60 normal elderly control subjects. Each subject was administered a short-term and a long-term memory task. On each task, high scores are better than low scores. How might the psychologist use the resulting task scores to determine if the theory is correct? (Respond to (a)–(e) above.)

10. According to an article in *Newsweek* (May 10, 2004, page 89), recent "studies have shown consistently that women are better than men at

reading and responding to subtle cues about mood and temperament." Some psychologists believe that such differences can be explained in part by biological differences between male and female brains. One such psychologist conducts a study in which day-old babies are shown three human faces and three mechanical objects. The time that the baby stares at each face/object is recorded. Of interest is how much time the baby spends staring at faces versus how much time the baby spends staring at objects. The psychologist's theory predicts that this comparison will differ by sex, with female babies preferring faces to objects to a greater extent than do male babies. How might the psychologist use his results to determine if his theory is correct? (Respond to (a)–(e) above.)

Problem Set C In the early 1960s, the Western Collaborative Group Study investigated the relation between behavior and risk of coronary heart disease in middle-aged men. Type A behavior is characterized by urgency, aggression and ambition; Type B behavior is noncompetitive, more relaxed and less hurried. The following data are the cholesterol measurements of 20 heavy men of each behavior type.[6] We consider whether or not they provide evidence that heavy Type A men have higher cholesterol levels than heavy Type B men.

Cholesterol Levels for Heavy Type A Men									
233	291	312	250	246	197	268	224	239	239
254	276	234	181	248	252	202	218	212	325

Cholesterol Levels for Heavy Type B Men									
344	185	263	246	224	212	188	250	148	169
226	175	242	252	153	183	137	202	194	213

1. Respond to (a)–(e) in Problem Set B.

2. Does it seem reasonable to assume that the samples $\vec{x}$ and $\vec{y}$, the observed values of $X_1, \ldots, X_{n_1}$ and $Y_1, \ldots, Y_{n_2}$, were drawn from normal distributions? Why or why not?

---

[6] S. Selvin (1991). *Statistical Analysis of Epidemiological Data.* Oxford University Press, New York, Table 2.1. These data appear as Data Set 47 in *A Handbook of Small Data Sets.* These 40 men were the heaviest in the study. Each weighed at least 225 pounds.

3. Assume that the X_i and the Y_j are normally distributed.

   (a) Test the null hypothesis derived above using Welch's approximate t-test. What is the significance probability? If we adopt a significance level of $\alpha = 0.05$, should we reject the null hypothesis?

   (b) Construct a (2-sided) confidence interval for Δ with a confidence coefficient of approximately 0.90.

4. Let P_1 denote the distribution of the X_i and let P_2 denote the distribution of the Y_j. Does it seem reasonable to assume that P_1 and P_2 belong to a shift family, i.e., that there exists a real number Δ such that $X_i' = X_i - \Delta \sim P_2$? Why or why not?

5. Assume that P_1 and P_2 do belong to a shift family. Let θ_1 and θ_2 denote the population medians of P_1 and P_2 and let $\Delta = \theta_1 - \theta_2$.

   (a) Modify the null hypothesis derived above by replacing $\Delta = \mu_1 - \mu_2$ with $\Delta = \theta_1 - \theta_2$. Use Wilcoxon's rank sum test to test the modified hypothesis. What is the significance probability? If we adopt a significance level of $\alpha = 0.05$, should we reject the null hypothesis?

   (b) Construct a (2-sided) confidence interval for Δ with a confidence coefficient of approximately 0.90.

Problem Set D Researchers obtained the following measurements of urinary β-thromboglobulin excretion in 12 diabetic patients and 12 normal control subjects.[7]

Normal	4.1	6.3	7.8	8.5	8.9	10.4
	11.5	12.0	13.8	17.6	24.3	37.2
Diabetic	11.5	12.1	16.1	17.8	24.0	28.8
	33.9	40.7	51.3	56.2	61.7	69.2

1. Do these measurements appear to be samples from symmetric distributions? Why or why not?

[7] B. A. van Oost, B. Veldhayzen, A. P. M. Timmermans, and J. J. Sixma (1983). Increased urinary β-thromboglobulin excretion in diabetes assayed with a modified RIA kit-technique. *Thrombosis and Haemostasis*, 9:18–20. These data appear as Data Set 313 in *A Handbook of Small Data Sets*.

2. Both samples of positive real numbers appear to be drawn from distributions that are skewed to the right; i.e., the upper tail of the distribution is longer than the lower tail of the distribution. Often, such distributions can be symmetrized by applying a suitable data transformation. Two popular candidates are:

   (a) The natural logarithm: $u_i = \log(x_i)$ and $v_j = \log(y_j)$.

   (b) The square root: $u_i = \sqrt{x_i}$ and $v_j = \sqrt{y_j}$.

   Investigate the effect of each of these transformations on the above measurements. Do the transformed measurements appear to be samples from symmetric distributions? Which transformation do you prefer?

3. Do the transformed measurements appear to be samples from normal distributions? Why or why not?

4. The researchers claimed that diabetic patients have increased urinary β-thromboglobulin excretion. Assuming that the transformed measurements are samples from normal distributions, how convincing do you find the evidence for their claim?

Problem Set E

1. Chemistry lab partners Arlen and Stuart collaborated on an experiment in which they measured the melting points of 20 specimens of two types of sealing wax. Twelve of the specimens were of one type (A); eight were of the other type (B). Each student then used Welch's approximate t-test to test the null hypothesis of no difference in mean melting point between the two methods:

   - Arlen applied Welch's approximate t-test to the original melting points, which were measured in degrees Fahrenheit.
   - Stuart first converted each melting point to degrees Celsius (by subtracting 32, then multiplying by 5/9), then applied Welch's approximate t-test to the converted melting points.

   Comment on the potential differences between these two analyses. In particular, is it *True* or *False* that (ignoring round-off error) Arlen and Stuart will obtain identical significance probabilities? Please justify your comments.

2. A graduate student in ornithology would like to determine if created marshes differ from natural marshes in their appeal to avian communities. He plans to observe $n_1 = 9$ natural marshes and $n_2 = 9$ created marshes, counting the number of red-winged blackbirds per acre that inhabit each marsh. His thesis committee wants to know how much he thinks he will be able to learn from this experiment.

   Let X_i denote the number of blackbirds per acre in natural marsh i and let Y_j denote the number of blackbirds per acre in created marsh j. In order to respond to his committee, the student makes the simplifying assumptions that $X_i \sim \text{Normal}(\mu_1, \sigma^2)$ and $Y_j \sim \text{Normal}(\mu_2, \sigma^2)$. He estimates that $\text{iqr}(X_i) = \text{iqr}(Y_j) = 10$. Calculate L, the length of the 0.90-level confidence interval for $\Delta = \mu_1 - \mu_2$ that he can expect to construct.

3. A film buff has formed the vague impression that movies tend to be longer than they used to be. Are they really longer? Or do they just *seem* longer? To investigate, he randomly samples U.S. feature films made in 1956 and U.S. feature films made in 1996, obtaining the data displayed in Table 11.2. Do these data provide convincing evidence that 1996 movies are longer than 1956 movies? Compute a significance probability that may be used to encourage or discourage the film buff's impression. Explain how this number should be interpreted. How did you obtain it? Identify and defend any assumptions that you made in your calculations.

Year	Title	Minutes
1956	*Accused of Murder*	74
	Away All Boats	114
	Baby Doll	114
	The Bold and the Brave	87
	Come Next Spring	92
	The Flaming Teen-Age	55
	Gun Girls	67
	Helen of Troy	118
	The Houston Story	79
	Patterns	83
	The Price of Fear	79
	The Revolt of Mamie Stover	92
	Written on the Wind	99
	The Young Guns	87
1996	*$40,000*	70
	Barb Wire	98
	Breathing Room	90
	Daddy's Girl	95
	Ed's Next Move	88
	From Dusk to Dawn	108
	Galgameth	110
	The Glass Cage	96
	Kissing a Dream	91
	Love & Sex etc.	88
	Love is All There Is	120
	Making the Rules	96
	Spirit Lost	90
	Work	90

Table 11.2: Running times of 14 feature films from 1956 and 14 feature films from 1996.

Chapter 12

The Analysis of Variance

Now we generalize our study of location problems from two to $k \geq 3$ populations. Again we are concerned with comparing the populations with respect to a measure of centrality. We designate the populations by $P_1, \ldots, P_k$ and the corresponding sample sizes by $n_1, \ldots, n_k$. Our bookkeeping will be facilitated by the use of double subscripts, e.g.,

$$\begin{array}{ccc} X_{11}, \ldots, X_{1n_1} & \sim & P_1, \\ X_{21}, \ldots, X_{2n_2} & \sim & P_2, \\ & \vdots & \\ X_{k1}, \ldots, X_{kn_k} & \sim & P_k. \end{array}$$

These expressions can be summarized succinctly by writing

$$X_{ij} \sim P_i.$$

We assume the following:

1. The X_{ij} are mutually independent continuous random variables.

2. P_i has location parameter θ_i, e.g., $\theta_i = \mu_i = EX_{ij}$ or $\theta_i = q_2(X_{ij})$.

3. We observe random samples $\vec{x}_i = \{x_{i1}, \ldots, x_{in_i}\}$, from which we attempt to draw inferences about $(\theta_1, \ldots, \theta_k)$. In general, we do *not* assume that $n_1 = \cdots = n_k$. However, certain procedures do require equal sample sizes. Furthermore, certain procedures that can be used with unequal sample sizes are greatly simplified when the sample sizes are equal.

The four questions that we posed at the beginning of Chapter 10 and asked in Chapters 10–11 can be asked here. What distinguishes k-sample problems from 1-sample and 2-sample problems is the number of populations from which the experimental units were drawn. The prototypical case of a k-sample problem is the case of several treatment populations.

One may wonder why we distinguish between $k = 2$ and $k \geq 3$ populations. In fact, many methods for k-sample problems can be applied to 2-sample problems, in which case they often simplify to methods studied in Chapter 11. However, many issues arise with $k \geq 3$ populations that do not arise with two populations, so the problem of comparing more than two location parameters is considerably more complicated than the problem of comparing only two. For this reason, our study of k-sample location problems will be less comprehensive than our previous studies of 1-sample and 2-sample location problems.

Throughout this chapter we assume that $P = \text{Normal}(\mu_i, \sigma^2)$. This is sometimes called the fixed effects model for the one-way analysis of variance (ANOVA). Notice that we are assuming that each normal population has the same variance. Recall that we criticized the assumption of equal variances for the normal 2-sample problem. In that setting, however, Welch's approximate t-test provides a viable alternative that is available in many popular statistical software packages. In the more complicated setting of k normal populations, the assumption of equal variances (sometimes called the assumption of *homoscedasticity*) is fairly standard, if only because it is less clear how to proceed when the variances are unequal.

12.1 The Fundamental Null Hypothesis

The fundamental problem of the analysis of variance is the problem of testing the null hypothesis that all of the population means are the same, i.e.,

$$H_0 : \mu_1 = \cdots = \mu_k, \tag{12.1}$$

against the alternative hypothesis that they are not all the same. Notice that the statement that the population means are not identical does *not* imply that each population mean is distinct. For example, if $\mu_1 = \mu_2 = 1.5$ and $\mu_3 = 2.2$, then H_0 is false. We stress that the analysis of variance is concerned with inferences about means, not variances.

To motivate our test of H_0, we formulate another null hypothesis that is equivalent to H_0. First, let

$$N = \sum_{i=1}^{k} n_i$$

denote the sum of the sample sizes and let

$$\bar{\mu}. = \sum_{i=1}^{k} \frac{n_i}{N} \mu_i$$

denote the *population grand mean*. The population grand mean is a weighted average of the individual population means, each population weighted in proportion to how many of the observations were drawn from it. If H_0 is true, then $\mu_1 = \cdots = \mu_k$ have a common value, say μ, and the population grand mean equals that common value:

$$\bar{\mu}. = \sum_{i=1}^{k} \frac{n_i}{N} \mu = \frac{\mu}{N} \sum_{i=1}^{k} n_i = \mu.$$

Next we introduce a quantity that measures how nearly the individual population means equal the population grand mean. Let

$$\gamma = \sum_{i=1}^{k} n_i \left(\mu_i - \bar{\mu}.\right)^2. \tag{12.2}$$

Notice that $\gamma \geq 0$ and that $\gamma = 0$ if and only if each $\mu_i = \bar{\mu}.$. But each $\mu_i = \bar{\mu}.$ if and only if each individual mean assumes a common value, which occurs if and only if the individual means are identical. Thus, H_0 is equivalent to the null hypothesis

$$H_0' : \gamma = 0,$$

which is to be tested against the alternative hypothesis

$$H_1' : \gamma > 0.$$

12.2 Testing the Fundamental Null Hypothesis

The idea that underlies our test is to estimate γ and reject H_0' when the estimate is sufficiently larger than zero. To estimate γ, we need only estimate the population means that appear in (12.2). The individual sample means,

$$\bar{X}_i = \frac{1}{n_i} \sum_{j=1}^{n_i} X_{ij},$$

are unbiased estimators of the individual population means, and the *sample grand mean*,

$$\bar{X}_{\cdot\cdot} = \sum_{i=1}^{k} \frac{n_i}{N} \bar{X}_{i\cdot} = \sum_{i=1}^{k} \frac{n_i}{N} \left(\frac{1}{n_i} \sum_{j=1}^{n_i} X_{ij} \right) = \frac{1}{N} \sum_{i=1}^{k} \sum_{j=1}^{n_i} X_{ij}$$

is an unbiased estimator of the population grand mean. Hence, a natural estimator of γ is the *between-groups* or *treatment* sum of squares,

$$SS_B = \sum_{i=1}^{k} n_i \left(\bar{X}_{i\cdot} - \bar{X}_{\cdot\cdot} \right)^2,$$

the variation of the individual sample means about the sample grand mean. A useful formula for computing the observed value of SS_B from the observed values of the individual sample means is

$$ss_B = \sum_{i=1}^{k} n_i \bar{x}_{i\cdot}^2 - \frac{1}{N} \left(\sum_{i=1}^{k} n_i \bar{x}_{i\cdot} \right)^2.$$

What remains is to determine when SS_B is "sufficiently larger than zero." We consider two cases, depending on whether or not the common population variance σ^2 is known.

12.2.1 Known Population Variance

Situations in which σ^2 is known are rarely encountered, but it is useful to consider how to proceed in this case. Here is the key fact that we require:

Theorem 12.1 *Under the fundamental null hypothesis (12.1), the random variable*

$$SS_B/\sigma^2 \sim \chi^2(k-1),$$

where $\chi^2(\nu)$ denotes the chi-squared distribution with ν degrees of freedom, introduced in Section 5.5. The quantity $k-1$ is the between-groups degrees of freedom.

Theorem 12.1 suggests a way to determine whether or not SS_B is "sufficiently larger than zero." Under H_0,

$$P\left(SS_B \geq q\right) = P\left(SS_B/\sigma^2 \geq q/\sigma^2\right) = P\left(Y \geq q/\sigma^2\right),$$

where $Y \sim \chi^2(k-1)$; hence, we can use the chi-squared distribution to compute significance probabilities and/or critical values.

Example 12.1 Suppose that we draw samples of $n_1 = 20$, $n_2 = 25$, and $n_3 = 30$ observations from normal populations with unknown means and common variance $\sigma^2 = 9$, obtaining sample means of $\bar{x}_1 = 1.489$, $\bar{x}_2 = 1.712$, and $\bar{x}_3 = 3.082$. To test the fundamental null hypothesis that the individual population means are identical, we first compute $N = 20+25+30 = 75$ and evaluate SS_B, obtaining

$$\begin{aligned} ss_B &= \left(20 \cdot 1.489^2 + 25 \cdot 1.712^2 + 30 \cdot 3.082^2\right) - \\ &\quad \left(20 \cdot 1.489 + 25 \cdot 1.712 + 30 \cdot 3.082\right)^2 / 75 \\ &\doteq 39.402. \end{aligned}$$

Now we use the R function pchisq to compute a significance probability **p**:

```
> 1-pchisq(39.402/9,df=2)
[1] 0.1120287
```

For conventional levels of significance, $\mathbf{p} > 0.10$ is too large to warrant rejecting the null hypothesis.

12.2.2 Unknown Population Variance

Now we consider the more realistic case of an unknown population variance. Our development will mimic the case of a known population variance, but it is complicated by the need to estimate σ^2. Recall that, in Section 11.1.2, we estimated the unknown common population variance of $k = 2$ normal populations with the pooled sample variance,

$$S_P^2 = \frac{(n_1 - 1)S_1^2 + (n_2 - 1)S_2^2}{(n_1 - 1) + (n_2 - 1)},$$

where S_i^2 is the sample variance for sample i. This procedure is easily extended to the present case of $k \geq 3$ by defining the pooled sample variance as

$$\begin{aligned} S_P^2 &= \frac{(n_1 - 1)S_1^2 + \cdots + (n_k - 1)S_k^2}{(n_1 - 1) + \cdots + (n_2 - 1)} \\ &= \frac{1}{n_1 + \cdots + n_k - k} \sum_{i=1}^{k} (n_i - 1)\, S_i^2 \\ &= \frac{1}{N - k} \sum_{i=1}^{k} \sum_{j=1}^{n_i} \left(X_{ij} - \bar{X}_{i\cdot}\right)^2. \end{aligned}$$

As in the case of $k = 2$,

$$\begin{aligned} ES_P^2 &= \frac{(n_1 - 1)ES_1^2 + \cdots + (n_k - 1)ES_k^2}{(n_1 - 1) + \cdots + (n_k - 1)} \\ &= \frac{(n_1 - 1)\sigma^2 + \cdots (n_k - 1)\sigma^2}{(n_1 - 1) + \cdots + (n_k - 1)} = \sigma^2, \end{aligned}$$

so the pooled sample variance is an unbiased estimator of a common population variance. It is also consistent and asymptotically efficient for estimating a common normal variance.

In the previous case of a known population variance, our statistic for testing the fundamental null hypothesis was SS_B/σ^2. In the present case of an unknown population variance, we estimate σ^2 with S_P^2. Our test statistic will turn out to be SS_B/S_P^2 multiplied by a constant.

In order to simplify the formulas that follow, we multiply S_P^2 by $N - k$, obtaining the *within-groups* or *error* sum of squares

$$SS_W = (N - k)S_P^2 = \sum_{i=1}^{k} (n_i - 1)\, S_i^2 = \sum_{i=1}^{k} \sum_{j=1}^{n_i} \left(X_{ij} - \bar{X}_{i\cdot}\right)^2.$$

In contrast to SS_B, which measures the variation of the individual sample means about the sample grand mean, SS_W measures the variations of the individual observations about the corresponding sample means. For completeness, we also define the *total* sum of squares,

$$SS_T = \sum_{i=1}^{k} \sum_{j=1}^{n_i} \left(X_{ij} - \bar{X}_{\cdot\cdot}\right)^2,$$

which measures the variation of the individual observations about the sample grand mean.

There is a beautiful relationship between SS_B, SS_W, and SS_T, viz.,

Theorem 12.2 $SS_B + SS_W = SS_T$

This formula turns out to be a corollary of the Pythagorean Theorem in N-dimensional Euclidean space! (In Section 15.2, we will explore a similar formula in greater detail.) The reason that our method for testing the fundamental null hypothesis is called the analysis of variance is that the method relies on decomposing total squared error into squared error between groups and squared error within groups. This elegant—and extremely useful—decomposition is only possible when we use *squared* error.

The quantities SS_B, SS_W, and SS_T are random variables. The following facts, which subsume Theorem 12.1, summarize the statistical behavior of these random variables.

Theorem 12.3 *The random variable*

$$SS_T/\sigma^2 \sim \chi^2(N-1).$$

The quantity $N-1$ is the total degrees of freedom.

Under the fundamental null hypothesis (12.1), SS_B and SS_W are independent random variables and

$$\begin{aligned} SS_B/\sigma^2 &\sim \chi^2(k-1), \\ SS_W/\sigma^2 &\sim \chi^2(N-k). \end{aligned}$$

The quantity $k-1$ is the between-groups degrees of freedom and the quantity $N-k$ is the within-groups degrees of freedom.

We have already remarked that the random variable

$$\frac{SS_B}{S_P^2} = \frac{SS_B}{SS_W/(N-k)}$$

would seem to be a natural statistic for testing the fundamental null hypothesis. Although sound in theory, this approach fails in practice because the distribution of SS_B/S_P^2 is not tractable. Fortunately, this approach can be salvaged by a trivial modification. Applying the definition of F distributions in Section 5.5 to the independent χ^2 random variables SS_B/σ^2 and SS_W/σ^2, we discover ...

Corollary 12.1 *Under the fundamental null hypothesis (12.1),*

$$F = \frac{\frac{SS_B}{\sigma^2}/(k-1)}{\frac{SS_W}{\sigma^2}/(N-k)} = \frac{SS_B/(k-1)}{SS_W/(N-k)} \sim F(k-1, N-k),$$

where $F(\nu_1, \nu_2)$ denotes an F distribution with ν_1 and ν_2 degrees of freedom.

The random variable F is the desired test statistic; notice that

$$F = \frac{SS_B/(k-1)}{SS_W/(N-k)} = \frac{1}{k-1}\frac{SS_B}{S_P^2}.$$

Appealing to Corollary 12.1, we see that the ANOVA F-test of the fundamental null hypothesis of equal population means is to reject H_0 at significance level α if and only if the significance probability

$$\mathbf{p} = P(Y \geq f) \leq \alpha,$$

where f denotes the observed value of F and $Y \sim F(k-1, N-k)$. Of course, we can also formulate the test using critical values instead of significance probabilities, in which case we reject H_0 at significance level α if and only if $f \geq q$, where q is the $1-\alpha$ quantile of the $F(k-1, N-k)$ distribution.

Example 12.2 Suppose that we draw samples of $n_1 = 25$, $n_2 = 20$, and $n_3 = 20$ observations from normal populations with unknown means and unknown common variance, obtaining the following sample quantities:

	$i=1$	$i=2$	$i=3$
n_i	25	20	20
$\bar{x}_{i\cdot}$	9.783685	10.908170	15.002820
s_i^2	29.89214	18.75800	51.41654

To test the null hypothesis of equal population means at significance level $\alpha = 0.05$, we begin by computing the observed values of SS_B and SS_W, obtaining $ss_B \doteq 322.4366$ and

$$ss_W = (25-1) \cdot 29.89214 + (20-1) \cdot 18.75800 + (20-1) \cdot 51.41654 \doteq 2050.7280.$$

It follows that the observed value of the test statistic is

$$f = \frac{ss_B/(k-1)}{ss_W/(N-k)} \doteq \frac{322.4366/2}{2050.7280/62} \doteq 4.874141.$$

Now we use the R function pf to compute a significance probability **p**:

```
> 1-pf(4.874141,df1=2,df2=62)
[1] 0.01081398
```

Because $\mathbf{p} < \alpha$, we reject the null hypothesis. Equivalently, we might use the R function qf to compute a critical value q:

```
> qf(1-.05,df1=2,df2=62)
[1] 3.145258
```

Because $f > q$, we reject the null hypothesis.

The information related to an ANOVA F-test is usually collected in an ANOVA table:

Source of Variation	Sum of Squares	Degrees of Freedom	Mean Squares	Test Statistic	Significance Probability
Between	SS_B	$k-1$	MS_B	F	**p**
Within	SS_W	$N-k$	$MS_W = S_P^2$		
Total	SS_T	$N-1$			

Notice that we have introduced new notation for the *mean squares*, $MS_B = SS_B/(k-1)$ and $MS_W = SS_W/(N-k)$, which allows us to write $F = MS_B/MS_W$. It is also helpful to examine $R^2 = SS_B/SS_T$, the proportion of total variation "explained" by differences in the sample means.

Example 12.2 (continued) For the ANOVA performed in Example 12.2, the ANOVA table is

Source	SS	df	MS	F	p
Between	322.4366	2	161.21830	4.874141	0.01081398
Within	2050.7280	62	33.07625		
Total	2373.1640	64			

The proportion of total variation explained by differences in the sample means is $322.4366/2373.1640 \doteq 0.1358678$. Thus, although there is sufficient variation between the sample means for us to infer that the population means are not identical, this variation accounts for a fairly small proportion of the total variation in the data.

12.3 Planned Comparisons

Rejecting the fundamental null hypothesis of equal population means leaves numerous alternatives. Typically, a scientist would like to say more than simply "$H_0 : \mu_1 = \cdots = \mu_k$ is false." Concluding that the population means are not identical naturally invites investigation of how they differ. Sections 12.3 and 12.4 describe several useful inferential procedures for performing more elaborate comparisons of population means. Section 12.3 describes two procedures that are appropriate when the scientist has determined specific comparisons of interest *in advance of the experiment*. For reasons that will become apparent, this is the preferred case. However, it is often the case that a specific comparison occurs to a scientist *after examining the results of the experiment*. Although statistical inference in such cases is rather tricky, a variety of procedures for *a posteriori* inference have been developed. Two such procedures are described in Section 12.4.

Inspired by a classic statistics text,[1] we motivate the concept of a *planned comparison* by considering an attempt to measure the constant of proportionality that appears in Newton's universal law of gravitation.

---

[1] K. A. Brownlee (1965). *Statistical Theory and Methodology in Science and Engineering*, Second Edition. John Wiley & Sons, New York.

Example 12.3 Newton's universal law of gravitation states that $F = Gm_1m_2/r^2$, where F is the gravitational force between points with masses m_1 and m_2 separated by distance r. The gravitational constant, G, is often represented in units of cubic meters per kilogram second squared. It can be measured using a torsion balance (also known as a torsion pendulum), the apparatus that Coulomb used in the 1780s to measure electrostatic force and that Cavendish used in 1798 to measure gravitational force.[2]

Cavendish attached small lead balls, each weighing about 1.61 pounds, to the ends of a six-foot rod. The rod was suspended from a wire with a known torsion coefficient. Two large lead balls, each weighing about 348 pounds, were independently suspended and positioned roughly 9 inches from the small balls. The attraction between the large and the small masses caused the rod to rotate, and therefore the wire to twist. By measuring the angle of rotation, Cavendish was able to deduce the force of attraction between the large and the small masses.

Subsequent experiments by other researchers used various types of torsion balances and refined the Cavendish experiment in various ways. In the 1890s, two independent researchers reported mean results of $G = 6.683 \times 10^{-11}$. C. V. Boys (1895) measured the deviation of a pendulum from its neutral position. His small masses weighed about 2.65 grams and his large masses weighed about 7.4 kilograms. C. V. Braun (1897) measured the time of a pendulum's swing. His small masses weighed about 54 grams and his large masses weighed about 9 kilograms.

In the 1920s, working at the Bureau of Standards, P. R. Heyl attempted to improve on Braun's experiment.[3] He made three series of measurements, using small masses (about 50 grams) of gold, platinum, and optical glass. The large masses were steel cylinders of about 66.3 and 66.4 kilograms. Let $i = 1$ denote gold, $i = 2$ denote platinum, and $i = 3$ denote glass. It is natural to ask not just if the three materials lead to identical determinations of G, by testing $H_0 : \mu_1 = \mu_2 = \mu_3$, but also to ask the following questions.

1. If glass differs from the two heavy metals, by testing

$$H_0 : \frac{\mu_1 + \mu_2}{2} = \mu_3 \quad \text{vs.} \quad H_1 : \frac{\mu_1 + \mu_2}{2} \neq \mu_3,$$

[2]Cavendish did not actually compute G, but the torsion balance has played a crucial role in the study of gravitation.

[3]P. R. Heyl (1927). A redetermination of the Newtonian constant of gravitation. *Proceedings of the National Academy of Science*, 13:601–605.

P. R. Heyl (1930). A redetermination of the constant of gravitation. *Bureau of Standards Journal of Research*, 5:1243–1250.

or, equivalently,

$$H_0 : \mu_1 + \mu_2 = 2\mu_3 \quad \text{vs.} \quad H_1 : \mu_1 + \mu_2 \neq 2\mu_3,$$

or, equivalently,

$$H_0 : \mu_1 + \mu_2 - 2\mu_3 = 0 \quad \text{vs.} \quad H_1 : \mu_1 + \mu_2 - 2\mu_3 \neq 0,$$

or, equivalently,

$$H_0 : \theta_1 = 0 \quad \text{vs.} \quad H_1 : \theta_1 \neq 0,$$

where $\theta_1 = \mu_1 + \mu_2 - 2\mu_3$.

2. If the two heavy metals differ from each other, by testing

$$H_0 : \mu_1 = \mu_2 \quad \text{vs.} \quad H_1 : \mu_1 \neq \mu_2,$$

or, equivalently,

$$H_0 : \mu_1 - \mu_2 = 0 \quad \text{vs.} \quad H_1 : \mu_1 - \mu_2 \neq 0,$$

or, equivalently,

$$H_0 : \theta_2 = 0 \quad \text{vs.} \quad H_1 : \theta_2 \neq 0,$$

where $\theta_2 = \mu_1 - \mu_2$.

Notice that both of the planned comparisons proposed in Example 12.3 have been massaged into testing a null hypothesis of the form $\theta = 0$. For this construction to make sense, θ must have a special structure, which statisticians identify as a *contrast*.

Definition 12.1 *A contrast is a linear combination (weighted sum) of the k population means,*

$$\theta = \sum_{i=1}^{k} c_i \mu_i,$$

for which $\sum_{i=1}^{k} c_i = 0$.

Example 12.3 (continued) In the contrasts suggested previously,

1. $\theta_1 = 1 \cdot \mu_1 + 1 \cdot \mu_2 + (-2) \cdot \mu_3$ and $1 + 1 - 2 = 0$; and
2. $\theta_2 = 1 \cdot \mu_1 + (-1) \cdot \mu_2 + 0 \cdot \mu_3$ and $1 - 1 + 0 = 0$.

We often identify different contrasts by their coefficients, e.g., $c = (1, 1, -2)$ or $c = (1, -1, 0)$.

The methods of Section 12.2 are easily extended to the problem of testing a single contrast, $H_0 : \theta = 0$ versus $H_1 : \theta \neq 0$. In Definition 12.1, each population mean μ_i can be estimated by the unbiased estimator $\bar{X}_{i\cdot}$; hence, an unbiased estimator of θ is

$$\hat{\theta} = \sum_{i=1}^{k} c_i \bar{X}_{i\cdot}.$$

We will reject H_0 if $\hat{\theta}$ is observed sufficiently far from zero.

Once again, we rely on a squared error criterion and ask if the observed quantity $(\hat{\theta})^2$ is sufficiently far from zero. However, the quantity $(\hat{\theta})^2$ is not a satisfactory measure of departure from $H_0 : \theta = 0$ because its magnitude depends on the magnitude of the coefficients in the contrast. To remove this dependency, we form a ratio that does not depend on how the coefficients were scaled. The sum of squares associated with the contrast θ is the random variable

$$SS_\theta = \frac{\left(\sum_{i=1}^{k} c_i \bar{X}_{i\cdot} \right)^2}{\sum_{i=1}^{k} c_i^2 / n_i}.$$

The following facts about the distribution of SS_θ lead to a test of $H_0 : \theta = 0$ versus $H_1 : \theta \neq 0$.

Theorem 12.4 *Under the fundamental null hypothesis* $H_0 : \mu_1 = \cdots = \mu_k$, SS_θ *is independent of* SS_W, $SS_\theta/\sigma^2 \sim \chi^2(1)$, *and*

$$F(\theta) = \frac{\frac{SS_\theta}{\sigma^2}/1}{\frac{SS_W}{\sigma^2}/(N-k)} = \frac{SS_\theta}{SS_W/(N-k)} \sim F(1, N-k).$$

The F-test of $H_0 : \theta = 0$ is to reject H_0 if and only if

$$\mathbf{p} = P_{H_0}\left(F(\theta) \geq f(\theta)\right) \leq \alpha,$$

i.e., if and only if

$$f(\theta) \geq q = \texttt{qf(1-}\alpha\texttt{,df1=1,df2=N-k)},$$

where $f(\theta)$ denotes the observed value of $F(\theta)$.

Example 12.3 (continued) Heyl reported the following x_{ij}, where Heyl's actual measurement of G is $(6600 + x_{ij}) \times 10^{-14}$.

Gold	83	81	76	78	79	72
Platinum	61	61	67	67	64	
Glass	78	71	75	72	74	

Applying the methods of Section 12.2, we obtain the following ANOVA table.

Source	SS	df	MS	F	p
Between	565.1	2	282.6	26.1	0.000028
Within	140.8	13	10.8		
Total	705.9	15			

To test $H_0 : \theta_1 = 0$ versus $H_1 : \theta_1 \neq 0$, we first compute

$$ss_{\theta_1} = \frac{[1 \cdot \bar{x}_{1\cdot} + 1 \cdot \bar{x}_{2\cdot} + (-2) \cdot \bar{x}_{3\cdot}]^2}{1^2/6 + 1^2/5 + (-2)^2/5} \doteq 29.16667,$$

then

$$f(\theta_1) = \frac{ss_\theta}{ss_W/(N-k)} \doteq \frac{29.16667}{140.8333/(16-3)} \doteq 2.692308.$$

Finally, we use the R function pf to compute a significance probability **p**:

```
> 1-pf(2.692308,df1=1,df2=13)
[1] 0.1247929
```

Because $\mathbf{p} > 0.05$, we decline to reject the null hypothesis at significance level $\alpha = 0.05$. Equivalently, we might use the R function qf to compute a critical value q:

```
> qf(1-.05,df1=1,df2=13)
[1] 4.667193
```

Because $f < q$, we decline to reject the null hypothesis.

In practice, one rarely tests a single contrast. However, testing multiple contrasts involves more than testing each contrast as though it was the only contrast. Entire books have been devoted to the problem of *multiple comparisons*. The remainder of this section describes two popular procedures for testing multiple contrasts that were specified *before* performing the experiment. The next section describes two popular procedures for testing multiple contrasts that were specified *after* examining the data.

12.3.1 Orthogonal Contrasts

When it can be used, the *method of orthogonal contrasts* is generally preferred. It is quite elegant, but has certain limitations. We begin by explaining what it means for contrasts to be orthogonal.

Definition 12.2 *Two contrasts with coefficient vectors* $(c_1, \ldots, c_k)$ *and* $(d_1, \ldots, d_k)$ *are orthogonal if and only if*

$$\sum_{i=1}^{k} \frac{c_i d_i}{n_i} = 0.$$

A collection of contrasts is mutually orthogonal if and only if each pair of contrasts in the collection is orthogonal.

Notice that, if $n_1 = \cdots = n_k$, then the orthogonality condition simplifies to

$$\sum_{i=1}^{k} c_i d_i = 0.$$

Readers who know some linear algebra should recognize that this condition states that the dot product between the vectors c and d vanishes, i.e., that the vectors c and d are orthogonal (perpendicular) to each other.

Example 12.3 (continued) Whether or not two contrasts are orthogonal depends not only on their coefficient vectors, but also on the size of the samples drawn from each population.

- Suppose that Heyl had collected samples of equal size for each of the three materials that he used. If $n_1 = n_2 = n_3$, then θ_1 and θ_2 are orthogonal because

$$1 \cdot 1 + 1 \cdot (-1) + (-2) \cdot 0 = 0.$$

- In fact, Heyl collected samples with sizes $n_1 = 6$ and $n_2 = n_3 = 5$. In this case, θ_1 and θ_2 are *not* orthogonal because

$$\frac{1 \cdot 1}{6} + \frac{1 \cdot (-1)}{5} + \frac{(-2) \cdot 0}{5} = \frac{1}{6} - \frac{1}{5} \neq 0.$$

  However, θ_1 is orthogonal to $\theta_3 = 18\mu_1 - 17\mu_2 - \mu_3$ because

$$\frac{1 \cdot 18}{6} + \frac{1 \cdot (-17)}{5} + \frac{(-2) \cdot (-1)}{5} = 3 - 3.2 + 0.2 = 0.$$

It turns out that the number of mutually orthogonal contrasts cannot exceed $k-1$. Obviously, this fact limits the practical utility of the method; however, families of mutually orthogonal contrasts have two wonderful properties that commend their use.

First, any family of $k-1$ mutually orthogonal contrasts partitions SS_B into $k-1$ separate components,

$$SS_B = SS_{\theta_1} + \cdots + SS_{\theta_{k-1}},$$

each with one degree of freedom. This information is usually incorporated into an expanded ANOVA table, as in...

Example 12.3 (continued) In the case of Heyl's data, the orthogonal contrasts θ_1 and θ_3 partition the between-groups sum-of-squares:

Source	SS	df	MS	F	p
Between	565.1	2	282.6	26.1	0.000028
θ_1	29.2	1	29.2	2.7	0.124793
θ_3	535.9	1	535.9	49.5	0.000009
Within	140.8	13	10.8		
Total	705.9	15			

Testing the fundamental null hypothesis, $H_0 : \mu_1 = \mu_2 = \mu_3$, results in a tiny signficance probability, leading us to conclude that the population means are not identical. The decomposition of the variation between groups into contrasts θ_1 and θ_3 provides insight into the differences between the population means. Testing the null hypothesis, $H_0 : \theta_1 = 0$, results in a large signficance probability, leading us to conclude that the heavy metals do not, in tandem, differ from glass. However, testing the null hypothesis, $H_0 : \theta_3 = 0$, results in a tiny signficance probability, leading us to conclude that the heavy metals do differ from each other. This is only possible if the glass mean lies between the gold and platinum means. For this simple example, our conclusions are easily checked by examining the raw data.

A second wonderful property of mutually orthogonal contrasts is that tests of mutually orthogonal contrasts are mutually independent. As we shall demonstrate, this property provides us with a powerful way to address a crucial difficulty that arises whenever we test multiple hypotheses. The difficulty is as follows. When testing a single null hypothesis that is true, there is a small chance (α) that we will falsely reject the null hypothesis and commit a Type I error. When testing multiple null hypotheses, each of which are true, there is a much larger chance that we will falsely reject at least one of them. We desire control of this *family-wide error rate*, often abbreviated FWER.

Definition 12.3 *The family-wide error rate (FWER) of a family of contrasts is the probability under the fundamental null hypothesis $H_0 : \mu_1 = \cdots = \mu_k$ of falsely rejecting at least one null hypothesis.*

The fact that tests of mutually orthogonal contrasts are mutually independent allows us to deduce a precise relation between the significance level(s) of the individual tests and the FWER.

1. Let E_r denote the event that $H_0 : \theta_r = 0$ is falsely rejected. Then $P(E_r) = \alpha$ is the rate of Type I error for an individual test.

2. Let E denote the event that at least one Type I error is committed, i.e.,
$$E = \bigcup_{r=1}^{k-1} E_r.$$
The family-wide rate of Type I error is FWER $= P(E)$.

3. The event that no Type I errors are committed is
$$E^c = \bigcap_{r=1}^{k-1} E_r^c,$$
and the probability of this event is $P(E^c) = 1 - \text{FWER}$.

4. By independence,
$$1 - \text{FWER} = P\left(E^c\right) = P\left(E_1^c\right) \times \cdots \times P\left(E_{k-1}^c\right) = (1-\alpha)^{k-1};$$
hence,
$$\text{FWER} = 1 - (1-\alpha)^{k-1}.$$

Notice that FWER $> \alpha$; i.e., the family rate of Type I error is greater than the error rate for an individual test. For example, if $k = 3$ and $\alpha = 0.05$, then
$$\text{FWER} = 1 - (1 - .05)^2 = 0.0975.$$

This phenomenon is sometimes called "alpha slippage." To protect against alpha slippage, we usually prefer to specify the family rate of Type I error that will be tolerated, then determine a significance level that will ensure the specified family rate. For example, if $k = 3$ and we desire FWER $= 0.05$, then we solve

$$0.05 = 1 - (1-\alpha)^2$$

to obtain a significance level of

$$\alpha = 1 - \sqrt{0.95} \doteq 0.0253.$$

12.3.2 Bonferroni t-Tests

It is often the case that one desires to test contrasts that are not mutually orthogonal. This can happen with a small family of contrasts. For example, suppose that we want to compare a control mean μ_1 to each of two treatment means, μ_2 and μ_3, in which case the natural contrasts have coefficient vectors $c = (1, -1, 0)$ and $d = (1, 0, -1)$. In this case, the orthogonality condition simplifies to $1/n_1 = 0$, which is impossible. Furthermore, as we have noted, families of more than $k - 1$ contrasts cannot be mutually orthogonal.

Statisticians have devised a plethora of procedures for testing multiple contrasts that are not mutually orthogonal. Many of these procedures address the case of multiple pairwise contrasts, i.e., contrasts for which each coefficient vector has exactly two nonzero components. We describe one such procedure that relies on *Bonferroni's inequality.*

Suppose that we plan m pairwise comparisons. These comparisons are defined by contrasts $\theta_1, \ldots, \theta_m$, each of the form $\mu_i - \mu_j$, not necessarily mutually orthogonal. Notice that each $H_0 : \theta_r = 0$ versus $H_1 : \theta_r \neq 0$ is a normal 2-sample location problem with equal variances. From this observation, the following facts can be deduced.

Theorem 12.5 *Under the fundamental null hypothesis* $H_0 : \mu_1 = \cdots = \mu_k$,

$$Z = \frac{\bar{X}_{i\cdot} - \bar{X}_{j\cdot}}{\sqrt{\left(\frac{1}{n_i} + \frac{1}{n_j}\right)\sigma^2}} \sim N(0,1)$$

and

$$T(\theta_r) = \frac{\bar{X}_{i\cdot} - \bar{X}_{j\cdot}}{\sqrt{\left(\frac{1}{n_i} + \frac{1}{n_j}\right) MS_W}} \sim t(N-k).$$

From Theorem 12.5, the t-test of $H_0 : \theta_r = 0$ is to reject if and only if

$$\mathbf{p} = P\left(|T(\theta_r)| \geq |t(\theta_r)|\right) \leq \alpha,$$

i.e., if and only if

$$|t(\theta_r)| \geq q = \texttt{qt(1-}\alpha\texttt{/2,df=N-k)},$$

where $t(\theta_r)$ denotes the observed value of $T(\theta_r)$. This t-test is virtually identical to Student's 2-sample t-test, described in Section 11.1.2, except that it pools all k samples to estimate the common variance instead of only pooling the two samples that are being compared.

At this point, you may recall that Section 11.1 strongly discouraged the use of Student's 2-sample t-test, which assumes a common population variance. Instead, we recommended Welch's approximate t-test. In the present case, our test of the fundamental null hypothesis $H_0 : \mu_1 = \cdots = \mu_k$ has already imposed the assumption of a common population variance, so our use of the T statistic in Theorem 12.5 is theoretically justified. But this justification is rather too glib, as it merely begs the question of why we assumed a common population variance in the first place. The general answer to this question is that the ANOVA methodology is extremely powerful and that comparable procedures in the case of unequal population variances may not exist. (Fortunately, ANOVA often provides useful insights even when its assumptions are violated. In such cases, however, one should interpret significance probabilities with extreme caution.) In the present case, a comparable procedure does exist, viz., the pairwise application of Welch's approximate t-test. The following discussion of how to control the family-wide error rate in such cases applies equally to either type of pairwise t-test.

Unless the pairwise contrasts are mutually orthogonal, we cannot use the multiplication rule for independent events to compute the family rate of Type I error. However, Bonferroni's inequality states that

$$\text{FWER} = P(E) = P\left(\bigcup_{r=1}^{m} E_r\right) \leq \sum_{r=1}^{m} P(E_r) = m\alpha;$$

hence, we can ensure that the family rate of Type I error is no greater than a specified FWER by testing each contrast at significance level $\alpha = \text{FWER}/m$.

Example 12.3 (continued) Instead of planning θ_1 and θ_3, suppose that we had planned θ_4 and θ_5, defined by coefficient vectors $c = (-1, 0, 1)$ and $d = (0, -1, 1)$ respectively. To test $H_0 : \theta_4 = 0$ and $H_0 : \theta_5 = 0$ with a family-wide error rate of FWER ≤ 0.10, we first compute

$$t(\theta_4) = \frac{\bar{x}_{3\cdot} - \bar{x}_{1\cdot}}{\sqrt{\left(\frac{1}{n_3} + \frac{1}{n_1}\right) ms_W}} \doteq -2.090605$$

and

$$t(\theta_5) = \frac{\bar{x}_{3\cdot} - \bar{x}_{2\cdot}}{\sqrt{\left(\frac{1}{n_3} + \frac{1}{n_2}\right) ms_W}} \doteq 4.803845,$$

resulting in the following significance probabilities:

```
> 2*pt(-2.090605,df=13)
[1] 0.0567719
> 2*pt(-4.803845,df=13)
[1] 0.0003444588
```

There are $m = 2$ pairwise comparisons. To ensure FWER ≤ 0.10, we compare the significance probabilities to $\alpha = 0.10/2 = 0.05$, which leads us to reject $H_0 : \theta_4 = 0$ and to decline to reject $H_0 : \theta_5 = 0$.

What do we lose by using Bonferroni's inequality instead of the multiplication rule? Without the assumption of independence, we must be slightly more conservative in choosing a significance level that will ensure a specified family-wide rate of error. For the same FWER, Bonferroni's inequality leads to a slightly smaller α than does the multiplication rule. The discrepancy grows as m increases.

12.4 Post Hoc Comparisons

We now consider situations in which we determine that a comparison is of interest *after* inspecting the data. For example, suppose that we had decided to compare gold to platinum *after* inspecting Heyl's data. This ought to strike you as a form of cheating. Almost every randomly generated data set will contain an appealing pattern that may draw the attention of an interested observer. To allow such patterns to determine what the scientist will investigate is to invite abuse. Fortunately, statisticians have devised procedures that protect ethical scientists from the heightened risk of Type I error when the null hypothesis was constructed after the data were examined. The present section describes two such procedures.

12.4.1 Bonferroni t-Tests

To fully appreciate the distinction between planned and *post hoc* comparisons, it is highly instructive to examine the method of Bonferroni t-tests. Suppose that only pairwise comparisons are of interest. Because we are testing *after* we have had the opportunity to inspect the data (and therefore to construct the contrasts that appear to be nonzero), we suppose that *all* pairwise contrasts were of interest *a priori*. Hence, whatever the number of pairwise contrasts actually tested *a posteriori*, we set

$$m = \binom{k}{2} = \frac{k(k-1)}{2}$$

and proceed as before.

The difference between planned and *post hoc* comparisons is especially sobering when k is large. For example, suppose that we desire that the family-wide error rate does not exceed 0.10 when testing two pairwise contrasts among $k = 10$ groups. If the comparisons were planned, then $m = 2$ and we can perform each test at signficance level $\alpha = 0.10/2 = 0.05$. However, if the comparisons were constructed after examining the data, then $m = 45$ and we must perform each test at signficance level $\alpha = 0.10/45 \doteq 0.0022$. Obviously, much stronger evidence is required to reject the same null hypothesis when the comparison is chosen after examining the data.

12.4.2 Scheffé F-Tests

The reasoning that underlies Scheffé F-tests for *post hoc* comparisons is analogous to the reasoning that underlies Bonferroni t-tests for *post hoc* comparisons. To accommodate the possibility that a general contrast was constructed after examining the data, Scheffé's procedure is predicated on the assumption that *all possible* contrasts were of interest *a priori*. This makes Scheffé's procedure the most conservative of all multiple comparison procedures.

Scheffé's F-test of $H_0 : \theta_r = 0$ versus $H_1 : \theta_r \neq 0$ is to reject H_0 if and only if

$$\mathbf{p} = \texttt{1-pf(}f(\theta)\texttt{/(k-1),df1=k-1,df2=N-k)} \leq \alpha,$$

i.e., if and only if

$$\frac{f(\theta_r)}{k-1} \geq q = \texttt{qf(1-}\alpha\texttt{,k-1,N-k)},$$

where $f(\theta_r)$ denotes the observed value of the $F(\theta_r)$ defined for the method of planned orthogonal contrasts. It can be shown that, no matter how many $H_0 : \theta_r = 0$ are tested by this procedure, the family-wide rate of Type I error is no greater than α.

Example 12.3 (continued) For the contrasts that we have considered, Scheffé's F-test produces the following results:

Source	$f(\theta_r)/2$	p
θ_1	1.3	0.29422
θ_2	25.3	0.00003
θ_3	24.7	0.00004
θ_4	2.2	0.15200
θ_5	11.5	0.00131

12.5 Case Study: Treatments of Anorexia

Table 12.1 displays weights of $N = 72$ anorexic girls, before and after treatment.[4] Each girl received one of three treatments: cognitive behavioral ($n_1 = 29$), standard ($n_2 = 26$), or family therapy ($n_3 = 17$). Quite a bit can be said about these data, to which we shall return in subsequent sections. Our present concern is with determining if either the cognitive behavioral treatment or the family therapy treatment tends to outperform the standard treatment. These are natural comparisons that undoubtedly concerned the experimenters before the data were collected.

Let B_{ij} denote the weight before treatment of girl j in sample i and let A_{ij} denote the weight after treatment of girl j in sample i. The effect of treatment on weight can be measured in various ways, e.g., $A_{ij} - B_{ij}$, A_{ij}/B_{ij}, or $\log(A_{ij}/B_{ij})$. It turns out that none of these choices is entirely satisfying. For ease of interpretation, we set $X_{ij} = A_{ij} - B_{ij}$. Box plots of $\vec{x}_1$, $\vec{x}_2$, and $\vec{x}_3$ are displayed in Figure 12.1.

The box plots reveal several interesting features of these data. First, the effect of standard treatment is highly variable and centered at zero (no effect). Second, there are quite a few outliers in the upper tail of $\vec{x}_1$; hence, the effect of cognitive behavioral treatment does not appear to be normally distributed. Third, family therapy treatment appears to be generally more effective than either standard treatment or cognitive behavioral treatment.

It turns out that several other features of these data—features well worth noting—are not revealed by the box plots, but let us proceed to perform an analysis of variance. The summary statistics for the three samples are as

[4]These data appear as Data Set 285 in *A Handbook of Small Data Sets*, which states that the weights are reported in kilograms. It appears more likely that the weights were actually reported in pounds.

Cognitive Behavioral		Standard		Family Therapy	
Before	After	Before	After	Before	After
80.5	82.2	80.7	80.2	83.8	95.2
84.9	85.6	89.4	80.1	83.3	94.3
81.5	81.4	91.8	86.4	86.0	91.5
82.6	81.9	74.0	86.3	82.5	91.9
79.9	76.4	78.1	76.1	86.7	100.3
88.7	103.6	88.3	78.1	79.6	76.7
94.9	98.4	87.3	75.1	76.9	76.8
76.3	93.4	75.1	86.7	94.2	101.6
81.0	73.4	80.6	73.5	73.4	94.9
80.5	82.1	78.4	84.6	80.5	75.2
85.0	96.7	77.6	77.4	81.6	77.8
89.2	95.3	88.7	79.5	82.1	95.5
81.3	82.4	81.3	89.6	77.6	90.7
76.5	72.5	78.1	81.4	83.5	92.5
70.0	90.9	70.5	81.8	89.9	93.8
80.4	71.3	77.3	77.3	86.0	91.7
83.3	85.4	85.2	84.2	87.3	98.0
83.0	81.6	86.0	75.4		
87.7	89.1	84.1	79.5		
84.2	83.9	79.7	73.0		
86.4	82.7	85.5	88.3		
76.5	75.7	84.4	84.7		
80.2	82.6	79.6	81.4		
87.8	100.4	77.5	81.2		
83.3	85.2	72.3	88.2		
79.7	83.6	89.0	78.8		
84.5	84.6				
80.8	96.2				
87.4	86.7				

Table 12.1: Weights of $N = 72$ anorexic girls before and after treatment.

follows:

	$i = 1$	$i = 2$	$i = 3$
n_i	29	26	17
$\bar{x}_{i\cdot}$	3.006897	−0.450000	7.264706
s_i^2	53.414240	63.819400	51.228680

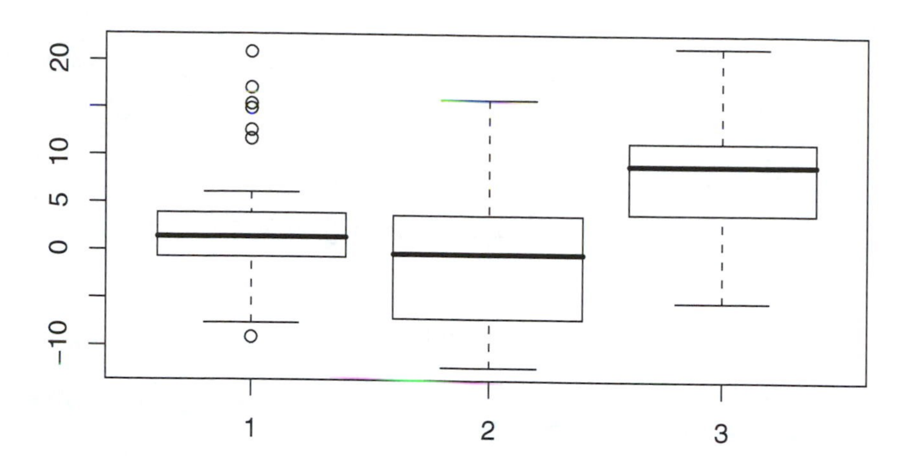

Figure 12.1: Box plots of weight gained by anorexic girls who received (1) cognitive behavioral treatment, (2) standard treatment, and (3) family therapy treatment.

Proceeding as in Example 12.2, we obtain the following ANOVA table:

Source	SS	df	MS	F	p
Between	614.6437	2	307.3218	5.4223	0.0065
Within	3910.7424	69	56.6774		
Total	4525.3861	71			

Although the assumptions of normality and equal population variances are suspect, the significance probability is so small that we feel comfortable rejecting the fundamental null hypothesis of equal population means.

Having concluded that the population means are not identical, we inquire how mean weight gain varies by treatment. The contrasts of interest are $\theta_1 = \mu_1 - \mu_2$ (cognitive behavioral versus standard) and $\theta_2 = \mu_3 - \mu_2$ (family therapy versus standard), both of which we suppose were constructed before the data were collected. Because

$$1 \cdot 0 + (-1) \cdot (-1) + 0 \cdot 1 = 1 \neq 0,$$

these contrasts are not orthogonal. However, both θ_1 and θ_2 are pairwise contrasts, so we can perform Bonferroni t-tests.

To test $H_0 : \theta_1 = 0$ and $H_0 : \theta_2 = 0$ with a family-wide error rate of FWER ≤ 0.05, we first compute

$$t(\theta_1) = \frac{\bar{x}_{1\cdot} - \bar{x}_{2\cdot}}{\sqrt{\left(\frac{1}{n_1} + \frac{1}{n_2}\right) ms_W}} \doteq 1.700144$$

and

$$t(\theta_2) = \frac{\bar{x}_{3\cdot} - \bar{x}_{2\cdot}}{\sqrt{\left(\frac{1}{n_3} + \frac{1}{n_2}\right) ms_W}} \doteq 3.285422,$$

resulting in the following significance probabilities:

```
> 2*pt(-1.700144,df=69)
[1] 0.09360765
> 2*pt(-3.285422,df=69)
[1] 0.001602338
```

There are $m = 2$ pairwise comparisons. To ensure FWER ≤ 0.05, we compare the significance probabilities to $\alpha = 0.05/2 = 0.025$, which leads us to decline to reject $H_0 : \theta_1 = 0$ and to reject $H_0 : \theta_2 = 0$.

In the present application, we might be searching for compelling evidence that either the cognitive behavioral treatment or the family therapy treatment is more effective than (as opposed to simply different from) the standard treatment. Thus, we might prefer to test $H_0 : \theta_1 \leq 0$ versus $H_1 : \theta_1 > 0$ and $H_0 : \theta_2 \leq 0$ versus $H_1 : \theta_2 > 0$. We do so by performing one-sided t-tests instead of two-sided t-tests. We decline to reject $H_0 : \theta_1 \leq 0$ and we reject $H_0 : \theta_2 \leq 0$ in favor of $H_1 : \theta_2 > 0$, concluding that mean weight gain under family therapy treatment is greater than mean weight gain under the standard treatment.

12.6 Exercises

Problem Set A Water was sampled at each of $k = 3$ sites (A,B,C) in the Bimini Lagoon, Bahamas. The salinity of each experimental unit was measured in parts per million (ppm). The resulting data are displayed in Table 12.2.[5]

[5] R. Till (1974). *Statistical Methods for the Earth Scientist.* Macmillan, London. These data appear as Data Set 253 in *A Handbook of Small Data Sets.*

A	37.54	37.01	36.71	37.03	37.32	37.01	37.03	37.70
	37.36	36.75	37.45	38.85				
B	40.17	40.80	39.76	39.70	40.79	40.44	39.79	39.38
C	39.04	39.21	39.05	38.24	38.53	38.71	38.89	38.66
	38.51	40.08						

Table 12.2: Salinity measurements (in ppm) in Bimini Lagoon.

1. Use side-by-side boxplots and normal probability plots to investigate the ANOVA assumptions of normality and homoscedasticity. Do these assumptions seem plausible? Why or why not?

2. Use ANOVA to test the null hypothesis that the three sites have the same mean salinity. Use a significance level of $\alpha = 0.05$ and organize your calculations in an ANOVA table.

Problem Set B Steady-state hemoglobin levels were measured on a total of $N = 41$ patients with $k = 3$ types of sickle cell disease. The $k = 3$ types are HB SS, HB ST (HB S/-thalassemia), and HB SC. The resulting data are displayed in Table 12.3.[6]

SS	7.2	7.7	8.0	8.1	8.3	8.4	8.4	8.5	8.6	8.7
	9.1	9.1	9.1	9.8	10.1	10.3				
ST	8.1	9.2	10.0	10.4	10.6	10.9	11.1	11.9	12.0	12.1
SC	10.7	11.3	11.5	11.6	11.7	11.8	12.0	12.1	12.3	12.6
	12.6	13.3	13.3	13.8	13.9					

Table 12.3: Hemoglobin levels of $N = 41$ patients with sickle cell disease.

1. Use side-by-side boxplots and normal probability plots to investigate the ANOVA assumptions of normality and homoscedasticity. Do these assumptions seem plausible? Why or why not?

[6] E. Anionwu, D. Watford, M. Brozovic, and B. Kirkwood (1981). Sickle cell disease in a British urban community. *British Medical Journal*, 282:283–286. These data appear as Data Set 310 in *A Handbook of Small Data Sets*.

2. Use ANOVA to test the null hypothesis that the three types of sickle cell disease have the same mean hemoglobin levels. Use a significance level of $\alpha = 0.05$ and organize your calculations in an ANOVA table.

Problem Set C A total of $N = 64$ patients with $k = 5$ types of advanced cancer were treated with ascorbate. The resulting survival times (in days) are displayed in Table 12.4.[7]

Stomach	Bronchus	Colon	Ovary	Breast
124	81	248	1234	1235
42	461	377	89	24
25	20	189	201	1581
45	450	1843	356	1166
412	246	180	2970	40
51	166	537	456	727
1112	63	519		3808
46	64	455		791
103	155	406		1804
876	859	365		3460
146	151	942		719
340	166	776		
396	37	372		
	223	163		
	138	101		
	72	20		
	245	283		

Table 12.4: Survival times (in days) of $N = 64$ cancer patients treated with ascorbate.

Let $\vec{x}_1, \ldots, \vec{x}_5$ denote the samples displayed in Table 12.4. Let $\vec{y}_1, \ldots, \vec{x}_5$ denote the corresponding samples of the logarithms of the survival times:

```
> y1 <- log(x1)
> y2 <- log(x2)
```

[7]E. Cameron and L. Pauling (1978). Supplemental ascorbate in the supportive treatment of cancer: re-evaluation of prolongation of survival times in terminal human cancer. *Proceedings of the National Academy of Science*, 75:4538–4542. These data appear as Data Set 323 in *A Handbook of Small Data Sets*.

```
> y3 <- log(x3)
> y4 <- log(x4)
> y5 <- log(x5)
```

1. Construct two side-by-side boxplots, one of $\vec{x}_1, \ldots, \vec{x}_5$, one of $\vec{y}_1, \ldots, \vec{x}_5$. Which data, the observed survival times or the transformed survival times, more nearly satisfy the ANOVA assumptions of normality and homoscedasticity? Explain.

2. Use ANOVA to investigate whether or not survival time differs with organ affected. Carefully state the null hypothesis that you are testing. Are you comparing population means of survival time or population means of log(survival time)? Use a significance level of $\alpha = 0.05$ and organize your calculations in an ANOVA table.

Problem Set D

1. For her master's thesis, a nutrition student at the University of Arizona decides to compare several weight loss strategies. Inspired by Chapter 14 of Jean Kerr's *Please Don't Eat the Daisies*, she recruits 140 moderately obese adult women and randomly assigns 20 women to each of the following diets: Rockefeller, Mayo, Atkins (high-protein), a low-protein diet, a blitz diet, a liquid diet, and—as a control—Aunt Jean's marshmallow fudge diet. Each woman is weighed before dieting, asked to follow the prescribed diet for eight weeks, then weighed again. The resulting data will be analyzed using the analysis of variance and related statistical techniques.

   (a) This is a k-sample problem. What is the value of k?

   (b) What null hypothesis is tested by an analysis of variance? (Your answer should specify relations between certain population parameters. Be sure to define these parameters!)

   (c) How many pairwise comparisons are possible?

   (d) The student is especially interested in three pairwise comparisons: Atkins versus low-protein, low-protein versus fudge, and fudge versus liquid. Specify contrasts that correspond to each of these comparisons.

   (e) Are the preceding contrasts orthogonal? Why or why not?

2. As part of her senior thesis, a William & Mary physics major decides to repeat Heyl's experiment for determining the gravitational constant using 4 different materials: silver, copper, topaz, and quartz. She plans to test 10 specimens of each material.

   (a) Three comparisons are planned:

      i. Metal (silver & copper) versus Gem (topaz & quartz)

      ii. Silver versus Copper

      iii. Topaz versus Quartz

   What contrasts correspond to these comparisons? Are they orthogonal? Why or why not? If the desired family rate of Type I error is 0.05, then what significance level should be used for testing the null hypotheses $H_0 : \theta_r = 0$?

   (b) After analyzing the data, an ANOVA table is constructed. Complete the table from the information provided.

Source	SS	df	MS	F	p
Between					
θ_1					0.001399
θ_2					0.815450
θ_3					0.188776
Within			9.418349		
Total					

   (c) Referring to the above table, explain what conclusion the student should draw about each of her planned comparisons.

   (d) Assuming that the ANOVA assumption of homoscedasticity is warranted, use the above table to estimate the common population variance.

3. The value of G, the gravitational constant, should not depend on the material (gold, platinum, glass) used for the small masses in a torsion balance. If the material mattered, then the contrasts that we proposed (gold & platinum vs. glass, gold vs. platinum) would be natural comparisons and we might guess that gold and platinum would behave similarly. In Heyl's experiment, however, gold and platinum behaved differently. Is there an explanation for this difference?

   In fact, Heyl never intended to compare different materials. After completing his first series of measurements, Heyl discovered that the gold balls had absorbed appreciable quantities of mercury during the seven-month experiment.[8] Heyl attempted to correct for the increase in weight in his calculation of G, but thought it prudent to repeat the experiment with platinum and glass balls, the former dipped in lacquer. These revelations might lead us to construct a different set of contrasts.

(a) Propose a contrast, θ_6, that compares gold to the average of platinum and glass.

(b) Propose a contrast, θ_7, that compares platinum to glass.

(c) Test $H_0 : \theta_6 = 0$ versus $H_0 : \theta_6 \neq 0$ and $H_0 : \theta_7 = 0$ versus $H_0 : \theta_7 \neq 0$ using the method of Bonferroni t-tests. Use a significance level of $\alpha = 0.05$ and assume that the contrasts were proposed before the data were examined.

(d) Heyl did not discard the results he obtained using gold balls, but perhaps he should have. If we agree to discard the gold data, then how should we test $H_0 : \theta_7 = 0$ versus $H_0 : \theta_7 \neq 0$?

Problem Set E R. R. Sokal observed 25 females of each of three genetic lines (RS, SS, NS) of the fruitfly *Drosophila melanogaster* and recorded the number of eggs laid per day by each female for the first 14 days of her life. The lines labelled RS and SS were selectively bred for resistance and for susceptibility to the insecticide DDT. A nonselected control line is labelled NS. The purpose of the experiment was to investigate the following research questions:

- Do the two selected lines (RS and SS) differ in fecundity from the nonselected line (NS)?

- Does the line selected for resistance (RS) differ in fecundity from the line selected for susceptibility (SS)?

The data are displayed in Table 12.5.[9]

1. Use side-by-side boxplots and normal probability plots to investigate the ANOVA assumptions of normality and homoscedasticity. Do these assumptions seem plausible? Why or why not?

---

[8] Heyl (1930) speculated that the mercury was "probably derived in vapor form from the manometer connected to the container."

[9] R. R. Sokal and F. J Rohlf (1981). *Biometry*, Second edition. W. H. Freeman, San Francisco. These data appear as Data Set 22 in *A Handbook of Small Data Sets.*

RS	12.8	21.6	14.8	23.1	34.6	19.7	22.6	29.6	16.4	20.3
	29.3	14.9	27.3	22.4	27.5	20.3	38.7	26.4	23.7	26.1
	29.5	38.6	44.4	23.2	23.6					
SS	38.4	32.9	48.5	20.9	11.6	22.3	30.2	33.4	26.7	39.0
	12.8	14.6	12.2	23.1	29.4	16.0	20.1	23.3	22.9	22.5
	15.1	31.0	16.9	16.1	10.8					
NS	35.4	27.4	19.3	41.8	20.3	37.6	36.9	37.3	28.2	23.4
	33.7	29.2	41.7	22.6	40.4	34.4	30.4	14.9	51.8	33.8
	37.9	29.5	42.4	36.6	47.4					

Table 12.5: Numbers of eggs laid by female fruitflies of $k = 3$ genetic lines.

2. Construct constrasts that correspond to the research questions framed above. Verify that these constrasts are orthogonal. At what significance level should the contrasts be tested in order to maintain a family rate of Type I error equal to 5%?

3. Use ANOVA and the method of orthogonal contrasts to construct an ANOVA table. State the null and alternative hypotheses that are tested by these methods. For each null hypothesis, state whether or not it should be rejected. (Use $\alpha = 0.05$ for the ANOVA hypothesis and the significance level calculated above for the contrast hypotheses.)

Problem Set F A number of Byzantine coins were discovered in Cyprus. These coins were minted during the reign of King Manuel I, Comnenus (1143–1180). It was determined that $n_1 = 9$ of these coins were minted in an early coinage, $n_2 = 7$ were minted several years later, $n_3 = 4$ were minted in a third coinage, and $n_4 = 7$ were minted in a fourth coinage.

The percent of silver content of each coin is displayed in Table 12.6.[10]

1. Investigate the ANOVA assumptions of normality and homoscedasticity. Do these assumptions seem plausible? Why or why not?

2. Construct an ANOVA table. State the null and alternative hypotheses tested by this method. Should the null hypothesis be rejected at the $\alpha = 0.10$ level?

[10] M. F. Hendy and J. A. Charles (1970). The production techniques, silver content and circulation history of the twelfth-century Byzantine Trachy. *Archaeometry*, 12:13–21. These data appear as Data Set 149 in *A Handbook of Small Data Sets.*

1	5.9	6.8	6.4	7.0	6.6	7.7	7.2	6.9	6.2
2	6.9	9.0	6.6	8.1	9.3	9.2	8.6		
3	4.9	5.5	4.6	4.5					
4	5.3	5.6	5.5	5.1	6.2	5.8	5.8		

Table 12.6: Percent of silver content of $N = 27$ Byzantine coins.

3. Examining these data, it appears that coins minted early in Manuel's reign (the first two coinages) tended to contain more silver than coins minted later in his reign (the last two coinages). Construct a contrast that is suitable for investigating if this is the case. State appropriate null and alternative hypotheses, then test them using Scheffé's F-test for multiple comparisons with a significance level of 5%.

Problem Set G Researchers compared antibody responses in normal and alloxan diabetic mice. Three groups of mice were studied: normal, alloxan diabetic, and alloxan diabetic treated with insulin. Several comparisons are of interest:

- Does the antibody response of alloxan diabetic mice differ from the antibody response of normal mice?
- Does treating alloxan diabetic mice with insulin affect their antibody response?
- Does the antibody response of alloxan diabetic mice treated with insulin differ from the antibody response of normal mice?

Table 12.7 displays the amounts of nitrogen-bound bovine serum albumin produced by the mice.[11]

1. Using the above data, investigate the ANOVA assumptions of normality and homoscedasticity. Do these assumptions seem plausible for these data? Why or why not?

[11]R. E. Dolkart, B. Halperin, and J. Perlman (1971). Comparison of antibody responses in normal and alloxan diabetic mice. *Diabetes*, 20:162–167. These data appear as Data Set 304 in *A Handbook of Small Data Sets*.

Normal	156	282	197	297	116	127	119	29	253	122
	349	110	143	64	26	86	122	455	655	14
Alloxan	391	46	469	86	174	133	13	499	168	62
	127	276	176	146	108	276	50	73		
Alloxan	82	100	98	150	243	68	228	131	73	18
+insulin	20	100	72	133	465	40	46	34	44	

Table 12.7: Amounts of nitrogen-bound bovine serum albumin produced by normal and alloxan diabetic mice.

2. Now transform the data by taking the square root of each measurement. Using the transformed data, investigate the ANOVA assumptions of normality and homoscedasticity. Do these assumptions seem plausible for the transformed data? Why or why not?

3. Using the transformed data, construct an ANOVA table. State the null and alternative hypotheses tested by this method. Should the null hypothesis be rejected at the $\alpha = 0.05$ level?

4. Using the transformed data, construct suitable contrasts for investigating the research questions framed above. State appropriate null and alternative hypotheses and test them using the method of Bonferroni t-tests. At what significance level should these hypotheses be tested in order to maintain a family rate of Type I error equal to 5%? Which null hypotheses should be rejected?

Chapter 13

Goodness-of-Fit

Chapters 9–12 discuss inferences about population means and medians. We now consider a rather different problem, in which we partition the sample space into k events and test various hypotheses about the probabilities of those events. Such hypotheses arise in a variety of applications.

13.1 Partitions

Let $E_1, \ldots, E_k$ be mutually disjoint events whose union is the sample space, S. We say that $E_1, \ldots, E_k$ *partition* S; the E_j are sometimes called *cells*.

Example 13.1 A single die is rolled. The possible outcomes are $S = \{1, 2, 3, 4, 5, 6\}$. Let $k = 6$ and $E_j = \{j\}$ for $j = 1, \ldots, k$.

Example 13.2 Let X be a discrete random variable with $X(S) = \{0, 1, 2, 3, \ldots\}$. Let $k = 5$, $E_j = \{j - 1\}$ for $j = 1, \ldots, 4$, and $E_5 = \{4, 5, 6, \ldots\}$.

Example 13.3 Let X be a continuous random variable with $X(S) = (-\infty, \infty)$. Let $k = 7$ and $E_1 = (-\infty, -5)$, $E_2 = [-5, -3)$, $E_3 = [-3, -1)$, $E_4 = [-1, 1)$, $E_5 = [1, 3)$, $E_6 = [3, 5)$, and $E_7 = [5, \infty)$.

Given $E_1, \ldots, E_k$, let $p_j = P(E_j)$. Our interest lies in testing hypotheses about the vector of cell probabilities, $\vec{p} = (p_1, \ldots, p_k)$. The set of all possible probability vectors is the *unit simplex* $\Pi \subset \Re^k$, defined by the requirement that $\vec{\pi} = (\pi_1, \ldots, \pi_k)$ if and only if $\pi_1, \ldots, \pi_k \geq 0$ and $\pi_1 + \cdots + \pi_k = 1$. We will test hypotheses of the form $H_0 : \vec{p} \in \Pi_0$ versus $H_1 : \vec{p} \in \Pi_1$, where Π_0 and Π_1 are disjoint sets of probability vectors whose union is Π.

In Example 13.1, the null hypothesis of a fair die can be written as $H_0 : \vec{p} \in \Pi_0$ by letting

$$\Pi_0 = \left\{ \left(\frac{1}{6}, \ldots, \frac{1}{6} \right) \right\} \subset \Re^6.$$

For simplicity, we might write

$$H_0 : p_1 = p_2 = p_3 = p_4 = p_5 = p_6 = 1/6.$$

To obtain data we repeat the experiment n times, each time recording which E_j occurs. We then tabulate o_j, the observed number of times that E_j occurs in n replications.

Example 13.1 (continued) In 1882, R. Wolf reported tossing a die $n = 20,000$ times, observing the following cell counts:[1]

j	1	2	3	4	5	6
o_j	3407	3631	3176	2916	3448	3422

Was Wolf tossing a fair die? The expected cell counts are $e_j = np_j$. Under the null hypothesis of a fair die, each $p_j = 1/6$ and each $e_j = 3333\frac{1}{3}$. The o_j evidently differ from the e_j, but perhaps the discrepancies are merely due to chance variation. Are the discrepancies sufficiently large to warrant rejecting the null hypothesis of a fair die?

13.2 Test Statistics

Goodness-of-fit tests compare observed cell counts to expected cell counts. The expected cell counts are computed under the assumption that the null hypothesis is correct; the null hypothesis is rejected if the discrepancy between observed and expected cell counts is sufficiently large. There are various ways to measure the discrepancy between observed and expected counts; perhaps the best known is *Pearson's chi-squared statistic*,

$$X^2 = \sum_{j=1}^{k} \frac{(o_j - e_j)^2}{e_j},$$

[1] Wolf's data are recorded as Data Set 131 in *A Handbook of Small Data Sets.*

which measures squared errors in relation to expected counts.

We will emphasize a slightly different approach to testing fit, one which has certain advantages in more complicated situations. Given cell probabilities $p_1, \ldots, p_k$, the probability, or *likelihood*, that we will observe $o_1, \ldots, o_k$ is

$$L(p_1, \ldots, p_k) = C \cdot p_1^{o_1} \cdots p_k^{o_k},$$

where C is the number of different ways that $X_1, \ldots, X_n$ might produce $o_1, \ldots, o_k$. (It will turn out that we need not compute the actual value of C.) The *maximum likelihood estimates* of $p_1, \ldots, p_k$ are the values of $p_1, \ldots, p_k$ that maximize $L(p_1, \ldots, p_k)$. In other words, the maximum likelihood estimates are the probabilities that afford the best chance of an experiment producing the data that actually were observed.

With no restrictions on $p_1, \ldots, p_k$ (other than $p_j \geq 0$ and $p_1 + \cdots + p_k = 1$), it turns out that the maximum likelihood estimate of p_j is $\hat{p}_j = o_j/n$, the observed frequency of cell j. In this case (but not always), the maximum likelihood estimate and the plug-in estimate are identical. The maximum value of the likelihood function is

$$L(\hat{p}_1, \ldots, \hat{p}_k).$$

The null hypothesis specifies restrictions on $p_1, \ldots, p_k$. For example, the null hypothesis of a fair die restricts $\vec{p}$ to one probability vector, $p_1 = \cdots = p_6 = 1/6$. In contrast, consider the null hypothesis that each die axis (1-6, 2-5, and 3-4) has equally likely faces. This hypothesis restricts $\vec{p}$ to probability vectors of the form

$$(p_{16}, p_{25}, p_{34}, p_{34}, p_{25}, p_{16}). \tag{13.1}$$

Let $\check{p}_1, \ldots, \check{p}_k$ denote the values of $p_1, \ldots, p_k$ that maximize the likelihood subject to the restrictions imposed by the null hypothesis. Depending on the nature of the restrictions, the $\check{p}_j$ may be easy or difficult to determine. If the null hypothesis specifies that $p_1 = \cdots = p_6 = 1/6$, then (obviously) $\check{p}_1 = \cdots = \check{p}_6 = 1/6$. If the null hypothesis specifies (13.1), then the maximum likelihood estimates turn out to be

$$\check{p}_{16} = \frac{(o_1 + o_6)/2}{n}, \ \check{p}_{25} = \frac{(o_2 + o_5)/2}{n}, \text{ and } \check{p}_{34} = \frac{(o_3 + o_4)/2}{n}.$$

Because the null hypothesis restricts the available choices for $p_1, \ldots, p_k$, it must be the case that

$$L(\check{p}_1, \ldots, \check{p}_k) \leq L(\hat{p}_1, \ldots, \hat{p}_k).$$

Hence, the *likelihood ratio statistic,*

$$\lambda = L\left(\check{p}_1, \ldots, \check{p}_k\right) / L\left(\hat{p}_1, \ldots, \hat{p}_k\right),$$

must lie in [0, 1]. If λ is nearly 1, then the restricted probabilities specified by the null hypothesis fit the observed data nearly as well as the unrestricted probabilities and H_0 seems plausible. If λ is much less than 1, then we may prefer to reject H_0.

Let $\check{e}_j = n\check{p}_j$, the expected cell counts if each $p_j = \check{p}_j$. Then

$$\begin{aligned} \lambda &= \frac{C\check{p}_1^{o_1} \cdots \check{p}_k^{o_k}}{C\hat{p}_1^{o_1} \cdots \hat{p}_k^{o_k}} \\ &= \frac{(\check{e}_1/n)^{o_1} \cdots (\check{e}_k/n)^{o_k}}{(o_1/n)^{o_1} \cdots (o_k/n)^{o_k}} \\ &= \left(\frac{\check{e}_1}{o_1}\right)^{o_1} \cdots \left(\frac{\check{e}_k}{o_k}\right)^{o_k} \end{aligned}$$

and the *likelihood ratio chi-squared statistic* is

$$G^2 = -2\log\lambda = 2\sum_{j=1}^{k} o_j \log\left(o_j/\check{e}_j\right).$$

Notice that $G^2 \geq 0$. Large values of G^2 are evidence against the null hypothesis. It turns out that the distribution of G^2 can usually be approximated by a chi-squared distribution. A popular rule of thumb is that the approximation will be adequate if each $\check{e}_j \geq 5$. Furthermore, the same chi-squared distribution that approximates the null distribution of G^2 also approximates the null distribution of X^2.

To determine the appropriate number of degrees of freedom for a particular G^2, one must examine the unrestricted and the restricted sets of possible $p_1, \ldots, p_k$. The correct degrees of freedom is the difference between the dimensions of these two sets. The unrestricted set has $k-1$ dimensions. (There are k probabilities, but they must sum to 1 so only $k-1$ are free to vary.) If the null hypothesis specifies a single point $\vec{p} \in \Re^k$, e.g., $H_0 : p_1 = \cdots = p_k = 1/k$, then the restricted set has dimension 0 and the correct degrees of freedom is $(k-1) - 0 = k-1$.

Example 13.1 (continued) To test the null hypothesis that Wolf's die was fair, we first compute the expected counts $\check{e}_j = 20,000/6$. Then

$$G^2 = 2\sum_{i=1}^{6} o_j \log\left(o_j/\check{e}_j\right) \doteq 95.8023$$

with $(6-1)-0=5$ degrees of freedom, resulting in an approximate significance probability of...

```
> 1-pchisq(95.8023,df=5)
[1] 0
```

The evidence that Wolf's die was not fair is overwhelming. Notice that

$$X^2 = \sum_{j=1}^{6} (o_j - \check{e}_j)^2 / \check{e}_j = 94.189,$$

not much different from G^2.

To test the null hypothesis that each axis of Wolf's die had equally likely faces, we first compute the following expected counts:

$$\begin{aligned} \check{e}_1 = \check{e}_6 &= (3407+3422)/2 = 3414.5, \\ \check{e}_2 = \check{e}_5 &= (3631+3448)/2 = 3539.5, \\ \check{e}_3 = \check{e}_4 &= (3176+2916)/2 = 3046.0. \end{aligned}$$

Using these expected counts, $G^2 = 15.8641$. The restricted set of probabilites specified by (13.1) has 2 dimensions (because $2(p_{16}+p_{25}+p_{34})=1$, only 2 of these probabilities are free to vary), so the correct degrees of freedom is $(6-1)-2=3$. Knowing G^2 and the correct degrees of freedom, we compute an approximate significance probability:

```
> 1-pchisq(15.8641,df=3)
[1] 0.001209101
```

At conventional significance levels, there is compelling evidence against the null hypothesis that each pair of opposing faces is equally likely. Compared to the fair die hypothesis, however, the hypothesis of equally likely opposing faces appears remarkably plausible.

13.3 Testing Independence

Suppose that $A_1, \ldots, A_r$ and $B_1, \ldots, B_c$ each partition S. We define a third partition by

$$E_{ij} = A_i \cap B_j$$

and use it to investigate the relation between the A_i and the B_j. The partitions $A_1, \ldots, A_r$ and $B_1, \ldots, B_c$ are mutually independent if and only if

$$P(E_{ij}) = P(A_i) \cdot P(B_j)$$

for each ij pair. We will use the methods of Section 13.2 to test the null hypothesis that these multiplication rules hold. Such chi-squared tests of independence are widely used.

Example 13.4 Karl Pearson studied the relation between two partitions of criminals, one by type of crime (arson, rape, violence, stealing, coining, fraud) and one by alcohol consumption (drinker, abstainer). The observed cell counts can be organized in the form of a *two-way contingency table*:[2]

	$B_1 =$ drink	$B_2 =$ abstain
$A_1 =$ arson	50	43
$A_2 =$ rape	88	62
$A_3 =$ violence	155	110
$A_4 =$ stealing	379	300
$A_5 =$ coining	18	14
$A_6 =$ fraud	63	144

Can we infer from these data that type of crime and drinking are related?

Let $p_{ij} = P(E_{ij})$ denote the cell probabilities, let

$$p_{i+} = p_{i1} + \cdots + p_{ic} = P(A_i)$$

denote the *row probabilities*, and let

$$p_{+j} = p_{1j} + \cdots + p_{rj} = P(B_j)$$

denote the *column probabilities*. The unrestricted maximum likelihood estimates of these quantities are

$$\hat{p}_{ij} = \frac{o_{ij}}{n}, \ \hat{p}_{i+} = \sum_{j=1}^{c} \hat{p}_{ij} = \frac{o_{i+}}{n}, \text{ and } \hat{p}_{+j} = \sum_{i=1}^{r} \hat{p}_{ij} = \frac{o_{+j}}{n}.$$

Under the null hypothesis of mutual independence, the restricted maximum likelihood estimates are

$$\check{p}_{i+} = \frac{o_{i+}}{n}, \ \check{p}_{+j} = \frac{o_{+j}}{n}, \text{ and } \check{p}_{ij} = \check{p}_{i+}\check{p}_{+j} = \frac{o_{i+}}{n}\frac{o_{+j}}{n}.$$

The expected cell counts are

$$\check{e}_{ij} = n\check{p}_{ij} = \frac{o_{i+}o_{+j}}{n}$$

[2]This table appears as Data Set 296 in *A Handbook of Small Data Sets.*

and the test statistic is

$$G^2 = 2\sum_{i=1}^{r}\sum_{j=1}^{c} o_{ij} \log\left(\frac{o_{ij}}{\check{e}_{ij}}\right) = 2\sum_{i=1}^{r}\sum_{j=1}^{c} o_{ij} \log\left(\frac{n o_{ij}}{o_{i+}o_{+j}}\right).$$

To compute degrees of freedom, we first note that Π, the unrestricted set of cell probability vectors, has dimension $k-1 = rc-1$. Under the null hypothesis, the individual cell probabilities are determined by the row and column probabilities. The r row probabilities must sum to 1, as must the c column probabilities. Hence, Π_0, the restricted set of cell probability vectors, has dimension $(r-1)+(c-1)$ and there are

$$(rc-1) - [(r-1)+(c-1)] = rc - r - c + 1 = (r-1)(c-1)$$

degrees of freedom.

Example 13.4 (continued) Under the null hypothesis of independence, the expected cell counts are (approximately)

	B_1 = drink	B_2 = abstain
A_1 = arson	49.11	43.89
A_2 = rape	79.21	70.79
A_3 = violence	139.93	125.07
A_4 = stealing	358.55	320.45
A_5 = coining	16.90	15.10
A_6 = fraud	109.31	97.69

,

resulting in $G^2 \doteq 50.5173$. With $(6-1)(2-1) = 5$ degrees of freedom, an approximate significance probability is

```
> 1-pchisq(50.5173,df=5)
[1] 1.085959e-09
```

There is compelling evidence that type of crime and drinking are related.

Examining the tables of the o_{ij} and the e_{ij}, we note several discrepancies. For example, more than half of the first five crimes were committed by drinkers, whereas more than 2/3 of the fraud crimes were committed by abstainers. What statistical analysis cannot do is explain *why* type of crime and drinking are related in this way.

13.4 Exercises

1. Does a horse's starting position affect the probability that it will win a race on a circular track? The following table[3] lists the numbers of wins from each starting position (1 is closest to the inside rail) in 144 8-horse races:

Starting position	1	2	3	4	5	6	7	8
Number of wins	29	19	18	25	17	10	15	11

   Assuming that starting positions were randomly assigned, use these data to test the null hypothesis that a horse's starting position does not affect its chance of winning.

2. M&M's Milk Chocolate Candies were first manufactured in 1940. In Spring 2006, the Mars company claimed that it was mixing M&M's in the following proportions:

Brown	Yellow	Red	Blue	Orange	Green
0.13	0.14	0.13	0.24	0.20	0.16

   To test this claim, I distributed two bags of M&M's to six students enrolled in Math 352 (Data Analysis) at the College of William & Mary, who observed the following counts:

Brown	Yellow	Red	Blue	Orange	Green
121	84	118	226	226	123

   Do you find the claimed proportions credible in light of these data?

3. According to Mendelian genetics, a recessive trait will appear in an offspring if and only if both parents contribute a recessive gene. If each parent has a dominant and a recessive gene, then the probability that their offspring will display the recessive trait is $1/4$. The following questions assume a population of offspring bred from parents with one dominant and one recessive gene for each of the two traits considered.

[3] These data were reported in the *New York Post* in 1955 and reprinted in the second edition of *Nonparametric Statistics for the Behavioral Sciences* in 1988. They also appear as Data Set 42 in *A Handbook of Small Data Sets*.

(a) A certain strain of tomato is either tall (dominant trait) or dwarf (recessive trait). The same strain has either cut leaves (dominant trait) or potato leaves (recessive trait). Let E_1 denote tall cut-leaf offspring, let E_2 denote tall potato-leaf offspring, let E_3 denote dwarf cut-leaf offspring, and let E_4 denote dwarf potato-leaf offspring. Assuming that the traits are independent, compute the probability of each E_j.

(b) In 1931, J. W. MacArthur reported experimental results for $n = 1611$ offspring.[4] MacArthur observed $o_1 = 926$, $o_2 = 288$, $o_3 = 293$, and $o_4 = 104$. Use these data to test the correctness of the cell probabilities computed in part (a).

4. A rectangular screen of 324×480 pixels was constructed for an experiment in visual perception.[5] The researchers devised a computer algorithm that was intended to color each pixel, independently and identically, black with probability $p = 0.29$ or white with probability $1 - p = 0.71$.

   To validate their coloring algorithm, the researchers randomly selected $n = 1000$ nonoverlapping squares of 4×4 pixels and counted the number of black pixels in each square. Let X_i denote the number of black pixels in square i. If the algorithm performed as intended, then $X_i \sim \text{Binomial}(16; 0.29)$ and the probabilities $P(X_i = j)$, $j \in X(S) = \{0, \ldots, 16\}$, can be computed by the R command `dbinom(0:16,16,.29)`.

   We partition $X(S)$ as follows: $E_1 = \{0, 1\}$, $E_j = \{j\}$ for $j = 2, \ldots, 8$, and $E_9 = \{9, \ldots, 16\}$. The following cell counts were observed:

j	1	2	3	4	5	6	7	8	9
o_j	30	93	159	184	195	171	92	45	31

   On the basis of these data, is there reason to doubt that the researchers' coloring algorithm performed as intended?

5. Let $X(S) = \{1, 2, \ldots, 9\}$ and consider the probability mass function

$$f(x) = P(X = x) = \log_{10}(1 + 1/x).$$

[4] These data were analyzed by R. S. Sokal and F. J. Rohlf in the second edition of *Biometry*, 1981. They appear as Data Set 39 in *A Handbook of Small Data Sets*.

[5] The screen was described by S. Laner, P. Morris, and R. C. Oldfield in the *Quarterly Journal of Experimental Psychology*, 9:105–108, 1957. Their results appear as Data Set 156 in *A Handbook of Small Data Sets*.

In 1881, S. Newcomb noted that this pmf approximates the distribution of the leading digit in the values of a logarithm table. In 1938, F. Benford noted the same phenomenon in a variety of different data sets. The phenomenon is now known as *Benford's law.*

(a) Verify that f is indeed a pmf.

(b) The following table[6] reports the leading digit of the populations of 305 towns on a randomly selected page of a gazeteer.

Leading digit	1	2	3	4	5	6	7	8	9
Number of towns	107	55	39	22	13	18	13	23	15

On the basis of these data, is it plausible that the leading digits of town populations follow Benford's law?

6. Let $X(S) = \{0, 1, 2, \ldots\}$. The random variable X is said to have a Poisson distribution with intensity parameter $\mu \in (0, \infty)$ if X has probability mass function

$$f(x) = P(X = x) = \lambda^x e^{-x}/x!.$$

The Poisson distribution frequently arises when counting arrivals in a fixed time interval. If $X_1, \ldots, X_n \sim \text{Poisson}(\mu)$, then $EX_i = \mu$ and the maximum likelihood estimator of μ is $\bar{X}_n$.

In 1910, E. Rutherford and M. Geiger counted the numbers of alpha-particle scintillations observed in each of $n = 2608$ 72-second intervals. The following table[7] reports the frequency of each count:

Count	0	1	2	3	4	5	6	7
Frequency	57	203	383	525	532	408	273	139

Count	8	9	10	11	12	13	14
Frequency	45	27	10	4	0	1	1

(a) Use the above data to compute $\bar{x}$, the average observed count.

(b) Now we partition $X(S)$ by setting $E_j = \{j - 1\}$ for $j = 1, \ldots, 10$ and $E_{11} = \{10, 11, 12, \ldots\}$. Test the null hypothesis that counts of alpha-particle scintillations follow a Poisson distribution.

[6] Data Set 174 in *A Handbook of Small Data Sets.*
[7] Data Set 279 in *A Handbook of Small Data Sets.*

7. In Example 13.4, test the null hypothesis that type of crime and drinking are independent using X^2 instead of G^2.

8. Using the data from Example 13.4, compute $o_{ij} \log(o_{ij}/\breve{e}_{ij})$ and $(o_{ij} - \breve{e}_{ij})^2/\breve{e}_{ij}$ for each cell. Both of these quantities measure the extent to which a cell's observed count deviates from its expected count. A value of 0 indicates a perfect match. For which two cells is the deviation greatest?

9. Discard the fraud row from the contingency table in Example 13.4. Does the remaining 5×2 contingency table provide compelling evidence of a relation between the remaining 5 types of crime and drinking?

10. A study of sandflies in Panama classified flies caught in light traps by sex (male, female) and height of trap (3 and 35 feet above ground), resulting in the following 2 × 2 contingency table:[8]

	3 ft	35 ft
Male	173	125
Female	150	73

Do these data provide evidence that the sex ratio of Panamanian sandflies varies with height above ground?

11. A study of Hodgkin's disease classified $n = 538$ patients by histological type (LP = lymphocyte predominance, NS = nodular sclerosis, MC = mixed cellularity, LD = lymphocyte depletion) and response to treatment (positive, partial, none), resulting in the following 4 × 3 contingency table:[9]

	Positive	Partial	None
LP	74	18	12
NS	68	16	12
MC	154	54	58
LD	18	10	44

Do these data provide evidence that a patient's response to treatment for Hodgkin's disease varies by histological type?

[8] These data were reported by H. A. Christensen, A. Herrer, and S. R. Telford in the *Annals of Tropical Medicine and Parasitology*, 66:55-66, 1972. They appear as Data Set 128 in *A Handbook of Small Data Sets*.

[9] Data Set 76 in *A Handbook of Small Data Sets*.

Chapter 14

Association

Section 13.3 introduced the concept of the relation between two different ways of partitioning the sample space. Given two random variables, X and Y, we could use the values of X and the values of Y to partition S, then use the methods of Section 13.3 to determine if there is compelling evidence against the null hypothesis that X and Y are independent. However, if X and Y have many possible values, then using these values to define small numbers of cells results in a fairly crude approximation of the relation between X and Y. Furthermore, this relation does not exploit various useful properties of real numbers, e.g., that $x_1, \ldots, x_r$ can be meaningfully ordered from smallest to largest.

In Chapters 14–15 we describe methods for assessing relationships between two continuous random variables, X and Y. Recall that $X : S \to \Re$ and $Y : S \to \Re$ are each functions that assign real numbers to experimental outcomes. For each $s \in S$, X assigns $X(s) \in \Re$ to s and Y assigns $Y(s) \in \Re$ to s. Thus, the pair (X, Y) assigns to each $s \in S$ a pair of real numbers, $(X(s), Y(s)) \in \Re^2$. A pair of random variables is sometimes called a *random vector*. Just as each random variable has a *univariate* probability distribution in $\Re$, so each pair of random variables has a *bivariate probability distribution* in $\Re^2$.

This chapter is organized as follows. In Section 14.1 we introduce the concept of a bivariate probability density function. In Section 14.2 we describe the special case of two normal random variables. The family of bivariate normal distributions is indexed by five parameters: the population means and variances of each random variable, and a fifth parameter that quantifies the relation between the random variables. In Section 14.3 we describe a different way of measuring association, one that does not require parametric assumptions about the distribution of (X, Y). In Section 14.4, we discuss the interpretation of association.

14.1 Bivariate Distributions

The continuous random variables (X, Y) define a function that assigns a pair of real numbers to each experimental outcome. Let

$$B = [a, b] \times [c, d] \subset \Re^2$$

be a rectangular set of such pairs and suppose that we want to compute

$$P((X, Y) \in B) = P(X \in [a, b], Y \in [c, d]).$$

Just as we compute $P(X \in [a, b])$ using the pdf of X, so we compute $P((X, Y) \in B)$ using the *joint probability density function* of (X, Y). To do so, we must extend the concept of area under the graph of a function of one variable to the concept of volume under the graph of a function of two variables.

Theorem 14.1 *Let X be a continuous random variable with pdf f_x and let Y be a continuous random variable with pdf f_y. In this context, f_x and f_y are called the* marginal pdfs *of (X, Y). Then there exists a function $f : \Re^2 \to \Re$, the* joint pdf *of (X, Y), such that*

$$P((X, Y) \in B) = \text{Volume}_B(f) = \int_a^b \int_c^d f(x, y)\, dy\, dx \tag{14.1}$$

for all rectangular subsets B. If X and Y are independent, then

$$f(x, y) = f_x(x) f_y(y).$$

Remark If (14.1) is true for all rectangular subsets of $\Re^2$, then it is true for all subsets in the sigma-field generated by the rectangular subsets.

We can think of the joint pdf as a function that assigns an elevation to a point identified by two coordinates, longitude (x) and latitude (y). Noting that topographic maps display elevations via contours of constant elevation, we can describe a joint pdf by identifying certain of its contours, i.e., subsets of $\Re^2$ on which $f(x, y)$ is contant.

Definition 14.1 *Let f denote the joint pdf of (X, Y) and fix $c > 0$. Then*

$$\left\{(x, y) \in \Re^2 \; : \; f(x, y) = c\right\}$$

is a contour *of f.*

14.2 Normal Random Variables

Suppose that $X \sim \text{Normal}(0, 1)$ and $Y \sim \text{Normal}(0, 1)$, not necessarily independent. To measure the degree of dependence between X and Y, we consider the quantity $E(XY)$.

- If there is a *positive association* between X and Y, then experimental outcomes that have
  - positive values of X will tend to have positive values of Y, so XY will tend to be positive;
  - negative values of X will tend to have negative values of Y, so XY will tend to be positive.

  Hence, $E(XY) > 0$ indicates positive association.

- If there is a *negative association* between X and Y, then experimental outcomes that have
  - positive values of X will tend to have negative values of Y, so XY will tend to be negative;
  - negative values of X will tend to have positive values of Y, so XY will tend to be negative.

  Hence, $E(XY) < 0$ indicates negative association.

If $X \sim \text{Normal}(\mu_x, \sigma_x^2)$ and $Y \sim \text{Normal}(\mu_y, \sigma_y^2)$, then we measure dependence after converting to standard units.

Definition 14.2 *Let $\mu_x = EX$ and $\sigma_x^2 = \text{Var}\, X < \infty$. Let $\mu_y = EY$ and $\sigma_y^2 = \text{Var}\, Y < \infty$. The* population product-moment correlation coefficient *of X and Y is*

$$\rho = \rho(X, Y) = E\left[\left(\frac{X - \mu_x}{\sigma_x}\right)\left(\frac{Y - \mu_y}{\sigma_y}\right)\right].$$

Notice that, although we motivated Definition 14.2 by examining the behavior of normal random variables, the formal definition of the product-moment correlation coefficient does not require X or Y to be normally distributed

Here are some properties of the product-moment correlation coefficient:

Theorem 14.2 *If X and Y have finite variances, then*

1. $-1 \leq \rho \leq 1$

2. $\rho = \pm 1$ *if and only if*

$$\frac{Y - \mu_y}{\sigma_y} = \pm \frac{X - \mu_x}{\sigma_x},$$

in which case Y is completely determined by X.

3. *If X and Y are independent, then $\rho = 0$.*

4. *If X and Y are normal random variables for which $\rho = 0$, then X and Y are independent.*

If $\rho = \pm 1$, then the values of (X, Y) fall on a straight line. If $|\rho| < 1$, then the five population parameters $(\mu_x, \mu_y, \sigma_x^2, \sigma_y^2, \rho)$ determine a unique bivariate normal pdf. The contours of this joint pdf are concentric ellipses centered at (μ_x, μ_y). We use one of these ellipses to display the basic features of the bivariate normal pdf in question.

Definition 14.3 *Let f denote a nondegenerate ($|\rho| < 1$) bivariate normal pdf. The* population concentration ellipse *is the contour of f that contains the four points*

$$(\mu_x \pm \sigma_x, \mu_y \pm \sigma_y).$$

It is not difficult to create an `R` function that plots concentration ellipses. The function `binorm.ellipse` is described in Appendix R and can be obtained from the web page for this book.

Example 14.1 The following `R` commands produce the population concentration ellipse for a bivariate normal distribution with parameters $(\mu_x, \mu_y, \sigma_x^2, \sigma_y^2, \rho) = (10, 20, 4, 16, 0.5)$:

```
> pop <- c(10,20,4,16,.5)
> binorm.ellipse(pop)
```

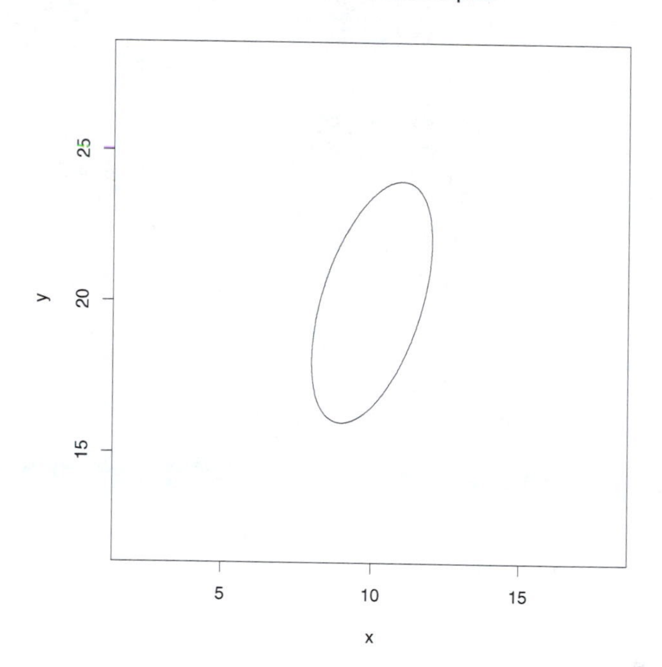

Figure 14.1: The population concentration ellipse for a bivariate normal distribution with parameters $(\mu_x, \mu_y, \sigma_x^2, \sigma_y^2, \rho) = (10, 20, 4, 16, 0.5)$.

The ellipse plotted by these commands is displayed in Figure 14.1.

Unless the population concentration ellipse is circular, it has a unique major axis. The line that coincides with this axis is the *first principal component* of the population and plays an important role in multivariate statistics. We will encounter this line again in Chapter 15.

14.2.1 Bivariate Normal Samples

A bivariate sample is a set of paired observations:

$$(x_1, y_1), (x_2, y_2), \ldots, (x_n, y_n).$$

We assume that each pair (x_i, y_i) was independently drawn from the same bivariate distribution. Bivariate samples are usually stored in an $n \times 2$ *data matrix*,

$$\begin{bmatrix} x_1 & y_1 \\ x_2 & y_2 \\ & \vdots & \\ x_n & y_n \end{bmatrix},$$

and are often displayed by plotting each (x_i, y_i) in the Cartesian plane. The resulting figure is called a *scatter plot* or *scatter diagram.*

x	y
87	87
25	57
76	91
84	67
91	67
82	66
94	86
89	74
92	92
76	85
84	75
99	92
92	55
74	74
84	74
94	69
99	98
63	81
82	80
91	85

Table 14.1: Scores on two statistics midterm tests.

Example 14.2 Twenty students enrolled in Math 351 (Applied Statistics) at the College of William & Mary produced scores on two midterm tests. The scores are displayed in Table 14.1 and plotted in Figure 14.2. Typically, it is easier to discern patterns by inspecting a scatter diagram than by inspecting a table of numbers. In particular, note the presence of an apparent outlier.

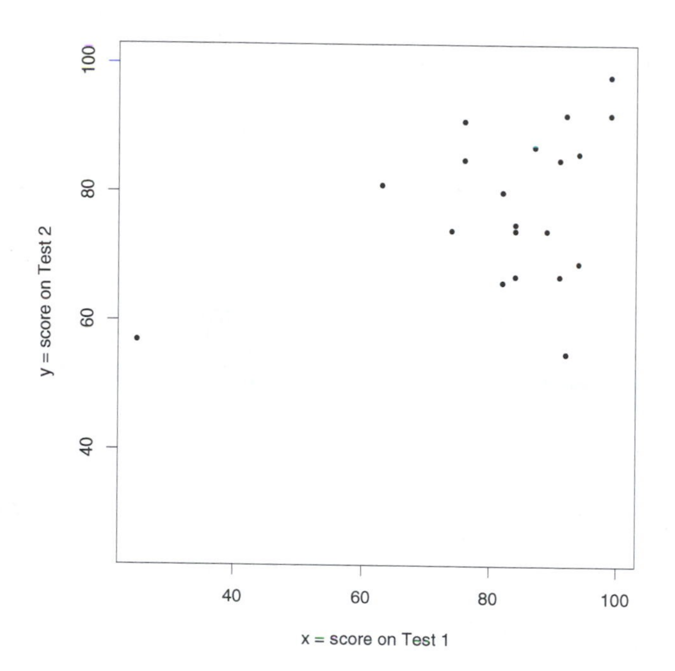

Figure 14.2: A scatter diagram of a bivariate sample. Each point corresponds to a student. The horizontal position of the point represents the student's score on the first midterm test; the vertical position of the point represents the student's score on the second midterm test.

The population from which the bivariate sample in Example 14.2 was drawn is not known, so this sample should not be interpreted as an example of a bivariate normal sample. However, it is not difficult to create an `R` function that simulates sampling from a specified bivariate normal population. The function `binorm.sample` is described in Appendix R and can be obtained from the web page for this book.

Example 14.1 (continued) The following `R` command draws $n = 5$ observations from the previously specified bivariate normal distribution:

```
> binorm.sample(pop,5)
          [,1]     [,2]
[1,] 12.293160 24.07643
[2,] 11.819520 24.13076
[3,] 11.529582 17.28637
[4,]  6.912459 23.39430
[5,] 11.043991 18.12538
```

Notice that `binorm.sample` returns the sample in the form of a data matrix.

Having observed a bivariate normal sample, we inquire how to estimate the five population parameters $(\mu_x, \mu_y, \sigma_x^2, \sigma_y^2, \rho)$. We have already discussed how to estimate the population means (μ_x, μ_y) with the sample means $(\bar{x}, \bar{y})$ and the population variances (σ_x^2, σ_y^2) with the sample variances (s_x^2, s_y^2). The plug-in estimate of ρ is

$$\begin{aligned}
\hat{\rho} &= \frac{1}{n}\sum_{i=1}^{n}\left[\left(\frac{x_i - \hat{\mu}_x}{\hat{\sigma}_x}\right)\left(\frac{y_i - \hat{\mu}_y}{\hat{\sigma}_y}\right)\right] \\
&= \frac{1}{n}\sum_{i=1}^{n}\left[\left(\frac{x_i - \bar{x}}{\sqrt{(n-1)s_x^2/n}}\right)\left(\frac{y_i - \bar{y}}{\sqrt{(n-1)s_y^2/n}}\right)\right] \\
&= \frac{1}{n-1}\sum_{i=1}^{n}\left[\left(\frac{x_i - \bar{x}}{s_x}\right)\left(\frac{y_i - \bar{y}}{s_y}\right)\right],
\end{aligned}$$

where

$$\hat{\sigma}_x = \sqrt{\widehat{\sigma_x^2}} \quad \text{and} \quad \hat{\sigma}_y = \sqrt{\widehat{\sigma_y^2}}.$$

This quantity is *Pearson's product-moment correlation coefficient*, usually denoted r.

It is not difficult to create an `R` function that computes the estimates $(\bar{x}, \bar{y}, s_x^2, s_y^2, r)$ from a bivariate data matrix. The function `binorm.estimate` is described in Appendix R and can be obtained from the web page for this book.

Example 14.1 (continued) The following `R` commands draw $n = 100$ observations from a bivariate normal distribution with parameters $\mu_x = 10$, $\mu_y = 20$, $\sigma_x^2 = 4$, $\sigma_y^2 = 16$ and $\rho = 0.5$, and then estimate the parameters from the sample:

```
> pop <- c(10,20,4,16,.5)
> Data <- binorm.sample(pop,100)
> binorm.estimate(Data)
[1]  9.8213430 20.3553502  4.2331147 16.7276819  0.5632622
```

Naturally, the estimates do not equal the estimands because of sampling variation.

Finally, it is not difficult to create an R function that plots a scatter diagram and overlays the *sample concentration ellipse*, i.e., the concentration ellipse constructed using the computed sample quantities $(\bar{x}, \bar{y}, s_x^2, s_y^2, r)$ instead of the unknown population quantities $(\mu_x, \mu_y, \sigma_x^2, \sigma_y^2, \rho)$. The function `binorm.scatter` is described in Appendix R and can be obtained from the web page for this book.

Example 14.1 (continued) The following R command creates the overlaid scatter diagram displayed in Figure 14.3:

```
> binorm.scatter(Data)
```

When analyzing bivariate data, it is good practice to examine both the scatter diagram and the sample concentration ellipse in order to ascertain how well the latter summarizes the former. A poor summary suggests that the sample may not have been drawn from a bivariate normal distribution, as in Figure 14.4.

14.2.2 Inferences about Correlation

We have already observed that $\hat{\rho} = r$ is the plug-in estimate of ρ. In this section, we consider how to test hypotheses about and construct confidence intervals for ρ.

Given normal random variables X and Y, an obvious question is whether or not they are uncorrelated. To answer this question, we test the null hypothesis $H_0 : \rho = 0$ against the alternative hypothesis $H_1 : \rho \neq 0$. (One might also be interested in one-sided hypotheses and ask, for example, whether or not there is convincing evidence of positive correlation.) We can derive a test from the following fact about the plug-in estimator of ρ.

Theorem 14.3 *Suppose that (X_i, Y_i), $i = 1, \ldots, n$, are independent pairs of random variables with a bivariate normal distribution. Let $\hat{\rho}$ denote the*

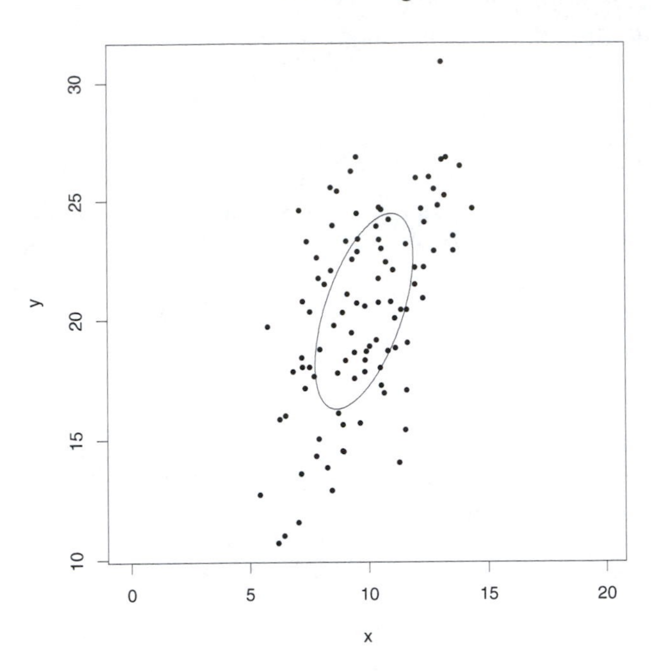

Figure 14.3: A scatter diagram of a bivariate normal sample, with the sample concentration ellipse overlaid.

plug-in estimator of ρ. *If* X_i *and* Y_i *are uncorrelated, i.e.,* $\rho = 0$, *then*

$$\frac{\hat{\rho}\sqrt{n-2}}{\sqrt{1-\hat{\rho}^2}} \sim t(n-2).$$

Assuming that (X_i, Y_i) have a bivariate normal distribution, Theorem 14.3 allows us to compute a significance probability for testing $H_0 : \rho = 0$ versus $H_1 : \rho \neq 0$. Let $T \sim t(n-2)$. Then the probability of observing $|\hat{\rho}| \geq |r|$ under H_0 is

$$\mathbf{p} = P\left(|T| \geq \frac{r\sqrt{n-2}}{\sqrt{1-r^2}}\right)$$

and we reject H_0 if and only if $\mathbf{p} \leq \alpha$. Equivalently, we reject H_0 if and

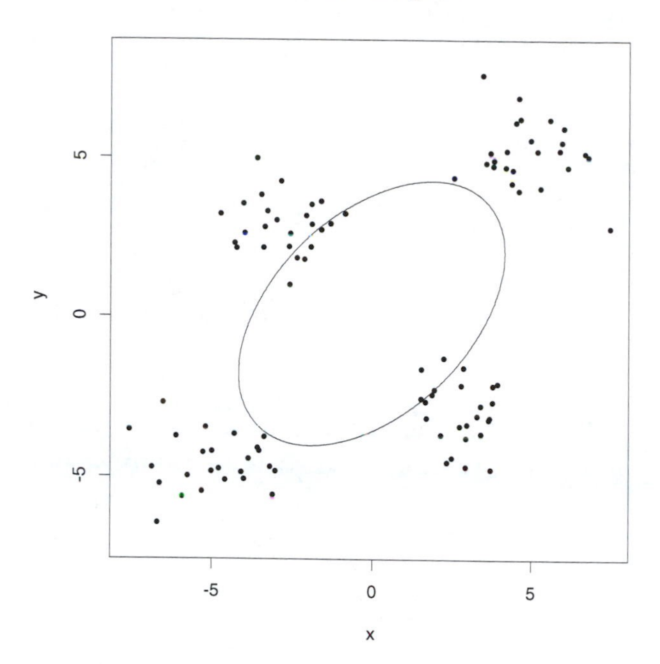

Figure 14.4: A scatter diagram for which the sample concentration ellipse is a poor summary. These data were not drawn from a bivariate normal distribution.

only if (iff)

$$\left| \frac{r\sqrt{n-2}}{\sqrt{1-r^2}} \right| \geq q_t \quad \text{iff} \quad \frac{r^2(n-2)}{1-r^2} \geq q_t^2 \quad \text{iff} \quad r^2 \geq \frac{q_t^2}{n-2+q_t^2},$$

where $q_t = \texttt{qt}(1-\alpha/2, n-2)$.

When testing hypotheses about correlation, it is important to appreciate the distinction between statistical significance and material significance. *Strong evidence that an association exists is not the same as evidence of a strong association.* The following examples illustrate the distinction.

Example 14.3 I used `binorm.sample` to draw a sample of $n = 300$ observations from a bivariate normal distribution with a population correlation coefficient of $\rho = 0.1$. This is a rather weak association. I then used `binorm.estimate` to compute a sample correlation coefficient of $r = 0.16225689$. The test statistic is

$$\frac{r\sqrt{n-2}}{\sqrt{1-r^2}} = 2.838604$$

and the significance probability is

$$\mathbf{p} = 2 * \texttt{pt}(-2.838604, 298) = 0.004842441.$$

This is fairly decisive evidence that $\rho \neq 0$, but concluding that X and Y are correlated does not warrant concluding that X and Y are strongly correlated.

Example 14.4 I used `binorm.sample` to draw a sample of $n = 10$ observations from a bivariate normal distribution with a population correlation coefficient of $\rho = 0.8$. This is a fairly strong association. I then used `binorm.estimate` to compute a sample correlation coefficient of $r = 0.3759933$. The test statistic is

$$\frac{r\sqrt{n-2}}{\sqrt{1-r^2}} = 1.147684$$

and the significance probability is

$$\mathbf{p} = 2 * \texttt{pt}(-1.147684, 8) = 0.2842594.$$

There is scant evidence that $\rho \neq 0$, despite the fact that X and Y are strongly correlated.

Although testing whether or not $\rho = 0$ may be important, other inferences about ρ may also be of interest. To construct confidence intervals for ρ, we must test $H_0 : \rho = \rho_0$ versus $H_1 : \rho \neq \rho_0$. To do so, we rely on an approximation due to Ronald Fisher. Let

$$\zeta = \frac{1}{2} \log\left(\frac{1+\rho}{1-\rho}\right)$$

and rewrite the hypotheses as $H_0 : \zeta = \zeta_0$ versus $H_1 : \zeta \neq \zeta_0$. This is sometimes called Fisher's z-transformation. Fisher discovered that

$$\hat{\zeta} = \frac{1}{2} \log\left(\frac{1+\hat{\rho}}{1-\hat{\rho}}\right) \approx \text{Normal}\left(\zeta, \frac{1}{n-3}\right),$$

which allows us to compute an approximate significance probability. Let $Z \sim \text{Normal}(0,1)$ and set

$$z = \frac{1}{2}\log\left(\frac{1+r}{1-r}\right).$$

Then

$$\mathbf{p} \approx P\left(|Z| \geq |z - \zeta_0|\sqrt{n-3}\right)$$

and we reject $H_0 : \zeta = \zeta_0$ if and only if $\mathbf{p} \leq \alpha$. Equivalently, we reject $H_0 : \zeta = \zeta_0$ if and only if

$$|z - \zeta_0|\sqrt{n-3} \geq q_z,$$

where $q_z = \texttt{qnorm}(1-\alpha/2)$.

To construct an approximate $(1-\alpha)$-level confidence interval for ρ, we first observe that

$$z \pm \frac{q_z}{\sqrt{n-3}} \tag{14.2}$$

is an approximate $(1-\alpha)$-level confidence interval for ζ. We then use the inverse of Fisher's z-transformation,

$$\rho = \frac{e^{2\zeta}-1}{e^{2\zeta}+1},$$

to transform (14.2) to a confidence interval for ρ.

Example 14.5 Suppose that we draw $n = 100$ observations from a bivariate normal distribution and observe $r = 0.5$. To construct a 0.95-level confidence interval, we use $q_z \doteq 1.96$. First we compute

$$z = \frac{1}{2}\log\left(\frac{1+0.5}{1-0.5}\right) = 0.5493061$$

and

$$z \pm \frac{q_z}{\sqrt{n-3}} \doteq 0.5493061 \pm \frac{1.96}{\sqrt{97}} = (0.350302, 0.7483103)$$

to obtain a confidence interval (a,b) for ζ. The corresponding confidence interval for ρ is

$$\left(\frac{e^{2a}-1}{e^{2a}+1}, \frac{e^{2b}-1}{e^{2b}+1}\right) = (0.3366433, 0.6341398).$$

Notice that the plug in estimate $\hat{\rho} = r = 0.5$ is *not* the midpoint of this interval.

14.3 Monotonic Association

The product-moment correlation coefficient, ρ, is but one measure of association. In Chapter 15 we will elaborate on the proper interpretation of ρ and discover that it measures a very specific type of association. Furthermore, the methods that we have presented for drawing inferences about ρ require that the random variables of interest have a bivariate normal distribution. In the present section we introduce a more general type of association for which inferences can be drawn without recourse to distributional assumptions.

One way to think about association is to consider the extreme case in which the random variable Y is completely determined by the random variable X, say $Y = \phi(X)$ for a function $\phi : \Re \to \Re$. In this case, the manner in which we might wish to quantify the association between X and Y depends on the nature of ϕ. The following definitions identify several useful types of functions.

Definition 14.4 *Let* $\phi : \Re \to \Re$ *and* $x_1, x_2 \in \Re$.

1. ϕ *is* nondecreasing *iff* $\phi(x_1) \leq \phi(x_2)$ *for every pair* $x_1 < x_2$.

2. ϕ *is* strictly increasing *iff* $\phi(x_1) < \phi(x_2)$ *for every pair* $x_1 < x_2$.

3. ϕ *is* nonincreasing *iff* $\phi(x_1) \geq \phi(x_2)$ *for every pair* $x_1 < x_2$.

4. ϕ *is* strictly decreasing *iff* $\phi(x_1) > \phi(x_2)$ *for every pair* $x_1 < x_2$.

5. ϕ *is* monotonic *iff* ϕ *is either nondecreasing or nonincreasing.*

6. ϕ *is* strictly monotonic *iff* ϕ *is either strictly increasing or strictly decreasing.*

7. ϕ *is* linear *iff there exist* $a, b \in \Re$ *such that* $\phi(x) = ax + b$ *for every* $x \in \Re$.

Notice that strictly increasing is a special case of nondecreasing, strictly decreasing is a special case of nonincreasing, and strictly monotonic is a special case of monotonic.

Example 14.6 Suppose that ϕ is linear.

- If $a > 0$, e.g., $\phi(x) = 2x + 1$, then ϕ is strictly increasing and therefore strictly monotonic.

- If $a < 0$, e.g., $\phi(x) = -3x + 1$, then ϕ is strictly decreasing and therefore strictly monotonic.

- If $a = 0$, e.g., $\phi(x) = 1$, then ϕ is constant. It is nondecreasing, but not strictly increasing. It is also nonincreasing, but not strictly decreasing. Thus, it is monotonic, but not strictly monotonic.

We conclude from Example 14.6 that every linear function is monotonic. However, most monotonic functions are not linear, e.g., $\phi(x) = x^3$, $\phi(x) = e^x$, and $\phi(x) = e^{-x}$.

Example 14.7 Consider the function $\phi(x) = x^2$.

- If $0 < x_1 < x_2$, then $\phi(x_1) = x_1^2 < x_2^2 = \phi(x_2)$. Hence, ϕ is strictly increasing on the positive real numbers.

- If $x_1 < x_2 < 0$, then $\phi(x_1) = x_1^2 > x_2^2 = \phi(x_2)$. Hence, ϕ is strictly decreasing on the negative real numbers.

Because ϕ is neither nondecreasing nor nonincreasing on all the real numbers, it is not monotonic.

In Chapter 15 we will see that the product-moment correlation coefficient measures the strength of linear association between X and Y. If there is a strong nonlinear relation between X and Y, then ρ may convey a misleading impression of how strongly X and Y are associated. In this section we explore a way of measuring the strength of monotonic association between X and Y.

Suppose that (X_1, Y_1) and (X_2, Y_2) are independent random vectors, each having the same bivariate pdf. Suppose that we observe (x_1, y_1) and (x_2, y_2). If either

$$x_1 < x_2 \quad \text{and} \quad y_1 < y_2$$

or

$$x_1 > x_2 \quad \text{and} \quad y_1 > y_2,$$

then we say that these observations are *concordant*.

Let κ denote the probability of concordance, i.e., the probability of observing a pair of concordant observations. If $Y = \phi(X)$ for a strictly increasing ϕ, then $\kappa = 1$. If $Y = \phi(X)$ for a strictly decreasing ϕ, then $\kappa = 0$.

Intermediate values of κ quantify the strength of monotonic association between X and Y. If X and Y are independent, then

$$\begin{aligned}
\kappa &= P\left(\left(X_1 < X_2 \bigcap Y_1 < Y_2\right) \bigcup \left(X_1 > X_2 \bigcap Y_1 > Y_2\right)\right) \\
&= P\left(X_1 < X_2 \bigcap Y_< Y_2\right) + P\left(X_1 > X_2 \bigcap Y_> Y_2\right) \\
&= P\left(X_1 < X_2\right)\left(Y_1 < Y_2\right) + \left(X_1 > X_2\right)\left(Y_1 > Y_2\right) \\
&= \frac{1}{2} \cdot \frac{1}{2} + \frac{1}{2} \cdot \frac{1}{2} = \frac{1}{2}.
\end{aligned}$$

If $\kappa > 1/2$, then there is a tendency for the joint distribution of (X, Y) to produce concordant pairs of observations and it follows that there is a positive monotonic association between X and Y. If $\kappa < 1/2$, then there is a tendency for the joint distribution of (X, Y) to produce discordant pairs of observations and it follows that there is a negative monotonic association between X and Y.

Notice that (x_1, y_1) and (x_2, y_2) are concordant if and only if $(x_1 - x_2)(y_1 - y_2) > 0$. Hence, the probability of concordance is

$$\kappa = P\left(\left[X_1 - X_2\right]\left[Y_1 - Y_2\right] > 0\right).$$

We say that (x_1, y_1) and (x_2, y_2) are *discordant* if and only if $(x_1 - x_2)(y_1 - y_2) < 0$. Because we have assumed that the (X_i, Y_i) are continuous, $P([X_1 - X_2][Y_1 - Y_2] = 0) = 0$ and the probability of discordance is

$$P\left(\left[X_1 - X_2\right]\left[Y_1 - Y_2\right] < 0\right) = 1 - \kappa.$$

By convention, correlation coefficients range from -1 (perfect negative correlation) to $+1$ (perfect positive correlation), with 0 corresponding to a complete lack of correlation. To transform κ from a probability to a correlation coefficient, we first subtract 1/2 to re-center the absence of monotonic association, then multiply by 2 to double the range. This results in

$$\tau = 2\left(\kappa - \frac{1}{2}\right) = 2\kappa - 1 = \kappa - (1 - \kappa),$$

the probability of concordance minus the probability of discordance, a correlation coefficient known as *Kendall's tau* (τ). Of course, the easiest way to interpret a value of τ is to compute $(\tau+1)/2$, the probability of concordance.

Having observed a bivariate random sample, $(x_1, y_1), \ldots, (x_n, y_n)$, how can we estimate κ (or τ)? If the sample contains n distinct values of x_i and n distinct values of y_i, then each comparison of an (x_i, y_i) and an (x_j, y_j)

is either concordant or discordant. There are $\binom{n}{2}$ such comparisons, so we estimate the probability of concordance to be the proportion of concordant comparisons in the sample:

$$\hat{\kappa} = \# \left\{ \text{concordant comparisons} \right\} / \binom{n}{2}.$$

Computing $\hat{\kappa}$ in the case of distinct values is easy if one begins by ordering the (x_i, y_i) pairs.

Example 14.8 Suppose that we observe

$$X = \begin{bmatrix} 1.0 & 3.1 \\ 6.1 & 5.4 \\ 3.8 & -2.8 \\ -7.1 & -5.1 \\ -1.9 & 0.8 \end{bmatrix}.$$

First, we reorder the data so that the x-values are increasing:

$$X' = \begin{bmatrix} -7.1 & -5.1 \\ -1.9 & 0.8 \\ 1.0 & 3.1 \\ 3.8 & -2.8 \\ 6.1 & 5.4 \end{bmatrix}.$$

For $(x, y) = (-7.1, -5.1)$, there are four comparisons to consider. Because $-5.1 < 0.8, 3.1, -2.8, 5.4$, each of these comparisons is concordant. For $(x, y) = (-1.9, 0.8)$, there are three comparisons to consider. (We have already compared $(-1.9, 0.8)$ and $(-7.1, -5.1)$.) Because $0.8 < 3.1, 5.4$, but $0.8 > -2.8$, two of these comparisons are concordant. For $(x, y) = (1.0, 3.1)$, there are two comparisons, one of which is concordant. For $(x, y) = (3.8, -2.8)$, there is one comparison. Because $-2.8 < 5.4$, it is concordant. Collecting these results, $\hat{\kappa} = (4 + 2 + 1 + 1)/\binom{5}{2} = 8/10 = 0.8$ and $\hat{\tau} = 2\hat{\kappa} - 1 = 0.6$.

I wrote an R function, `kappa.distinct`, that performs the calculations illustrated in Example 14.8. This function is described in Appendix R and can be obtained from the web page for this book. However, `kappa.distinct` is not intended for general use, as it assumes that both the observed x-values and the observed y-values are distinct.

What if the observed x-values and/or the observed y-values are not distinct? Our probability model says that ties occur with zero probability, yet ties do occur in experimental data. Suppose, for example, that $x_1 < x_2$ but that $y_1 = y_2$. In this case, comparison of (x_1, y_1) and (x_2, y_2) provides no information about the probability of concordance. One way to proceed is to count each uninformative comparison as half concordant and half discordant, resulting in

$$\hat{\kappa} = \left(\# \{\text{concordant comparisons}\} + \frac{1}{2} \{\text{uninformative comparisons}\} \right) \Big/ \binom{n}{2}.$$

As before, $\hat{\tau} = 2\hat{\kappa} - 1$.

I wrote an `R` function, `kappa`, that computes $\hat{\kappa}$ in the presence of ties. This function is described in Appendix R and can be obtained from the web page for this book.

Having derived a point estimator of τ, it is natural to inquire how one might test hypotheses about or construct confidence intervals for τ. If one could test $H_0 : \tau = \tau_0$ versus $H_0 : \tau \neq \tau_0$ at significance level α, then one could construct a $(1-\alpha)$-level confidence interval for τ consisting of all τ_0 for which $H_0 : \tau = \tau_0$ is not rejected. Unfortunately, no such test is readily available. However, one can test the important special case of $H_0 : \tau = 0$, the null hypothesis that X and Y are not monotonically associated.

Somewhat remarkably, when $\tau = 0$ the distribution of $\hat{\tau}$ does not depend on any attributes of the bivariate distribution of (X, Y) other than τ. Hence, there is a distribution-free test of $H_0 : \tau = 0$, if one can compute (or approximate) the significance probability.

If $\tau = 0$ and no ties are present, then the exact distribution of τ is known and exact significance probabilities can be computed. The calculations become more complicated as n increases, so exact significance probabilities are rarely computed for large sample sizes, say $n > 50$. If $\tau = 0$ and n is large, then the exact distribution of τ can be approximated by a normal distribution with zero mean. If no ties are present, then the variance of the approximating normal distribution is $(4n + 10)/[9(n^2 - n)]$; if ties are present, then this formula must be modified.

Instead of computing exact significance probabilities and/or normal approximations of them, we will estimate significance probabilities by simulation. Under $H_0 : \tau = 0$, we can generate a random value of $\hat{\tau}$ (or, equivalently, of $\hat{\kappa}$) by computing the monotonic association between $\{1, 2, \ldots, n\}$ and a random permutation of $\{1, 2, \ldots, n\}$. If we generate a large number of random values of $\hat{\kappa}$, then the proportion of these values that lie at least as far from 0.5 as the observed value of $\hat{\kappa}$ estimates the significance probability. I wrote an R function `kappa/p/sim`, that does precisely that. It is described in Appendix R and can be obtained from the web page for this book.

x	y
24.90	33.41
38.01	26.77
27.82	29.94
64.55	73.04
82.21	74.95
71.99	70.45
46.51	49.23
41.99	67.21
35.38	49.36
−4.14	28.77
79.04	63.24
54.97	30.51
28.89	23.50
3.62	40.09
73.15	82.68

Table 14.2: The bivariate sample used in Example 14.9.

Example 14.9 The bivariate data in Table 14.2 were drawn for the purpose of testing the null hypothesis of no monotonic association at significance level $\alpha = 0.05$: The estimated probability of concordance, $\hat{\kappa} \doteq 0.7619048$, and an approximate significance probability, $\mathbf{p} = 0.00605$, were obtained as follows:

```
> kappa(X)
[1] 0.7619048
> kappa.p.sim(X,20000)
[1] 0.00605
```

Because $\mathbf{p} < \alpha$, we reject the null hypothesis of no monotonic association.

14.4 Explaining Association

When we infer that two phenomena are associated, we invariably proceed to inquire why that association exists. Constructing causal explanations is ultimately the province of science, not statistics, but statisticians can help scientists think clearly about association.

We begin with a famous example. For a population of father-son pairs, let X measure the adult height of the father and let Y measure the adult height of the son. The positive association between X and Y is relatively easy to explain. On the one hand, the value of X is determined years before the value of Y. Thus, it is hard to conceive of a mechanism through which the value of Y might influence the value of X. On the other hand, fathers contribute genetic material to their children. Thus, there is an obvious mechanism through which the value of X influences the value of Y.

Notice that we say "X influences Y," not "X causes Y." The association between X and Y is imperfect: a son's height is affected not just by his father's genes, but also by his mother's genes and a host of environmental factors. The value of X directly influences the value of Y because a father is partly responsible for his son's genes, but there are other influences that are not related to X. Furthermore, there may be other ways in which X indirectly influences Y. For example, if tall men tend to marry tall women, then X also influences Y by influencing what genes the mother may possess.

Now we change the example. Instead of father-son pairs, consider sister-brother pairs. Let X measure the adult height of the sister and let Y measure the adult height of the brother. Again, there is a positive association between X and Y, but not because either variable directly influences the other. Siblings share the same parents, and tall parents tend to have tall children. This pattern tends to induce a second pattern, a tendency for the siblings of tall individuals to be tall themselves.

When two variables are associated, but neither directly influences the other, the induced pattern of association is sometimes described as spurious. Unfortunately, the popular phrase "spurious correlation" is potentially misleading. If X and Y are truly associated, then science demands an explanation. The explanation may be obvious, as for the positive association between sister and brother heights, or it may be quite subtle, but a pattern of association should not be dismissed as an illusion simply because the explanation of that pattern is difficult to discern.

It is absolutely crucial to appreciate that association reveals statistical patterns, nothing more. Sometimes the patterns revealed may seem counterintuitive. To illustrate, suppose that we sample students who take Math

351 (Applied Statistics) at the College of William & Mary. For each student, we measure the number of hours in the semester that s/he studied statistics (X) and the semester average that s/he achieved in Math 351 (Y). To our considerable surprise, we observe a negative association between X and Y. How can this be? Imagine the headline in the student newspaper: "Studying Causes Bad Grades!" Imagine an editorial that exhorts students to improve their GPAs by studying less!

Of course we know perfectly well that, within reason, studying improves academic performance. A student who increases the time that she spends studying will likely improve her grade. Why, then, a pattern of negative association between X and Y? To answer this question, we need to understand what influences the number of hours that students study statistics and what influences how well they perform in Math 351. The overwhelming majority of students who take Math 351 are highly motivated. If they struggle with statistics, then they tend to study more. Thus, talented students who master material quickly tend to study less and perform better than students who struggle, resulting in a statistical pattern of negative correlation.[1]

14.5 Case Study: Anorexia Treatments Revisited

In Section 12.5, we performed an analysis of variance on weight gains by anorexic girls under three treatments. The results verified the impressions, conveyed by Figure 12.1, that mean weight gain varies by treatment. In particular, the family therapy treatment was found to be more effective than the standard treatment. It must be emphasized, however, that this is a statement about an *average* effect. In the present section, we explore the effects of treating anorexia in greater depth.

The weight gains in question were computed by subtracting the weights before treatment from the weights after treatment. It is instructive to examine the original (before,after) pairs reported in Table 12.1. Figure 14.5 displays two scatter diagrams constructed from the (before,after) pairs for girls who received the standard treatment. In each scatter diagram, weight before treatment is plotted on the horizontal axis. On the left, weight after treatment is plotted on the vertical axis; on the right, weight gain is plotted on the vertical axis.

The second scatter diagram in Figure 14.5 reveals a striking pattern. On average, girls who received the standard treatment did not gain weight.

[1] Although this example is hypothetical, it was based on a study at Penn State University that found that B students tended to study more than A students.

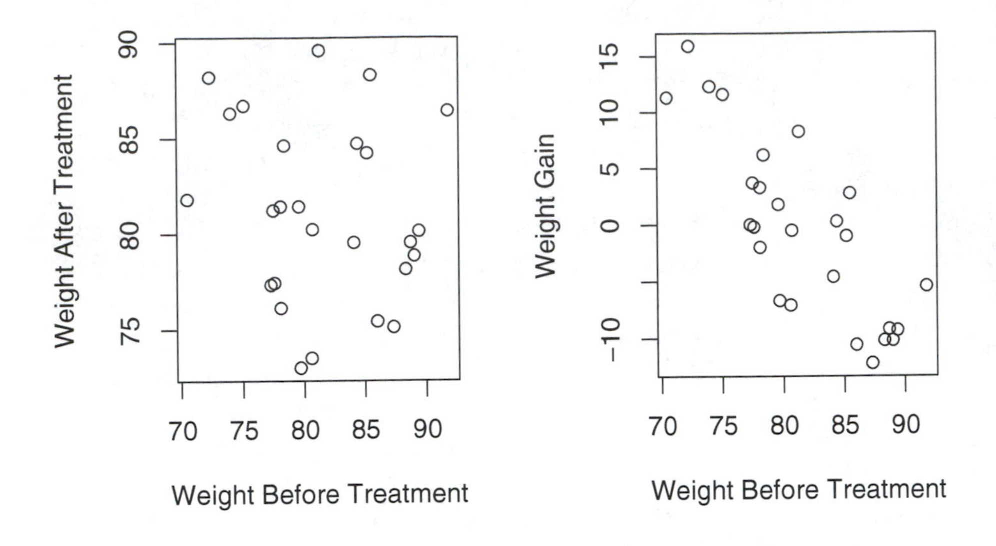

Figure 14.5: Two scatter diagrams for anorexic girls who received the standard treatment. The raw data are reported in Table 12.1.

The sample mean of the observed weight gains is actually negative, $\bar{x}_2 = -0.45$. However, each of the 4 girls who weighed the least before treatment gained weight during treatment, whereas each of the 7 girls who weighed the most before treatment lost weight during treatment. This pattern can be quantified in various ways. Pearson's product-moment correlation coefficient (for weight gain versus weight before treatment) is $r \doteq -0.81$ and the sample probability of concordance is $\hat{\kappa} \doteq 0.22$. Thus, there appears to be a strong negative association between weight before treatment and weight gain during treatment.

Absent further information, we cannot explain *why* weight before treatment and weight gain during treatment are negatively associated. We might *speculate* that the girls who weighed the least before treatment are the girls who were most severely anorexic, and that the girls who weighed the most before treatment are the girls who were least severely anorexic. However, absent information about each girl's age, height, body type, etc., such speculation is entirely fanciful. Furthermore, even if it were true, what might we then conclude? It *might* be that the standard treatment is effective for severely anorexic girls and counterproductive for mildly anorexic girls. Or, it *might* be that each girl is progressing through a cycle on which the standard treatment has no effect. Did the girls who were mildly anorexic at

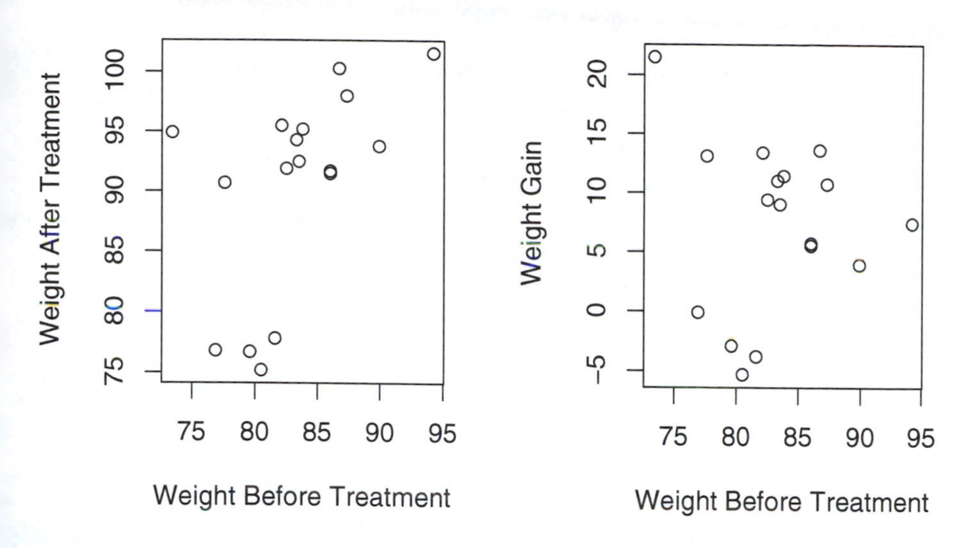

Figure 14.6: Two scatter diagrams for anorexic girls who received the family therapy treatment. The raw data are reported in Table 12.1.

the beginning of the study inevitably continue to decline, while the girls who were severely anorexic inevitably hit bottom and rebounded? It is not possible to construct a scientific explanation from the information provided.

Let us now turn from the standard treatment to the family therapy treatment, which analysis of variance revealed (Section 12.5) to be superior (on average) to the standard treatment. Figure 14.6 displays two scatter diagrams constructed from the (before,after) pairs for girls who received the standard treatment. As in Figure 14.5, weight before treatment is plotted on the horizontal axis. On the left, weight after treatment is plotted on the vertical axis; on the right, weight gain is plotted on the vertical axis.

The scatter diagrams in Figure 14.6 reveal another striking pattern. After treatment, 4 girls weighed less than 80. The rest weighed more than 90. These 4 girls are the only girls who did not gain weight during family therapy treatment. Before treatment they were among the lighter girls assigned to family therapy treatment, but other girls of comparable initial weight gained weight during family therapy treatment. What is interesting about this pattern is that it does *not* resemble chance variation. Rather, there appear to be two groups of anorexic girls: those who gained weight during family therapy treatment and those who did not. Without additional information, we cannot attempt to explain this phenomenon and identify promising candidates for this type of treatment.

Finally, let us drop the 4 girls who did not gain weight and focus on the 13 who did. We see the same pattern of negative association that we observed for the standard treatment. For family therapy treatment, Pearson's product-moment correlation coefficient (for weight gain versus weight before treatment) is $r \doteq -0.76$ and the sample probability of concordance is $\hat{\kappa} \doteq 0.26$. Again, we cannot explain this pattern without more information; however, the fact that both treatments exhibit the same pattern should cause us to prefer explanations that do not depend on the type of treatment.

14.6 Exercises

1. For each of the situations described below, (i) state whether you would expect to observe a positive or a negative pattern of association, and (ii) provide an explanation of the proposed pattern.

   (a) Consider a sample of houses. Let x denote the size of the house (in square feet) and let y denote the selling price of the house (in thousands of dollars).

   (b) Consider a sample of men at least 25 years of age who competed in a 5 km race in June 2008. Let x denote the runner's age (in years) and let y denote the runner's time (in minutes).

   (c) Consider a sample of men at least 25 years of age who competed in a 5 km race in June 2008. Let x denote the runner's income (in thousands of dollars) and let y denote the runner's time (in minutes).

   (d) Consider a sample of Ohio Division I high school football programs for the year 2008. Let x denote the head coach's salary (in thousands of dollars) and let y denote the team's final OHSAA points average (higher numbers are better).

   (e) Consider a sample of boys who are attending an Indiana public elementary school. Let x denote the weight of the student (in pounds) and let y denote how fast the student can run 50 yards (in seconds).

2. After returning home from school, Quinn and Nick like to play World of Warcraft (WoW), an extremely popular massively multiplayer online game. They usually play for about two hours, until their mother, Dana, returns from work. Concerned that her boys are not spending

enough time on their science homework, Dana surreptitiously monitors their activities. For each of 40 school days, she measures x, the number of minutes that Quinn and Nick play WoW, and y, the number of minutes that Quinn and Nick spend doing science homework. To her considerable consternation, she discovers a negative association between x and y.

(a) Confronted by their mother, Quinn and Nick argue that small values of y are associated with large values of x because they intentionally play more WoW on the days that they are assigned less science homework. What do you think of this explanation? Can you suggest another?

(b) Knowing their mother's reverence for science, Quinn and Nick also argue that playing WoW is training them to become scientists, citing a study that playing WoW fosters scientific habits of mind.[2] They suggest that, in the interest of their scientific education, Dana allow them to play more WoW and do less science homework. What do you think of this suggestion?

(c) Suppose the existence of an unobservable third variable, z, that measures the daily increase in Quinn's and Nick's scientific prowess. No one knows how to measure z, but suppose that z is positively associated with y and negatively associated with x. Does the negative association between x and z contradict the study cited by Quinn and Nick? Why or why not?

3. Consider the bivariate data displayed in Table 14.3.

(a) Construct a normal probability plot of the x values. Do the x values appear to have been drawn from a normal distribution? Why or why not?

(b) Construct a normal probability plot of the y values. Do the y values appear to have been drawn from a normal distribution? Why or why not?

(c) Construct a scatter diagram of the (x, y) values. Do the (x, y) values appear to have been drawn from a bivariate normal distribution? Why or why not?

---

[2]C. Steinkuehler and S. Duncan. Scientific habits of mind in virtual worlds. To appear in *Journal of Science Education & Technology*. A 2008 preprint is available on Constance Steinkuhler's web page. The authors found that 86 percent of a random sample of 1984 posts to class related forums were engaged in social knowledge construction rather than social banter.

(d) Suggest an explanation for the phenomena observed in (a)–(c). Is this a paradox? How do you think that these (x, y) pairs were obtained?

x	y	x	y	x	y	x	y
4.813	5.505	3.208	3.235	2.034	1.575	3.804	4.078
3.449	3.576	4.025	4.391	4.189	4.623	1.578	0.930
2.558	2.316	3.302	3.368	3.493	3.639	3.848	4.140
1.657	1.042	2.017	1.551	3.268	3.320	3.211	3.240
3.988	4.339	3.394	3.498	3.985	4.334	2.925	2.835
3.250	2.988	4.482	1.246	2.492	4.060	2.549	3.980
3.568	2.538	2.583	3.931	3.006	3.333	3.598	2.495
3.248	2.991	3.271	2.959	3.613	2.475	3.985	1.949
2.921	3.453	2.107	4.604	3.203	3.055	1.843	4.978
3.116	3.178	3.551	2.563	3.415	2.756	2.651	3.836

Table 14.3: An unusual bivariate data set.

4. Consider the before and after weights for cognitive behavioral therapy displayed in Table 12.1. Let b_i denote the weight of girl i before treatment, let a_i denote the weight of girl i after treatment, and let $d_i = a_i - b_i$.

   (a) Create a scatter diagram of the (b_i, a_i) pairs.

   (b) Create a scatter diagram of the (b_i, d_i) pairs.

   (c) Comment on any patterns you discern in the preceding scatter diagrams.

   (d) Do the (b_i, a_i) pairs appear to have been drawn from a bivariate normal distribution? Why or why not?

5. Consider the test score data reported in Example 14.2.

   (a) Quantify the association between midterm test scores by computing Pearson's product-moment correlation coefficient. Is the association positive or negative?

   (b) Examining the scatter diagram displayed in Figure 14.2, one student appears to be an outlier. Omitting the corresponding row of the data matrix, recompute Pearson's product-moment correlation coefficient. How does the outlier affect the value of r? Hint: If `Data` is a complete data matrix, then `Data[-17,]` is the same data matrix without row 17.

6. In Section 14.5 we measured the strength of association between weight before treatment (B_i) and weight gained during treatment (D_i). For the standard treatment, we obtained $r \doteq -0.81$.

   Now assume that (B_i, D_i) has a bivariate normal distribution with population correlation coefficient ρ.

   (a) Construct a confidence interval for ρ that has a confidence coefficient of approximately 0.95.

   (b) Using a significance level of $\alpha = 0.05$, can we reject $H_0 : \rho = 0$ in favor of $H_1 : \rho \neq 0$?

7. K. Pearson and A. Lee analyzed the heights (in inches) of $n = 11$ pairs of siblings.[3] The heights are displayed in Table 14.4. Assuming that the pairs of heights were drawn from a bivariate normal population, construct a confidence interval for ρ that has a confidence level of approximately 0.90.

sister	brother
69	71
64	68
65	66
63	67
65	70
62	71
65	70
64	73
66	72
59	65
62	66

Table 14.4: Heights of $n = 11$ pairs of siblings.

[3] K. Pearson and A. Lee (1902–3). On the laws of inheritance in man. *Biometrika*, 2:357. These data appear as Data Set 373 in *A Handbook of Small Data Sets*.

8. Let $\alpha = 0.05$.

   (a) Suppose that we sample from a bivariate normal distribution with $\rho = 0.5$. Assuming that we observe $r = 0.5$, how large a sample will be needed to reject $H_0 : \rho = 0$ in favor of $H_0 : \rho \neq 0$?

   (b) Suppose that we sample from a bivariate normal distribution with $\rho = 0.1$. Assuming that we observe $r = 0.1$, how large a sample will be needed to reject $H_0 : \rho = 0$ in favor of $H_0 : \rho \neq 0$?

9. Table 14.5 displays the heights (cm) and resting pulse measurements (beats per minute) for a sample of $n = 50$ hospital patients.[4]

   (a) Use `binorm.scatter` to create a scatter diagram of these data. Does it seem plausible that this sample was drawn from a bivariate normal distribution? Why or why not?

   (b) Use `binorm.estimate` to compute r. Use `kappa` to compute $\hat{\kappa}$, then compute $\hat{\tau}$.

   (c) Assuming that the sample was drawn from a bivariate normal distribution, test $H_0 : \rho = 0$ versus $H_1 : \rho \neq 0$ at a significance level of $\alpha = 0.05$. What is **p**? Can H_0 be rejected?

   (d) Use `kappa.p.sim` to test $H_0 : \tau = 0$ versus $H_1 : \tau \neq 0$ at a significance level of $\alpha = 0.05$. What is **p**? Can H_0 be rejected?

x	y	x	y	x	y	x	y	x	y
160	68	172	116	167	80	185	80	162	84
163	95	175	80	177	80	185	80	165	76
162	80	182	100	173	92	162	88	167	92
172	90	170	80	177	90	170	80	168	90
163	80	178	80	158	80	182	76	157	80
167	80	160	78	170	84	170	90	160	80
177	80	182	80	166	72	168	80	170	80
155	80	148	82	175	104	175	76	168	80
160	84	180	68	153	70	175	84	185	80
145	64	165	82	170	84	165	84	175	72

Table 14.5: Height (x) and resting pulse (y) of $n = 50$ hospital patients.

[4]Data Set 416 in *A Handbook of Small Data Sets*.

10. Table 14.6 displays the final examination scores (out of a maximum possible score of 75) and completion times (seconds) for 134 students.[5]

    (a) Let x denote the completion time in minutes and let y denote the score. Use `binorm.scatter` to create a scatter diagram of the (x, y) pairs. Does it seem plausible that these data were drawn from a bivariate normal distribution? Why or why not?

    (b) Do these data provide compelling evidence of a monotonic association between completion time and score? Perform an appropriate test and report a significance probability.

    (c) Notice that 9 students completed the exam in ≤ 20 minutes. Their scores ranged from 40 to 72. Suggest a reason for so much variation.

    (d) Notice that 11 students scored ≤ 40 points. Their completion times ranged from 20 minutes to 40 minutes and 35 seconds. Suggest a reason why these students decided to stop working so early and turn in their exam papers.

    (e) Do you believe that how long you work on the final exam in this course will affect your final exam score? Why or why not?

[5] I. Basak, W. R. Balch, and P. Basak (1992). Skewness: asymptotic critical values for a test related to Pearson's measure. *Journal of Applied Statistics*, 19:479–487. These data appear as Data Set 272 in *A Handbook of Small Data Sets*.

$60x$	y	$60x$	y	$60x$	y	$60x$	y	$60x$	y
2860	49	1278	56	1910	60	2105	66	1604	46
2063	49	1677	41	2730	53	1496	42	1475	62
2013	70	1945	40	2235	51	1301	67	1106	68
2000	55	1754	42	1993	51	2467	48	2040	58
1420	52	1200	40	1613	60	1265	56	1594	47
1934	55	1307	51	1532	64	3813	47	1215	66
1519	61	1895	53	2339	66	1216	68	1418	61
2735	65	1798	62	2109	52	1167	58	1828	58
2329	57	1375	61	1649	45	1767	59	2305	45
1590	71	2665	49	2238	48	1683	45	1902	55
1699	49	1743	54	1733	51	1648	31	2013	54
1816	48	1722	57	1981	73	1144	47	2026	54
1824	49	2562	71	1440	63	1162	56	1875	54
1899	69	2277	45	1482	32	1460	38	2227	41
1714	44	1579	70	1758	59	1726	47	2325	65
1741	53	1785	58	2540	68	1862	65	1674	66
1968	49	1068	62	1637	35	3284	61	2435	38
1721	52	1411	28	1779	64	1683	45	2715	51
2120	53	1162	72	1069	62	1654	63	1773	49
1435	36	1646	37	1929	51	2725	66	1656	49
1909	61	1489	67	2605	52	1992	44	2320	51
1707	68	1769	51	1491	44	1332	57	1908	42
1431	67	1550	55	1321	64	1840	56	1853	61
2024	53	1313	68	1326	65	1704	56	1302	69
1725	33	2472	58	1797	56	1510	54	2161	42
1634	64	2036	61	1158	52	3000	61	1715	53
1949	57	1914	43	1595	59	1758	58		

Table 14.6: Examination completion time in seconds ($60x$) and score (y) for $n = 134$ students.

Chapter 15

Simple Linear Regression

This chapter continues our study of the relationship between two random variables, X and Y. In Chapter 14, we quantified association by measures of correlation, e.g., Pearson's product-moment correlation coefficient. Another way to quantify the association between X and Y is to quantify the extent to which knowledge of X allows one to predict values of Y. Notice that this approach to association is asymmetric: one variable (conventionally denoted X) is the *predictor variable* and the other variable (conventionally denoted Y) is the *response variable*.[1]

Given a value, $x \in \Re$, of the predictor random variable, we restrict attention to the experimental outcomes that can possibly result in this value:

$$S(x) = X^{-1}(x) = \{x \in S : X(s) = x\}.$$

Restricting Y to $S(x)$ results in a conditional random variable, $Y|X = x$. We write the expected value of this random variable as

$$\mu(x) = E(Y|X = x)$$

and note that $\mu(\cdot)$ varies with x. Thus, $\mu(\cdot)$ is a function that assigns mean values of various conditional random variables to values of x. This function is the conditional mean function, also called the *prediction function* or the *regression function*. Given $X = x$, the predicted value of Y is $\hat{y}(x) = \mu(x)$.

The nature of the regression function reveals much about the relation between X and Y. For example, if larger values of x result in larger values of $\mu(x)$, then there is some kind of positive association between X and

[1] The predictor variable is often called the *independent variable* and the response variable is often called the *dependent variable*. We will eschew this terminology, as it has nothing to do with the probabilistic (in)dependence of events and random variables.

Y. Whether or not this kind of association has anything to do with the correlation of X and Y remains to be seen.

15.1 The Regression Line

Suppose that $Y \sim \text{Normal}(\mu_y, \sigma_y^2)$ and that we want to predict the outcome of an experiment in which we observe Y. If we know μ_y, then an obvious value of Y to predict is $EY = \mu_y$. The expected value of the squared error of this prediction is $E(Y - \mu_y)^2 = \text{Var}\, Y = \sigma_y^2$.

Now suppose that $X \sim \text{Normal}(\mu_x, \sigma_x^2)$ and that we observe $X = x$. Again we want to predict Y. Does knowing $X = x$ allow us to predict Y more accurately? The answer depends on the association between X and Y. If X and Y are independent, then knowing $X = x$ will not help us predict Y. If X and Y are dependent, then knowing $X = x$ should help us predict Y.

Example 15.1 Suppose that we want to predict the adult height to which a male baby will grow. Knowing only that adult male heights are normally distributed, we would predict the average height of this population. However, if we knew that the baby's father had attained a height of 6′-11″, then we surely would be inclined to revise our prediction and predict that the baby will grow to a greater-than-average height.

When X and Y are normally distributed, the key to predicting Y from $X = x$ is the following result.

Theorem 15.1 *Suppose that (X, Y) have a bivariate normal distribution with parameters $(\mu_x, \mu_y, \sigma_x^2, \sigma_y^2, \rho)$. Then the conditional distribution of Y given $X = x$ is*

$$Y|X = x \sim \text{Normal}\left(\mu_y + \rho\frac{\sigma_y}{\sigma_x}(x - \mu_x), \left(1 - \rho^2\right)\sigma_y^2\right).$$

Because $Y|X = x$ is normally distributed, the obvious value of Y to predict when $X = x$ is

$$\hat{y}(x) = E(Y|X = x) = \mu_y + \rho\frac{\sigma_y}{\sigma_x}(x - \mu_x). \tag{15.1}$$

Interpreting (15.1) as a function that assigns a predicted value of Y to each value of x, we see that the prediction function (15.1) corresponds to a line that passes through the point (μ_x, μ_y) with slope $\rho\sigma_y/\sigma_x$. The prediction function (15.1) is the *population regression function* and the corresponding line is the *population regression line.*

The expected squared error of the prediction (15.1) is

$$\mathrm{Var}(Y|X = x) = (1 - \rho^2)\sigma_y^2.$$

Notice that this quantity does not depend on the value of x. If X and Y are strongly correlated, then $\rho \approx \pm 1$, $(1-\rho^2)\sigma_y^2 \approx 0$, and prediction is extremely accurate. If X and Y are uncorrelated, then $\rho = 0$, $(1 - \rho^2)\sigma_y^2 = \sigma_y^2$, and the accuracy of prediction is not improved by knowing $X = x$. These remarks suggest a natural way of interpreting what ρ actually measures: the proportion by which the expected squared error of prediction is reduced by virtue of knowing $X = x$ is

$$\frac{\sigma_y^2 - (1 - \rho^2)\,\sigma_y^2}{\sigma_y^2} = \rho^2,$$

the *population coefficient of determination.* Statisticians often express this interpretation by saying that ρ^2 is "the proportion of variation explained by linear regression." Of course, as we emphasized in Section 14.4, this is not an explanation in the sense of articulating a causal mechanism.

Example 15.2 Suppose that $(\mu_x, \mu_y, \sigma_x^2, \sigma_y^2, \rho) = (10, 20, 2^2, 4^2, 0.5)$. Then

$$\hat{y}(x) = 20 + 0.5 \cdot \frac{4}{2}(x - 10) = x + 10$$

and $\rho^2 = 0.25$.

Rewriting (15.1), the equation for the population regression line, as

$$\frac{\hat{y}(x) - \mu_y}{\sigma_y} = \rho\,\frac{x - \mu_x}{\sigma_x},$$

we discern an important fact:

Corollary 15.1 *Suppose that (x, y) lies on the population regression line. If x lies z standard deviations above μ_x, then y lies ρz standard deviations above μ_y.*

Example 15.2 (continued) The value $x = 12$ lies $(12 - 10)/2 = 1$ standard deviations above the X-population mean, $\mu_x = 10$. The predicted y value that corresponds to $x = 12$, $\hat{y}(12) = 12 + 10 = 22$, lies $(22 - 20)/4 = 0.5$ standard deviations above the Y-population mean, $\mu_y = 20$.

Example 15.2 (continued) The 0.90 quantile of X is

$$x = \texttt{qnorm(.9,mean=10,sd=2)} = 12.5631.$$

The predicted y-value that corresponds to $x = 12.5631$ is $\hat{y}(12.5631) = 22.5631$. At what quantile of Y does the predicted y-value lie? The answer is

$$P\left(Y \leq \hat{y}(x)\right) = \texttt{pnorm(22.5631,mean=20,sd=4)} = 0.7391658.$$

At first, most students find the preceding example counterintuitive. If x lies at the 0.90 quantile of X, then should we not predict $\hat{y}(x)$ to lie at the 0.90 quantile of Y? This is a natural first impression, but one that must be dispelled. We begin by considering two familiar situations:

1. Consider the case of a young boy whose father is extremely tall, at the 0.995 quantile of adult male heights. We surely would predict that the boy will grow to be quite tall. But precisely how tall? A father's height does not completely determine his son's height. Height is also affected by myriad other factors, considered here as chance variation. Statistically speaking, it's more likely that the boy will grow to an adult height slightly shorter than his extremely tall father than that he will grow to be even taller.

2. Consider the case of two college freshmen, William and Mary, who are enrolled in an introductory chemistry class of 250 students. On the first midterm examination, Mary attains the 5th highest score and William obtains the 245th highest (5th lowest) score. How should we predict their respective performances on the second midterm examination? There is undoubtedly a strong, positive correlation between scores on the two tests. We surely will predict that Mary will do quite well on the second test and that William will do rather badly. But how well and how badly? One test score does not completely determine another—if it did, then computing semester grades would be easy! Mary can't do much better on the second test than she did on the first, but she might easily do worse. Statistically speaking, it's likely that she'll rank slightly below 5th on the second test. Likewise, William can't do much worse on the second test than he did on the first. Statistically speaking, it's likely that he'll rank slightly above 245th on the second test.

The phenomenon that we have just described, that experimental units with extreme X quantiles will tend to have less extreme Y quantiles, is

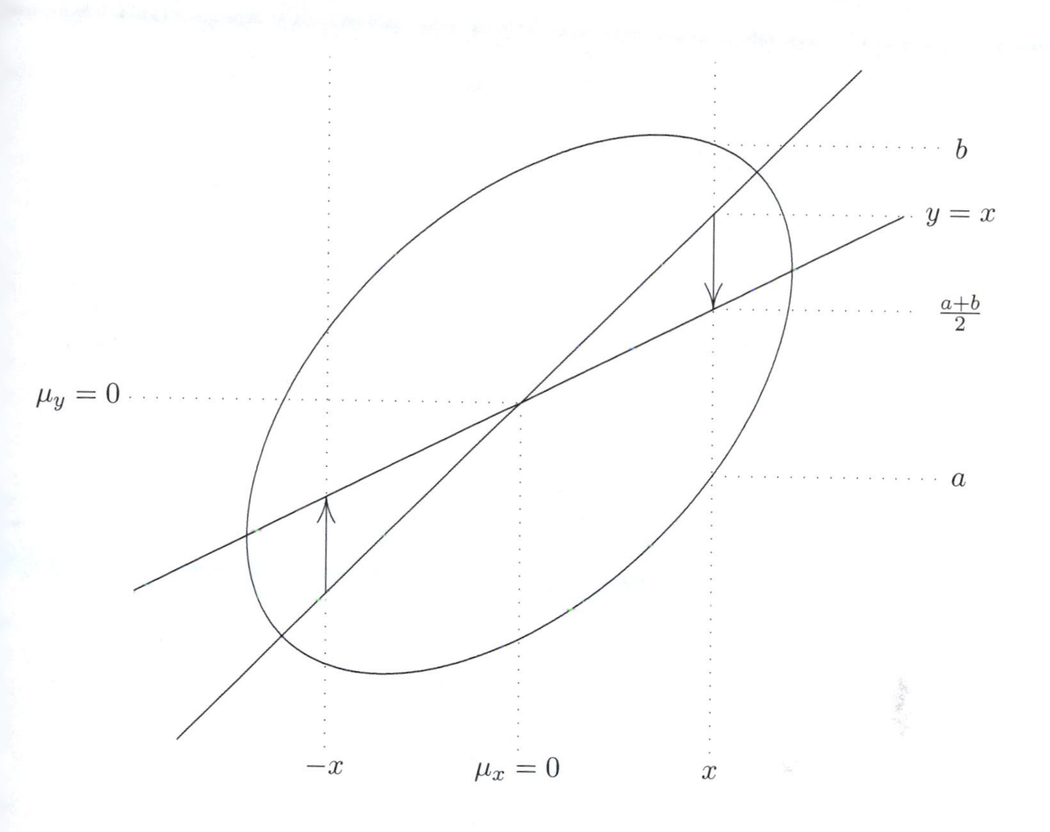

Figure 15.1: The regression effect.

purely statistical. It was first discerned by Sir Francis Galton, who called it "regression to mediocrity." Modern statisticians call it *regression to the mean*, or simply *the regression effect.*

Having refined our intuition, we can now explain the regression effect by examining the population concentration ellipse in Figure 15.1. For simplicity, we assume that X and Y have been converted to standard units. The bivariate normal population represented in Figure 15.1 has population parameters $\mu_x = \mu_y = 0$, $\sigma_x^2 = \sigma_y^2 = 1$, and $\rho = 0.5$. Recall that the line that coincides with the major axis of the ellipse is called the first principal component. In Figure 15.1, the first principal component is the line $y = x$ and the regression line is the line $y = x/2$. Both lines pass through the point $(\mu_x, \mu_y) = (0, 0)$, but their slopes differ by a factor of $|\rho| = 0.5$.

Let us explore the implications of the fact that, if $|\rho| < 1$, then *the regression line does not coincide with the major axis of the concentration ellipse.* Given $X = x$, it might seem tempting to predict $Y = x$. But this would be a mistake! Here, $x > \mu_x$ and clearly

$$P(Y > x|X = x) < \frac{1}{2},$$

so $\hat{y}(x) = x$ overpredicts $Y|X = x$. Similarly, $\hat{y}(-x) = -x$ underpredicts $Y|X = -x$.

The population regression line is the line of conditional expected values, $y = E(Y|X = x)$. Let (x, a) and (x, b) denote the lower and upper points at which the vertical line $X = x$ intersects the population concentration ellipse. As one might guess, it turns out that

$$\hat{y}(x) = E(Y|X = x) = \frac{a+b}{2}.$$

However, the midpoint of the vertical line segment that connects (x, a) and (x, b) is *not* (x, x). The discrepancy between using the first principal component to predict $\hat{y}(x) = x$ and using the regression line to predict $\hat{y}(x) = (a + b)/2$, indicated by an arrow in Figure 15.1, is the regression effect.

The correlation coefficient ρ mediates the strength of the regression effect. If $\rho = \pm 1$, then

$$\frac{Y - \mu_x}{\sigma_y} = \pm \frac{X - \mu_x}{\sigma_x}$$

and Y is completely determined by X. In this case there is no regression effect: if x lies z standard deviations above μ_x, then we know that y lies z standard deviations above μ_y. At the other extreme, if $\rho = 0$, then knowing $X = x$ does not reduce the expected squared error of prediction at all. In this case, we regress all the way to the mean: regardless of where x lies, we predict $\hat{y} = \mu_y$.

Thus far, we have focused on predicting Y from $X = x$ in the case that the population concentration ellipse is known. We have done so in order to emphasize that the regression effect is an inherent property of prediction, not a statistical anomaly caused by chance variation. In practice, however, the population concentration ellipse typically is not known and we must rely on the sample concentration ellipse, estimated from bivariate data. This means that we must substitute $(\bar{x}, \bar{y}, s_x^2, s_y^2, r)$ for $(\mu_x, \mu_y, \sigma_x^2, \sigma_y^2, \rho)$. The *sample regression function* is

$$\hat{y}(x) = \bar{y} + r\frac{s_y}{s_x}(x - \bar{x}) \qquad (15.2)$$

and the corresponding line is the *sample regression line*. Notice that the slope of the sample regression line does not depend on whether we use plug-in or unbiased estimates of the population variances. The variances affect the regression line through the (square root of) their ratio,

$$\frac{\widehat{\sigma_y^2}}{\widehat{\sigma_x^2}} = \frac{\frac{1}{n}\sum_{i=1}^n (y_i - \bar{y})^2}{\frac{1}{n}\sum_{i=1}^n (x_i - \bar{x})^2} = \frac{\frac{1}{n-1}\sum_{i=1}^n (y_i - \bar{y})^2}{\frac{1}{n-1}\sum_{i=1}^n (x_i - \bar{x})^2} = \frac{s_y^2}{s_x^2},$$

which is not affected by the choice of plug-in or unbiased.

Example 15.2 (continued) I used `binorm.sample` to draw a sample of $n = 100$ observations from a bivariate normal distribution with parameters

$$\texttt{pop} = \left(\mu_x, \mu_y, \sigma_x^2, \sigma_y^2, \rho\right) = \left(10, 20, 2^2, 4^2, 0.5\right).$$

I then used `binorm.estimate` to compute sample estimates of `pop`, obtaining

$$\begin{aligned} \texttt{est} &= \left(\bar{x}, \bar{y}, s_x^2, s_y^2, r\right) \\ &= (10.0006837, 19.3985929, 4.4512393, 14.1754248, 0.4707309). \end{aligned}$$

The resulting formula for the sample regression line is

$$\hat{y}(x) = \bar{y} + r\frac{s_y}{s_x}(x - \bar{x}) = \bar{y} + 1.784545\,(x - \bar{x}) = 1.55192 + 1.784545x.$$

It is not difficult to create an R function that plots a scatter diagram of the sample and overlays both the sample concentration ellipse and the sample regression line. The function `binorm.regress` is described in Appendix R and can be obtained from the web page for this book. The commands used in this example are as follows:

```
> pop <- c(10,20,4,16,.5)
> Data <- binorm.sample(pop,100)
> est <- binorm.estimate(Data)
> binorm.regress(Data)
```

The scatter diagram created by `binorm.regress` is displayed in Figure 15.2.

15.2 The Method of Least Squares

In Section 15.1 we derived the regression line from properties of bivariate normal distributions. Having derived it, we now note that the sample regression line can be computed from *any* set of $n \geq 2$ points $(x_i, y_i) \in \Re^2$ for

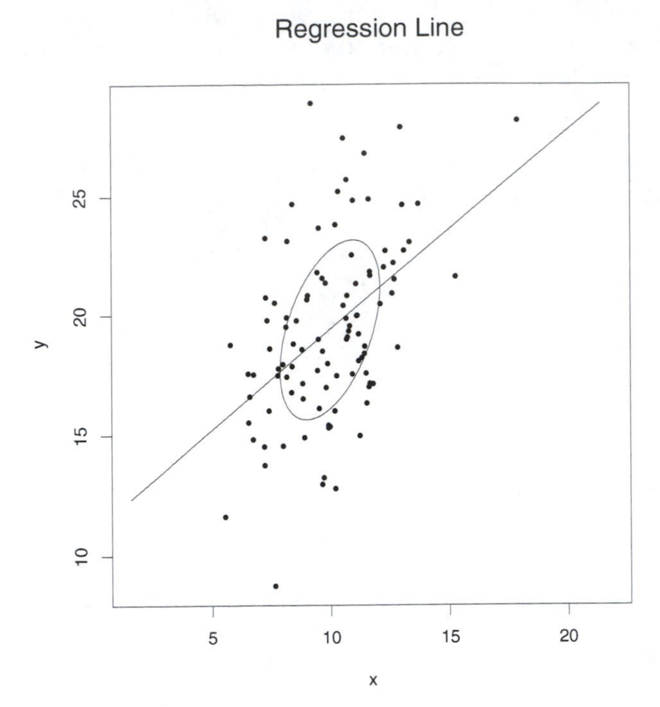

Figure 15.2: Scatter diagram, sample concentration ellipse, and sample regression line of $n = 100$ observations sampled from a bivariate normal distribution. Notice that the sample regression line is *not* the major axis of the sample concentration ellipse.

which the x_i assume more than one distinct value (and therefore $s_x > 0$). In this section, we derive the regression line in this more general setting.

Given points $(x_i, y_i) \in \Re^2$, $i = 1, \ldots, n$, we ask two conceptually distinct questions:

1. What line best *summarizes* the (x, y) pairs?

2. What line best *predicts* values of y from values of x?

We will answer each of these questions by applying the method of least squares. The possible lines are of the form $y = a + bx$. Given a candidate line, we measure the error between the line and each (x_i, y_i), then sum the squared errors from $i = 1, \ldots, n$. The best line is the one that minimizes

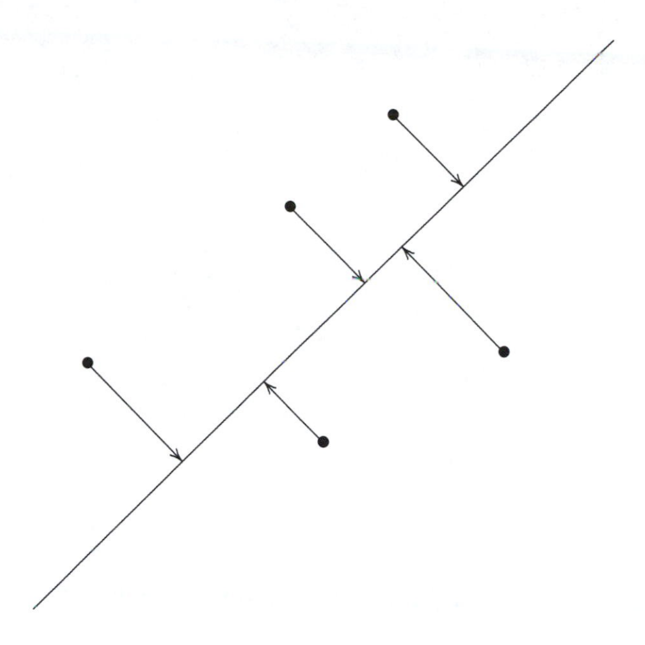

Figure 15.3: Perpendicular errors for summary.

this sum of squared errors:

$$\min_{a,b} \sum_{i=1}^{n} \left[\text{error} \left(\begin{array}{c} (x_i, y_i) \\ y = a + bx \end{array} \right) \right]^2 \tag{15.3}$$

The distinction between (1) summary and (2) prediction lies in how we define error.

To define the line that best summarizes the (x, y) pairs, it is natural to define the error between a point and a line as the Euclidean distance from the point to the line. This is found by measuring the length of the perpendicular line segment that connects them, as in Figure 15.3. Thus,

$$\begin{array}{c}\text{summary}\\ \text{error}\end{array} \left(\begin{array}{c} (x_i, y_i) \\ y = a + bx \end{array} \right) = \begin{array}{c}\text{perpendicular}\\ \text{distance}\end{array} \left(\begin{array}{c} (x_i, y_i) \\ y = a + bx \end{array} \right).$$

Using this definition of error, the solution of Problem 15.3 is the major axis of the sample concentration ellipse, the first principal component of the sample. We emphasize: *the first principal component is used for summary, not prediction.*

In contrast, to define the line that best predicts y values from x values, it is natural to define the error between a point (x_i, y_i) and a line $y = a + bx$

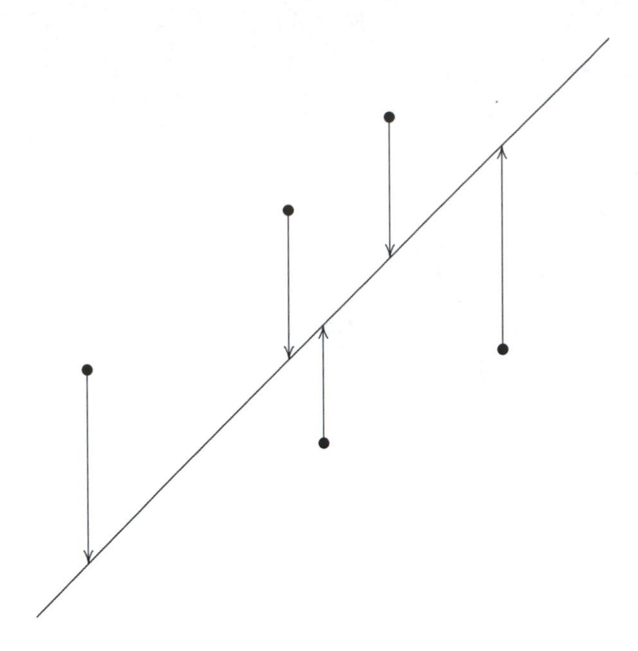

Figure 15.4: Vertical errors for prediction.

as the difference between the observed value $y = y_i$ and the predicted value

$$y = \hat{y}(x_i) = a + bx_i.$$

The difference $y_i - \hat{y}(x_i)$ is a *residual error* and the absolute difference $|y_i - \hat{y}(x_i)|$ is the length of the vertical line segment that connects (x_i, y_i) and $y = a + bx$, as in Figure 15.4. Using this definition of error, the solution of Problem 15.3 is the sample regression line. We emphasize: *the regression line is used for prediction, not summary.*

The remainder of this section provides a more detailed exposition of the squared error approach to prediction. Let

$$\text{SS}(a, b) = \sum_{i=1}^{n} (y_i - a - bx_i)^2,$$

the sum of the squared residual errors that result from the prediction function $\hat{y}(x) = a + bx$. The method of least squares chooses (a, b) to minimize $\text{SS}(a, b)$. Before analyzing this problem, we first consider an easier problem. If we knew $\{y_1, \ldots, y_n\}$ but not the corresponding $\{x_1, \ldots, x_n\}$, then it would be impossible to measure errors associated with prediction functions that involve x. In this situation we would be forced to restrict attention to prediction functions of the form $\hat{y} = a$, which corresponds to restricting attention to lines with zero slope. The method of least squares then chooses a to minimize

$$\sum_{i=1}^{n} (y_i - a)^2 = \mathrm{SS}(a, 0).$$

Theorem 15.2 *The value of a that minimizes* $\mathrm{SS}(a,0)$ *is* $a = \bar{y}$.

Proof We can conclude that $\mathrm{SS}(a,0)/n$ is minimal when $a = \bar{y}$ by applying part (2) of Theorem 6.1 to the empirical distribution of $\{y_1, \ldots, y_n\}$; however, it is instructive to verify this conclusion by direct calculation:

$$\begin{aligned}
\mathrm{SS}(a,0) &= \sum_{i=1}^{n} (y_i - a)^2 = \sum_{i=1}^{n} (y_i - \bar{y} + \bar{y} - a)^2 \\
&= \sum_{i=1}^{n} (y_i - \bar{y})^2 + \sum_{i=1}^{n} 2\,(y_i - \bar{y})\,(\bar{y} - a) + \sum_{i=1}^{n} (\bar{y} - a)^2 \\
&= (n-1)s_y^2 + 2\,(\bar{y} - a)\left[\sum_{i=1}^{n} y_i - n\bar{y}\right] + n\,(\bar{y} - a)^2 \\
&= (n-1)s_y^2 + n\,(\bar{y} - a)^2
\end{aligned}$$

The second term in this expression is the only term that involves a. It achieves its minimal value of zero when $a = \bar{y}$. □

For future reference, we define the *total sum of squares* to be

$$\mathrm{SS}_T = \mathrm{SS}\,(\bar{y}, 0) = \sum_{i=1}^{n} (y_i - \bar{y})^2 = (n-1)s_y^2.$$

This is the smallest squared error possible when predicting y without information about x.

Now we consider the problem of finding the line $y = a + bx$ that best predicts values of y from values of x. The method of least squares chooses (a,b) to minimize $\mathrm{SS}(a,b)$. Let (a^*, b^*) denote the minimizing values of (a,b) and define the *error sum of squares* to be

$$\mathrm{SS}_E = \mathrm{SS}\,(a^*, b^*)\,.$$

Because we have not restricted attention to $b = 0$, $\hat{y}(x) = a^* + b^* x$ must predict at least as well as $\hat{y} = \bar{y}$. Thus,

$$\mathrm{SS}_E = \mathrm{SS}\,(a^*, b^*) \le \mathrm{SS}\,(\bar{y}, 0) = \mathrm{SS}_T.$$

We have already stated that $y = a^* + b^*x$ is the sample regression line. We can verify that statement by a calculation that resembles the proof of Theorem 15.2.

Theorem 15.3 *Let $(x_i, y_i) \in \Re^2$, $i = 1, \ldots, n$, be a set of (x, y) pairs with at least two distinct values of x. Let*

$$b^* = r\frac{s_y}{s_x} \quad \text{and} \quad a^* = \bar{y} - b^*\bar{x}.$$

Then

$$\mathrm{SS}\left(a^*, b^*\right) \leq \mathrm{SS}(a, b)$$

for all choices of (a, b).

Proof First, write

$$\begin{aligned} \mathrm{SS}(a,b) &= \sum_{i=1}^{n} (y_i - a - bx_i)^2 = \sum_{i=1}^{n} (y_i - \bar{y} + \bar{y} - b\bar{x} + b\bar{x} - a - bx_i)^2 \\ &= \sum_{i=1}^{n} \left[(y_i - \bar{y}) + (\bar{y} - b\bar{x} - a) - b\left(x_i - \bar{x}\right)\right]^2. \end{aligned}$$

Expanding the square in this expression results in six terms. The three squared terms are:

$$\begin{aligned} \sum_{i=1}^{n} (y_i - \bar{y})^2 &= (n-1)s_y^2, \\ \sum_{i=1}^{n} \left(\bar{y} - b\bar{x} - a\right)^2 &= n\left(\bar{y} - b\bar{x} - a\right)^2, \\ \sum_{i=1}^{n} (-b)^2 \left(x_i - \bar{x}\right)^2 &= b^2 \sum_{i=1}^{n} \left(x_i - \bar{x}\right)^2 = b^2(n-1)s_x^2. \end{aligned}$$

The three cross-product terms are:

$$\begin{aligned} &\sum_{i=1}^{n} 2\left(y_i - \bar{y}\right)\left(\bar{y} - b\bar{x} - a\right) = 2\left(\bar{y} - b\bar{x} - a\right) \sum_{i=1}^{n} \left(y_i - \bar{y}\right) \\ &= 2\left(\bar{y} - b\bar{x} - a\right)\left[\sum_{i=1}^{n} y_i - n\bar{y}\right] = 0, \\ &\sum_{i=1}^{n} 2\left(y_i - \bar{y}\right)(-b)\left(x_i - \bar{x}\right) = -2b \sum_{i=1}^{n} \left(x_i - \bar{x}\right)\left(y_i - \bar{y}\right) \end{aligned}$$

$$= -2b(n-1)s_x s_y \frac{\frac{1}{n-1}\sum_{i=1}^{n}(y_i - \bar{y})(x_i - \bar{x})}{s_x s_y} = -2b(n-1)s_x s_y r,$$

$$\sum_{i=1}^{n} 2(\bar{y} - b\bar{x} - a)(-b)(x_i - \bar{x}) = -2b(\bar{y} - b\bar{x} - a)\sum_{i=1}^{n}(x_i - \bar{x}) = 0.$$

Hence,

$$\begin{aligned} \text{SS}(a,b) &= (n-1)s_y^2 + n(\bar{y} - b\bar{x} - a)^2 + b^2(n-1)s_x^2 - 2b(n-1)s_x s_y r \\ &= n(\bar{y} - b\bar{x} - a)^2 + (n-1)\left[b^2 s_x^2 - 2bs_x r s_y + r^2 s_y^2\right] \\ &\quad -(n-1)r^2 s_y^2 + (n-1)s_y^2 \\ &= n(\bar{y} - b\bar{x} - a)^2 + (n-1)[bs_x - rs_y]^2 + \left(1 - r^2\right)(n-1)s_y^2. \end{aligned}$$

The third term in this expression does not involve b or a. The second term achieves its minimal value of zero when $b = rs_y/s_x = b^*$. The first term is the only term that involves a. Whatever the value of b, the first term achieves its minimal value of zero when $a = \bar{y} - b\bar{x}$. Hence, for $b = b^*$, the minimizing value of a is $a = \bar{y} - b^*\bar{x} = a^*$. □

The total sum of squares, SS_T, measures the prediction error from $\hat{y} = \bar{y}$. The error sum of squares,

$$\begin{aligned} \text{SS}_E &= \text{SS}(a^*, b^*) = \sum_{i=1}^{n}[y_i - (\bar{y} - b^*\bar{x}) - b^* x_i]^2 \\ &= \sum_{i=1}^{n}[y_i - \bar{y} - b^*(x_i - \bar{x})]^2 = \sum_{i=1}^{n}\left[(y_i - \bar{y}) - r\frac{s_y}{s_x}(x_i - \bar{x})\right]^2 \\ &= \sum_{i=1}^{n}(y_i - \bar{y})^2 - 2r\frac{s_y}{s_x}\sum_{i=1}^{n}(x_i - \bar{x})(y_i - \bar{y}) + r^2\frac{s_y^2}{s_x^2}\sum_{i=1}^{n}(x_i - \bar{x})^2 \\ &= (n-1)s_y^2 - 2rs_y^2(n-1)\frac{\frac{1}{n-1}\sum_{i=1}^{n}(x_i - \bar{x})(y_i - \bar{y})}{s_x s_y} + r^2 s_y^2(n-1) \\ &= (n-1)s_y^2 - 2(n-1)s_y^2 r^2 + r^2(n-1)s_y^2 \\ &= (n-1)s_y^2\left(1 - r^2\right) \\ &= \left(1 - r^2\right)\text{SS}_T, \end{aligned}$$

measures the prediction error from the sample regression line. Now we define the *regression sum of squares* to be the sum of the squared differences between the two predictions,

$$\begin{aligned}
\mathrm{SS}_R &= \sum_{i=1}^{n} \left[\hat{y} - \hat{y}\left(x_i\right)\right]^2 = \sum_{i=1}^{n} \left[\bar{y} - \bar{y} - r\frac{s_y}{s_x}\left(x_i - \bar{x}\right)\right]^2 \\
&= r^2 \frac{s_y^2}{s_x^2} \sum_{i=1}^{n} \left(x_i - \bar{x}\right)^2 = r^2 s_y^2 (n-1) = r^2 \mathrm{SS}_T.
\end{aligned}$$

The three sums of squares $(\mathrm{SS}_R, \mathrm{SS}_E, \mathrm{SS}_T)$ are precisely analogous to the three sums of squares $(\mathrm{SS}_B, \mathrm{SS}_W, \mathrm{SS}_T)$ that arise in the analysis of variance and they enjoy an identical property:

$$\mathrm{SS}_R + \mathrm{SS}_E = r^2 \mathrm{SS}_T + \left(1 - r^2\right) \mathrm{SS}_T = \mathrm{SS}_T$$

This is the Pythagorean Theorem in n-dimensional Euclidean space! The points

$$A = \begin{bmatrix} \bar{y} \\ \vdots \\ \bar{y} \end{bmatrix}, \quad B = \begin{bmatrix} \bar{y} - r\frac{s_y}{s_x}\left(x_1 - \bar{x}\right) \\ \vdots \\ \bar{y} - r\frac{s_y}{s_x}\left(x_n - \bar{x}\right) \end{bmatrix}, \quad C = \begin{bmatrix} y_1 \\ \vdots \\ y_n \end{bmatrix}$$

are the vertices of a right triangle in $\Re^n$. The right angle occurs at vertex B. The squared Euclidean distances of the sides that meet at B are

$$d^2(A,B) = \mathrm{SS}_R \quad \text{and} \quad d^2(B,C) = \mathrm{SS}_E$$

and the squared Euclidean distance of the hypotenuse is

$$d^2(A,C) = \mathrm{SS}_T,$$

so

$$d^2(A,B) + d^2(B,C) = \mathrm{SS}_R + \mathrm{SS}_E = \mathrm{SS}_T = d^2(A,C).$$

To quantify the extent to which knowledge of x improves our ability to predict y, we measure the proportion by which the squared error of prediction is reduced when we use the sample regression line instead of the constant prediction $\hat{y} = \bar{y}$. This proportion is just

$$\frac{\mathrm{SS}\left(\bar{y}, 0\right) - \mathrm{SS}(a,b)}{\mathrm{SS}\left(\bar{y}, 0\right)} = \frac{\mathrm{SS}_T - \mathrm{SS}_E}{\mathrm{SS}_T} = \frac{\mathrm{SS}_R}{\mathrm{SS}_T} = \frac{r^2 \mathrm{SS}_T}{\mathrm{SS}_T} = r^2,$$

the sample coefficient of determination. Again, we conclude that the square of Pearson's product-moment correlation coefficient measures the proportion of variation "explained" by simple linear regression.

Example 15.2 (continued) For the bivariate sample displayed in Figure 15.2, the total sum of squares is

$$\mathrm{SS}_T = (n-1)s_y^2 = 99 \cdot 14.1754248 = 1403.3671$$

and the coefficient of determination is

$$r^2 = 0.4707309^2 = 0.2215876.$$

Hence, the regression sum of squares is

$$\mathrm{SS}_R = r^2 \mathrm{SS}_T = 0.2215876 \cdot 1403.367 = 310.9688$$

and the error sum of squares is

$$\mathrm{SS}_E = \mathrm{SS}_T - \mathrm{SS}_R = 1403.3671 - 310.9688 = 1092.3983.$$

15.3 Computation

A bivariate sample consists of $2n$ numbers. However, all of the quantities used in the preceding sections can be computed from just six fundamental quantities:

$$n \quad \textstyle\sum_{i=1}^n x_i \quad \sum_{i=1}^n y_i \quad \sum_{i=1}^n x_i^2 \quad \sum_{i=1}^n y_i^2 \quad \sum_{i=1}^n x_i y_i$$

These quantities are used by many calculators. One reason that they are so convenient is that they are easily incremented as new (x, y) pairs are observed.

Example 15.2 (continued) For the bivariate sample displayed in Figure 15.2, the six fundamental quantities are as follows:

$$n = 100 \quad \textstyle\sum_{i=1}^n x_i = 1000.068 \quad \displaystyle\sum_{i=1}^n y_i = 1939.859$$

$$\sum_{i=1}^n x_i^2 = 10442.04 \quad \textstyle\sum_{i=1}^n y_i^2 = 39033.01 \quad \displaystyle\sum_{i=1}^n x_i y_i = 19770.1$$

Now suppose that we draw another (x, y) pair from the same population, say $(8.9, 13.5)$. Then the new sample has the following fundamental quantities:

$$
\begin{array}{rr}
n = 100 + 1 & \sum_{i=1}^{n} x_i^2 = 10442.04 + 8.9^2 \\
\sum_{i=1}^{n} x_i = 1000.068 + 8.9 & \sum_{i=1}^{n} y_i^2 = 39033.91 + 13.5^2 \\
\sum_{i=1}^{n} y_i = 1939.859 + 13.5 & \sum_{i=1}^{n} x_i y_i = 19770.1 + 8.9 \cdot 13.5
\end{array}
$$

Three useful quantities are easily computed from the six fundamental quantities:

$$
\begin{aligned}
t_{xx} &= \sum_{i=1}^{n} (x_i - \bar{x})(x_i - \bar{x}) = \sum_{i=1}^{n} \left(x_i^2 - 2\bar{x}x_i + \bar{x}^2 \right) \\
&= \sum_{i=1}^{n} x_i^2 - 2\bar{x} \sum_{i=1}^{n} x_i + n\bar{x}^2 = \sum_{i=1}^{n} x_i^2 - 2n\bar{x}^2 + n\bar{x}^2 \\
&= \sum_{i=1}^{n} x_i^2 - \frac{1}{n} \left(\sum_{i=1}^{n} x_i \right)^2 \\
t_{yy} &= \sum_{i=1}^{n} (y_i - \bar{y})(y_i - \bar{y}) = \sum_{i=1}^{n} y_i^2 - \frac{1}{n} \left(\sum_{i=1}^{n} y_i \right)^2 \\
t_{xy} &= \sum_{i=1}^{n} (x_i - \bar{x})(y_i - \bar{y}) = \sum_{i=1}^{n} \left(y_i^2 - \bar{y}x_i - \bar{x}y_i + \bar{x}\bar{y} \right) \\
&= \sum_{i=1}^{n} x_i y_i - \bar{y} \sum_{i=1}^{n} x_i - \bar{x} \sum_{i=1}^{n} y_i + n\bar{x}\bar{y} = \sum_{i=1}^{n} x_i y_i - n\bar{x}\bar{y} \\
&= \sum_{i=1}^{n} x_i y_i - \frac{1}{n} \left(\sum_{i=1}^{n} x_i \right) \left(\sum_{i=1}^{n} y_i \right)
\end{aligned}
$$

These quantities are useful because all of the important quantities derived in the preceding sections are easily computed from them. Here are the formulas:

1. Sample variances:

$$
s_x^2 = \frac{t_{xx}}{n-1} \qquad s_y^2 = \frac{t_{yy}}{n-1}
$$

2. Pearson's product-moment correlation coefficient:

$$\begin{aligned} r &= \frac{\frac{1}{n-1}\sum_{i=1}^{n}(x_i-\bar{x})(y_i-\bar{y})}{s_x s_y} = \frac{t_{xy}}{\sqrt{t_{xx}}\sqrt{t_{yy}}} \\ r^2 &= \frac{t_{xy}^2}{t_{xx}t_{yy}} \end{aligned}$$

3. Sample regression coefficients:

$$\begin{aligned} b^* &= r\frac{s_y}{s_x} = \frac{\frac{1}{n-1}\sum_{i=1}^{n}(x_i-\bar{x})(y_i-\bar{y})}{s_x^2} = \frac{t_{xy}}{t_{xx}} \\ a^* &= \bar{y} - b^*\bar{x} = \frac{1}{n}\sum_{i=1}^{n} y_i - \frac{t_{xy}}{t_{xx}}\frac{1}{n}\sum_{i=1}^{n} x_i \end{aligned}$$

4. Sums of squares:

$$\begin{aligned} \text{SS}_T &= \sum_{i=1}^{n}(y_i-\bar{y})^2 = t_{yy} \\ \text{SS}_R &= r^2\text{SS}_T = \frac{t_{xy}^2}{t_{xx}t_{yy}}t_{yy} = \frac{t_{xy}^2}{t_{xx}} \\ \text{SS}_E &= \text{SS}_T - \text{SS}_R = t_{yy} - \frac{t_{xy}^2}{t_{xx}} \end{aligned}$$

15.4 The Simple Linear Regression Model

Let $x_1, \ldots, x_n$ be a list of real numbers for which $s_x > 0$. Suppose that:

1. Associated with each x_i is a random variable

$$Y_i \sim \text{Normal}\left(\mu_i, \sigma^2\right).$$

Notice that the Y_i have a common population variance $\sigma^2 > 0$. This is analogous to the homoscedasticity assumption of the analysis of variance.

2. The population means μ_i satisfy the linear relation

$$\mu_i = \beta_0 + \beta_1 x_i$$

for some $\beta_0, \beta_1 \in \Re$. The population parameters (β_0, β_1) are called the population regression coeffcients.

These assumptions define the *simple linear regression model.* Suppose that we sample from a bivariate normal distribution, then condition on the observed values $x_1, \ldots, x_n$. It follows from Theorem 15.1 that this is a special case of the simple linear regression model in which

$$\begin{aligned}
\beta_1 &= \rho \frac{\sigma_y}{\sigma_x}, \\
\beta_0 &= \mu_y - \rho \frac{\sigma_y}{\sigma_x} \mu_x = \mu_y - \beta_1 \mu_x, \\
\sigma^2 &= \left(1 - \rho^2\right) \sigma_y^2.
\end{aligned}$$

The simple linear regression model has three unknown parameters. The method of least squares estimates (β_0, β_1) by

$$\begin{aligned}
\hat{\beta}_1 &= b^* = r \frac{s_y}{s_x} = \frac{t_{xy}}{t_{xx}}, \\
\hat{\beta}_0 &= a^* = \bar{y} - \hat{\beta}_1 \bar{x}.
\end{aligned}$$

These are also the plug-in estimates of (β_0, β_1), and the plug-in estimate of σ^2 is

$$\widehat{\sigma^2} = \frac{1}{n} \sum_{i=1}^{n} \left(y_i - \hat{\beta}_0 - \hat{\beta}_1 x_i\right)^2 = \frac{1}{n} \mathrm{SS}_E.$$

We proceed to explore some properties of the corresponding estimators. These properties are consequences of the following key facts:

Theorem 15.4 *Assume the simple linear regression model. Then the random variables $\hat{\beta}_1$ and SS_E are independent and satisfy*

$$\hat{\beta}_1 \sim \text{Normal}\left(\beta_1, \frac{\sigma^2}{t_{xx}}\right) \tag{15.4}$$

$$\frac{SS_E}{\sigma^2} \sim \chi^2(n-2). \tag{15.5}$$

It follows from (15.4) that $E\hat{\beta}_1 = \beta_1$, and consequently that

$$\begin{aligned}
E\hat{\beta}_0 &= E\left(\frac{1}{n}\sum_{i=1}^{n} Y_i - \hat{\beta}_1 \frac{1}{n}\sum_{i=1}^{n} x_i\right) = \frac{1}{n}\sum_{i=1}^{n} E\left(Y_i - \hat{\beta}_1 x_i\right) \\
&= \frac{1}{n}\sum_{i=1}^{n} \left(\beta_0 + \beta_1 x_i - \beta_1 x_i\right) = \frac{1}{n}\sum_{i=1}^{n} \beta_0 = \beta_0.
\end{aligned}$$

Thus, $(\hat{\beta}_0, \hat{\beta}_1)$ are unbiased estimators of (β_0, β_1). Furthermore, it follows from (15.5) and Corollary 5.1 that $E(\text{SS}_E/\sigma^2) = n - 2$. Hence, $E[\text{SS}_E/(n-2)] = \sigma^2$ and

$$\text{MS}_E = \frac{1}{n-2}\text{SS}_E$$

is an unbiased estimator of σ^2.

Converting (15.4) to standard units results in

$$\frac{\hat{\beta}_1 - \beta_1}{\sqrt{\sigma^2/t_{xx}}} \sim \text{Normal}(0,1). \tag{15.6}$$

Dividing (15.6) by (15.5), it follows from Definition 5.7 that

$$\frac{\left(\hat{\beta}_1 - \beta_1\right)/\sqrt{\sigma^2/t_{xx}}}{\sqrt{\frac{\text{SS}_E}{\sigma^2}/(n-2)}} = \frac{\hat{\beta}_1 - \beta_1}{\sqrt{\text{MS}_E/t_{xx}}} \sim t(n-2).$$

This fact allows us to construct confidence intervals for β_1. Given α, we first compute the critical value

$$q_t = \texttt{qt}(1 - \alpha/2, n-2).$$

Then

$$\hat{\beta}_1 \pm q_t\sqrt{\frac{\text{MS}_E}{t_{xx}}}$$

is a $(1-\alpha)$-level confidence interval for β_1.

Remark It may be helpful to write

$$\begin{aligned}\frac{\text{MS}_E}{t_{xx}} &= \frac{(1-r^2)\,\text{SS}_T/(n-2)}{(n-1)s_x^2} = \frac{(1-r^2)\,(n-1)s_y^2/(n-2)}{(n-1)s_x^2} \\ &= \left(1-r^2\right)\frac{s_y^2}{s_x^2}/(n-2).\end{aligned}$$

Example 15.3 Suppose that $n = 100$ bivariate observations produce the following estimates:

$$\begin{aligned}\bar{x} &= 97.255564 \\ \bar{y} &= 103.872210 \\ s_x^2 &= 425.062476 \\ s_y^2 &= 872.229230 \\ r &= -0.485857\end{aligned}$$

To construct a 0.95-level confidence interval for β_1, we first compute

$$\hat{\beta}_1 = r\frac{s_y}{s_x} = -0.5070697 \cdot \sqrt{\frac{414.7388683}{434.9825540}} = -0.695981,$$

$q_t = \texttt{qt}(.975, \texttt{df} = 98) = 1.984467$, and

$$\frac{\text{MS}_E}{t_{xx}} = \frac{1-r^2}{n-2} \cdot \frac{s_y^2}{s_x^2} = \frac{1-0.485857^2}{98} \cdot \frac{872.229230^2}{425.062476^2} = 0.01599605.$$

The desired confidence interval is then

$$\begin{aligned} \hat{\beta}_1 \pm q_t\sqrt{\frac{\text{MS}_E}{t_{xx}}} &= -0.695981 \pm 1.984467 \cdot \sqrt{0.01599605} \\ &= (-0.9469675, -0.4449945). \end{aligned}$$

Next we consider how to test $H_0 : \beta_1 = 0$ versus $H_1 : \beta_1 \neq 0$. This is an important decision because rejecting $H_0 : \beta_1 = 0$ means that we are convinced that values of x help us predict values of y. Furthermore, if we sampled from a bivariate normal population, then

$$\beta_1 = \rho\frac{\sigma_y}{\sigma_x} = 0$$

if and only if $\rho = 0$. Because normal random variables X and Y are independent if and only if they are uncorrelated, the null hypothesis $H_0 : \beta_1 = 0$ is equivalent to the null hypothesis that X and Y are independent.

If $\beta_1 = 0$, then

$$\frac{\hat{\beta}_1}{\sqrt{\text{MS}_E/t_{xx}}} \sim t(n-2).$$

Hence, the significance probability for testing $H_0 : \beta_1 = 0$ is

$$\mathbf{p} = P\left(|T| \geq \left|\frac{\hat{\beta}_1}{\sqrt{\text{MS}_E/t_{xx}}}\right|\right),$$

where the random variable $T \sim t(n-2)$, and we reject $H_0 : \beta_1 = 0$ if and only if $\mathbf{p} \leq \alpha$. Equivalently, we reject $H_0 : \beta_1 = 0$ if and only if we observe

$$\left|\frac{\hat{\beta}_1}{\sqrt{\text{MS}_E/t_{xx}}}\right| \geq q_t,$$

where q_t is the critical value defined above. Notice that

$$\begin{aligned}
\frac{\hat{\beta}_1}{\sqrt{\text{MS}_E/t_{xx}}} &= \frac{t_{xy}/t_{xx}}{\sqrt{\text{MS}_E/t_{xx}}} = \frac{t_{xy}}{\sqrt{t_{xx}}} \frac{1}{\sqrt{\text{SS}_E/(n-2)}} \\
&= \frac{t_{xy}}{\sqrt{t_{xx}}\sqrt{t_{yy}}} \frac{\sqrt{t_{yy}}\sqrt{n-2}}{\sqrt{t_{yy} - t_{xy}^2/t_{xx}}} \\
&= r \frac{\sqrt{n-2}}{\sqrt{1 - t_{xy}^2/(t_{xx}t_{yy})}} = \frac{r\sqrt{n-2}}{\sqrt{1-r^2}},
\end{aligned}$$

so this is the same t-test that we described in Section 14.2.2 for testing $H_0 : \rho = 0$ versus $H_1 : \rho \neq 0$.

It follows from Theorem 5.5 that

$$\left(\frac{\hat{\beta}_1}{\sqrt{\text{MS}_E/t_{xx}}} \right)^2 \sim F(1, n-2).$$

Hence, an F-test that is equivalent to the t-test derived in the preceding paragraph rejects $H_0 : \beta_1 = 0$ if and only if we observe

$$(n-2)\frac{r^2}{1-r^2} \geq q_F,$$

where the critical value q_F is defined by

$$q_F = \texttt{qf}(1-\alpha, 1, n-2).$$

Equivalently, we reject $H_0 : \beta_1 = 0$ if and only if the significance probability

$$\mathbf{p} = P\left(F \geq (n-2)\frac{r^2}{1-r^2} \right) \leq \alpha,$$

where the random variable $F \sim F(1, n-2)$. The results of the F-test of $H_0 : \beta_1 = 0$ are traditionally presented in the form of an ANOVA table:

Source of Variation	Sum of Squares	Degrees of Freedom	Mean Square	F-Test Statistic	p-Value
Regression	$r^2\text{SS}_T$	1	$r^2\text{SS}_T$	$(n-2)\frac{r^2}{1-r^2}$	$\mathbf{p}$
Error	$(1-r^2)\text{SS}_T$	$n-2$	$\frac{1-r^2}{n-2}\text{SS}_T$		
Total	SS_T				

Example 15.3 (continued) Let us now test $H_0 : \beta_1 = 0$ versus $H_1 : \beta_1 \neq 0$ at a significance level of $\alpha = 0.05$. Of course, we know that we will reject H_0 because the 0.95-level confidence interval constructed from these data did not contain the hypothesized slope $\beta_1 = 0$.

The t-test statistic is

$$t = \frac{\hat{\beta}_1}{\sqrt{\text{MS}_E / t_{xx}}} = \frac{-0.695981}{\sqrt{0.01599605}} = -5.502893,$$

which results in a significance probability of

$$\mathbf{p} = 2 * \texttt{pt}(-5.502893, \texttt{df} = 98) = 2.989589 \times 10^{-7}.$$

Because $\mathbf{p} < \alpha$, we reject $H_0 : \beta_1 = 0$.

Equivalently, we can compute $SS_T = (n-1)s_y^2 = 99 \cdot 425.062476 \doteq 42081.19$ and $r^2 = 0.236057$, then construct the following ANOVA table:

Source of Variation	Sum of Squares	DF	Mean Square	F-Test Statistic	p-Value
Regression	20383.689	1	20383.6885	30.28183	2.989589×10^{-7}
Error	65967.005	98	673.1327		
Total	86350.694				

Again, we reject $H_0 : \beta_1 = 0$ because $\mathbf{p} < \alpha$. Notice that we obtain the same significance probability with either test.

Although equivalent, the t-test and F-test of $H_0 : \beta_1 = 0$ each enjoy certain advantages. The former is more flexible, as it is easily adapted to test one-sided hypotheses. The F-test is more readily generalized to testing a variety of hypotheses that naturally arise when studying more complicated regression models.

15.5 Assessing Linearity

The simple linear regression model posits that expected values of the random variables Y_i satisfy a linear relation,

$$EY_i = \beta_0 + \beta_1 x_i.$$

We now consider how to assess the plausibility of this assumption in light of the observed data.

Let $Z_i = Y_i - \beta_0 - \beta_1 x_i$, the amount by which Y_i exceeds $\beta_0 + \beta_1 x_i$. If $EY_i = \beta_0 + \beta_1 x_i$, then $EZ_i = 0$. Because β_0 and β_1 are unknown population parameters, we cannot observe Z_i. However, we can observe $\hat{Z}_i = Y_i - \hat{\beta}_0 - \hat{\beta}_1 x_i$, the amount by which Y_i exceeds the sample regression line. These quantities,

$$z_i = y_i - \hat{\beta}_0 - \hat{\beta}_1 x_i \qquad \text{for } i = 1, \ldots, n,$$

are called *residuals*. Notice that the average residual is

$$\bar{z} = \bar{y} - \hat{\beta}_0 - \hat{\beta}_1 \bar{x} = \bar{y} - \left(\bar{y} - \hat{\beta}_1 \bar{x} \right) - \hat{\beta}_1 \bar{x} = 0.$$

Residuals can be analyzed in various ways, leading to various insights. We will examine scatter diagrams of the pairs (x_i, z_i). These pairs are computed by the function `binorm.resid`, described in Appendix R and available from the web page for this book.

Recall that $\hat{\beta}_0$ and $\hat{\beta}_1$ are unbiased estimators of β_0 and β_1. Hence, if $EY_i = \beta_0 + \beta_1 x_i$, then

$$EZ_i = EY_i - E\hat{\beta}_0 - \left(E\hat{\beta}_1 \right) x_i = \beta_0 + \beta_1 x_i - \beta_0 - \beta_1 x_i = 0.$$

We will examine scatter diagrams of the (x_i, z_i) to try to determine if, in fact, $EZ_i = 0$. If we observe systematic patterns that suggest otherwise, then we must question the assumption of linearity and the propriety of simple linear regression. (More advanced regression techniques, e.g., multiple linear regression, nonlinear regression, and nonparametric regression, are beyond the scope of this book.)

Example 15.4 Suppose that Y_i measures the concentration of a drug at time $x_i = 0.05, 0.10, \ldots, 5.00$, with

$$Y_i \sim \text{Normal}\left(\mu_i, 0.04 \right)$$

and

$$\mu_i = \exp\left(-x_i/5 \right) + 3 \exp\left(-2x_i/5 \right).$$

The relation between the μ_i and the x_i is not linear, but let us proceed as though it were. Bivariate data was generated by the following `R` commands:

```
> x <- seq(from=.05,to=5,by=.05)
> y <- exp(-.2*x) + 3*exp(-.4*x) + rnorm(100,sd=.2)
> Data <- cbind(x,y)
```

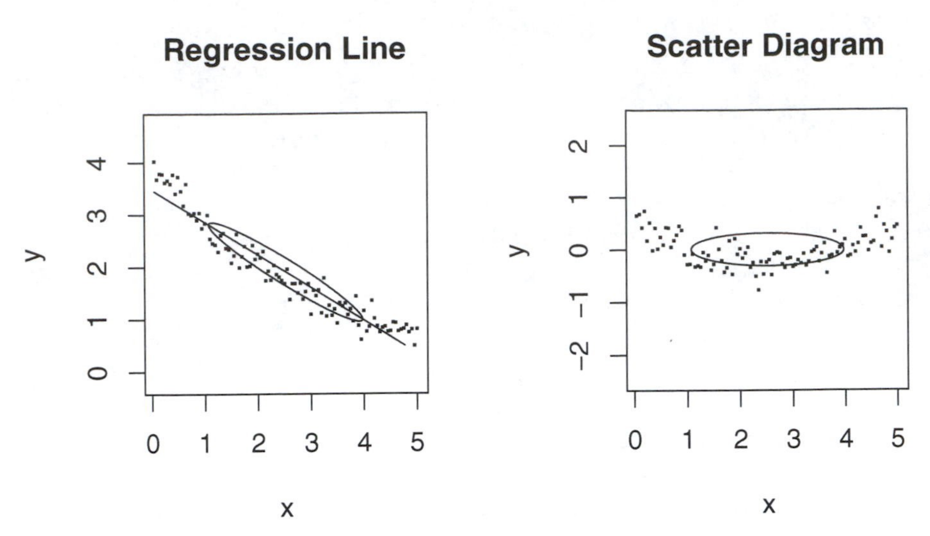

Figure 15.5: Simulated data from a nonlinear regression model. The observed (x_i, y_i) pairs are plotted on the left. The regression line approximates the monotonic relation between x and y, but that relation is not linear. The systematic departure from linearity is more clearly revealed in the right plot, where the residuals are plotted against the x_i.

Figure 15.5 contains two scatter diagrams. The plot on the left displays the (x_i, y_i) pairs, with the concentration ellipse and regression line superimposed. It was created by the following command:

```
> binorm.regress(Data)
```

The plot on the right displays the (x_i, z_i) pairs, with the concentration ellipse corresponding to those pairs superimposed. It was created by the following commands:

```
> Resid <- binorm.resid(Data)
> binorm.scatter(Resid)
```

Notice that the function `binorm.scatter` automatically labels the horizontal axis as `x` and the vertical axis as `y`, although in this case we are actually plotting (x_i, z_i).

It is easy to see that the relation between time (x) and concentration (y) is not linear. This nonlinearity is more clearly revealed by removing the linear trend in the (x_i, y_i) pairs and examining the (x_i, z_i) pairs. If the relation between x and y was linear, then removing the linear trend should result in residuals (z) that have no systematic relation to x. In fact, the residuals tend to be positive for extreme values of x and negative for intermediate values. Because it is highly improbable that such a pronounced pattern resulted by chance, we conclude that the simple linear regression model is not adequate for these data.

So far, we have treated the x_i as n distinct values. Henceforth, assume that there are k distinct values of x_i and that we observe n_i values of

$$Y_i \sim \text{Normal}\left(\beta_0 + \beta_1 x_i, \sigma^2\right)$$

for $i = 1, \ldots, k$. As in Chapter 12, we represent this situation using double subscripts:

$$Y_{ij} \sim \text{Normal}\left(\mu_i, \sigma^2\right),$$

where $\mu_i = \beta_0 + \beta_1 x_i$, $i = 1, \ldots, k$, $j = 1, \ldots, n_i$, and $N = n_1 + \cdots + n_k = n$. This is the model for a one-way analysis of variance, with the additional restriction that the population means are linearly related. In this setting, the regression sum of squares is

$$SS_R = \sum_{i=1}^{k} \sum_{j=1}^{n_i} \left(\hat{\beta}_0 + \hat{\beta}_1 x_i - \bar{y}_{\cdot\cdot}\right)^2 = \sum_{i=1}^{k} n_i \left(\hat{\beta}_0 + \hat{\beta}_1 x_i - \bar{y}_{\cdot\cdot}\right)^2$$

and the error sum of squares is

$$SS_E = \sum_{i=1}^{k} \sum_{j=1}^{n_i} \left(y_{ij} - \hat{\beta}_0 - \hat{\beta}_1 x_i\right)^2 = \sum_{i=1}^{k} \sum_{j=1}^{n_i} z_{ij}^2.$$

Because $\bar{z}_{\cdot\cdot} = 0$, we can write

$$SS_E = \sum_{i=1}^{k} \sum_{j=1}^{n_i} \left(y_{ij} - \hat{\beta}_0 - \hat{\beta}_1 x_i\right)^2 = \sum_{i=1}^{k} \sum_{j=1}^{n_i} \left(z_{ij} - \bar{z}_{\cdot\cdot}\right)^2,$$

the total sum of squares were we to perform an analysis of variance on the residuals. Such an analysis decomposes SS_E into

$$\begin{aligned} SS_B &= \sum_{i=1}^{k} \sum_{j=1}^{n_i} \left(\bar{z}_{i\cdot} - \bar{z}_{\cdot\cdot}\right)^2 = \sum_{i=1}^{k} n_i \bar{z}_{i\cdot}^2 \\ &= \sum_{i=1}^{k} n_i \left(\frac{1}{n_i} \sum_{j=1}^{n_i} \left(y_{ij} - \hat{\beta}_0 - \hat{\beta}_1 x_i\right) \right)^2 \\ &= \sum_{i=1}^{k} n_i \left(\bar{y}_i - \hat{\beta}_0 - \hat{\beta}_1 x_i\right)^2 \end{aligned}$$

and

$$\begin{aligned} SS_W &= \sum_{i=1}^{k}\sum_{j=1}^{n_i}(z_{ij}-\bar{z}_{i\cdot})^2 \\ &= \sum_{i=1}^{k}\sum_{j=1}^{n_i}\left(y_{ij}-\hat{\beta}_0-\hat{\beta}_1 x_i-\frac{1}{n_i}\sum_{\ell=1}^{n_i}\left(y_{i\ell}-\hat{\beta}_0-\hat{\beta}_1 x_i\right)\right)^2 \\ &= \sum_{i=1}^{k}\sum_{j=1}^{n_i}\left(y_{ij}-\hat{\beta}_0-\hat{\beta}_1 x_i-\left(\bar{y}_{i\cdot}+\hat{\beta}_0+\hat{\beta}_1 x_i\right)\right)^2 \\ &= \sum_{i=1}^{k}\sum_{j=1}^{n_i}(y_{ij}-\bar{y}_{i\cdot})^2 . \end{aligned}$$

We see that SS_B measures the squared deviations of the sample means at $x_1,\ldots,x_k$ from the corresponding means predicted by simple linear regression. Hence, this SS_B is sometimes called the sum of squares due to lack of fit (SS_{lof}). Moreover, SS_W measures the squared deviations of the individual observations at $x_1,\ldots,x_k$ from the corresponding sample means at $x_1,\ldots,x_k$. Hence, this SS_W is sometimes called the sum of squares due to pure error (SS_{pe}).

The fundamental null hypothesis of the analysis of variance is that the population residual means, in this case

$$E\hat{Z}_{ij} = E\left[Y_{ij}-\hat{\beta}_0-\hat{\beta}_1 x_1\right],$$

have a common value. Under the simple linear regression model, each $E\hat{Z}_{ij}=0$ and the null hypothesis of equal population residual means is correct. Conversely, if the population residual means differ, then the $\mu_i = EY_{ij}$ are not linearly related. These observations suggest the possibility of testing the assumption of linearity by performing an analysis of variance.

There are two subtleties that must be considered before proceeding. First, ANOVA requires equality of population variances. It turns out that, under the simple linear regression model, the variances of the $\hat{Z}_{ij}$ are indeed equal. (If a scatter diagram of (x_{ij}, z_{ij}) suggests otherwise, then one should question the assumption of equal variances in the simple linear regression model.) Second, we know that the grand mean of the residuals must equal zero. It turns out that this constraint changes the degrees of freedom in the analysis of variance.

In an ordinary ANOVA, the total sum of squares has $N-1$ degrees of freedom. Here, the analogous SS_E has $n-2=N-2$ degrees of freedom. In an ordinary ANOVA, the between-groups sum of squares has $k-1$ degrees of freedom. Here, the analogous SS_{lof} has $k-2$ degrees of freedom. In an ordinary ANOVA, the within-groups sum of squares has $(N-1)-(k-1) = N-k$ degrees of freedom. Here, the analogous SS_{pe} also has $(n-2)-(k-2) = n-k = N-k$ degrees of freedom. Thus, the ANOVA table for simple linear regression can be expanded as follows:

Source of Variation	Sum of Squares	Degrees of Freedom	Mean Square	F-Test Statistic	p-Value
Regression	$r^2\text{SS}_T$	1	MS_R	$\text{MS}_R/\text{MS}_{pe}$	$\mathbf{p}_R$
Total Error	$(1-r^2)\text{SS}_T$	$n-2$			
Lack of Fit	SS_{lof}	$k-2$	MS_{lof}	$\text{MS}_{lof}/\text{MS}_{pe}$	$\mathbf{p}_{lof}$
Pure Error	SS_{pe}	$n-k$	MS_{pe}		
Total	SS_T				

Notice that we now test $H_0 : \beta_1 = 0$ using $F = \text{MS}_R/\text{MS}_{pe}$ instead of $F = \text{MS}_R/\text{MS}_E$.

Example 15.4 (continued) Now suppose that Y_{ij} measures the concentration of a drug at time $x_i = 0.5, 1.0, \ldots, 5.0$, with

$$Y_{ij} \sim \text{Normal}\,(\mu_i, 0.04)$$

for $j = 1, \ldots, 10$. Bivariate data was generated by the following `R` commands:

```
> x <- seq(from=.5,to=5,by=.5)
> x <- matrix(x,byrow=TRUE,nrow=10,ncol=10)
> x <- as.vector(x)
> y <- exp(-.2*x) + 3*exp(-.4*x) + rnorm(100,sd=.2)
> Data <- cbind(x,y)
```

To construct an ANOVA table, first we compute $\text{SS}_T = (n-1)s_x^2 \doteq 99 \cdot 0.799773 \doteq 79.17752$ and $r^2 = 0.8954554$. To decompose $\text{SS}_E = (1-r^2)\text{SS}_T \doteq 8.27758$ into SS_{lof} and SS_{pe}, we must first compute the regression line.

The regression coefficients are

$$\hat{\beta}_1 = rs_y/s_x \doteq -0.946285 \cdot 0.8943003/1.443376 = -0.5863082$$

and

$$\hat{\beta}_0 = \bar{y} - \hat{\beta}_1\bar{x} \doteq 1.750056 + 0.5863082 \cdot 2.75 \doteq 3.362404.$$

Knowing the regression line allows us to compute

$$\begin{aligned} \text{SS}_{lof} &= \sum_{i=1}^{k} n_i \left(\bar{y}_{i\cdot} - \hat{\beta}_0 - \hat{\beta}_1 x_i \right)^2 \\ &\doteq \sum_{i=1}^{10} 10 \left(\bar{y}_{i\cdot} - 3.362404 + 0.5863082 i/2 \right)^2 \doteq 3.862893 \end{aligned}$$

and

$$\text{SS}_{pe} = \text{SS}_E - \text{SS}_{lof} \doteq 8.27758 - 3.862893 \doteq 4.414687.$$

Completing the rest of the ANOVA table is easy:

Source of Variation	Sum of Squares	Degrees of Freedom	Mean Square	F-Test Statistic	p-Value
Regression	70.89994	1	70.89994	1445.401	0
Error	8.27758	98			
Lack of Fit	3.86289	8	0.4828616	9.844	0
Pure Error	4.41469	90	0.04905208		
Total	79.17752				

Despite the fact that nearly 90 percent of the variation in concentration is "explained" by simple linear regression of concentration on time, there is overwhelming evidence that the relation between concentration and time is not linear.

15.6 Case Study: Are Thick Books More Valuable?

Data Set 142 in *A Handbook of Small Data Sets* contains the replacement values (in pence) and the width (in millimeters) of a sample of $n = 100$ books. We convert pence to pounds (1 pound equals 100 pence) and explore the question of how well one can predict replacement value from width.[2]

Figure 15.6 contains two scatter diagrams, each with a concentration ellipse and regression line superimposed. The observed data arre plotted on the left. The horizontal axis (x) measures width and the vertical axis (y) measures replacement value. Because both variables are evidently skewed,

[2]There is a concern here that we will ignore. The sample was drawn from a collection of 1554 books. Is the population the collection, or is the collection itself a sample from a larger population? If the latter, then how can we describe that population? Clearly, the relations that hold for a (private?) English collection are not universal.

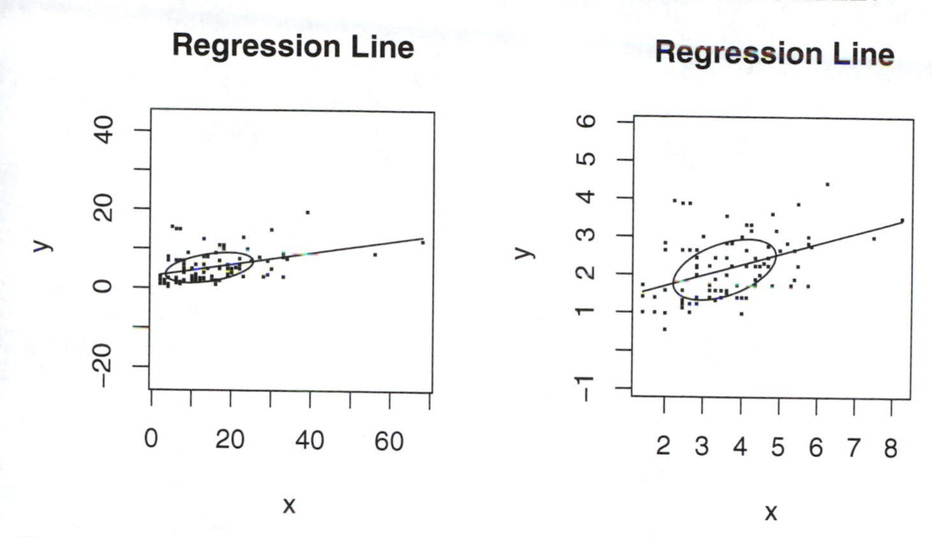

Figure 15.6: Two scatter diagrams for a sample of $n = 100$ books. On the left, replacement value (y) is plotted against width (x); on the right, $\sqrt{y}$ is plotted against $\sqrt{x}$.

we are reluctant to assume that the sample was drawn from a bivariate normal distribution. Let x_i and y_i denote the square roots of the width and replacement value of book i. The (x_i, y_i) pairs are plotted on the right, and it seems somewhat more reasonable to assume that these pairs were drawn from a bivariate normal distribution. A scatter diagram (not shown) of the residuals against the x_i is unremarkable.

The $n = 100$ (x_i, y_i) pairs produce the following estimates:

$$\begin{aligned} \bar{x} &= 3.5490964 \\ \bar{y} &= 2.1296724 \\ s_x^2 &= 1.8019344 \\ s_y^2 &= 0.6429247 \\ r &= 0.4665971 \end{aligned}$$

The proportion of variation "explained" by simple linear regression is $r^2 \doteq 0.22$. The magnitude of this quantity is hardly surprising. Thicker books do tend to cost more, but the price of a book is determined by much more than its width.

Is the relation between width and replacement value statistically significant? We answer that question by testing $H_0 : \rho = 0$ versus $H_1 : \rho \neq 0$, which is equivalent to testing $H_0 : \beta_1 = 0$ versus $H_1 : \beta_1 \neq 0$. The results are summarized in the following ANOVA table:

Source of Variation	Sum of Squares	DF	Mean Square	F-Test Statistic	p-Value
Regression	13.85732	1	13.85732	27.27369	9.92651×10^{-7}
Error	49.79223	98	0.5080839		
Total	63.64955				

Evidently, there is compelling evidence of a relation.

Now suppose that we draw another book from the same population and observe a width of 25 millimeters. To predict the replacement value of this book, we need the sample regression line:

$$y = \hat{\beta}_0 + \hat{\beta}_1 x = \bar{y} + r\frac{s_y}{s_x}(x - \bar{x}) = 1.140504 + 0.27871x.$$

For $x = \sqrt{25} = 5$, we predict

$$\hat{y}(5) = 1.140504 + 0.27871 \cdot 5 = 2.534054.$$

The predicted replacement value is $\hat{y}(5)^2 \doteq 6.42$ pounds.

15.7 Exercises

1. According to Stanford University Professor Claude M. Steele,

   > The SAT, for example, correlates .42 with freshman grades... This means that it measures about 18 percent of the characteristics, whatever they are, that determine freshman grades.[3]

   Comment on this passage. Do you agree with Professor Steele's interpretation of what $r = 0.42$ means?

2. In the athletics event known as the shot put, male competitors "put" the "shot," a 16-pound metal ball. (Female competitors use a smaller shot.) In the United States, high school male competitors put a 12-pound shot, then graduate to the 16-pound shot used in NCAA, USATF, and IAAF competition. In its August 2002 "Stat Corner," the respected athletics periodical *Track & Field News* proclaimed an "Inverse Relationship Between 12 & 16lb Shots":[4]

---

[3]Not just a test, *The Nation*, May 3, 2004, p. 40.

[4]The quoted passage and Table 15.1 are reproduced with the permission of *Track and Field News*.

> A look at the accompanying all-time Top 11 lists for high schoolers with the 12lb shot—11 because there have been 11 of them over 70 [feet]—and for U.S. men with the 16 sends two messages to aspiring prep putters:
>
> - If you're not very good in high school, don't worry about it; few of the big guys were either.
> - If you're great in high school, that may be about as good as you'll ever get.
>
> The numbers are astounding. We'll leave it to a technical expert to figure out why...

The numbers are displayed in Table 15.1.[5] Comment on the inverse relationship noted by *Track & Field News*. Do you agree with their two messages?

3. Suppose that (X, Y) have a bivariate normal distribution with parameters $(5, 3, 1, 4, 0.5)$. Compute the following quantities:

   (a) $P(Y > 6)$

   (b) $E(Y|X = 6.5)$

   (c) $P(Y > 6|X = 6.5)$

4. Assume that the population of all sister-brother heights has a bivariate normal distribution and that the data in Table 14.4 were sampled from this population.

   (a) Consider the population of all sister-brother heights. Estimate the proportion of all brothers who are at least $5'\ 10''$.

   (b) Suppose that Carol is $5'\ 1''$. Predict her brother's height.

   (c) Consider the population of all sister-brother heights for which the sister is $5'\ 1''$. Estimate the proportion of these brothers who are at least $5'\ 10''$.

5. Assume that the population of all sister-brother heights has a bivariate normal distribution and that the data in Table 14.4 were sampled from this population.

---

[5] Perhaps the most astounding number is Michael Carter's prodigious heave of 81-3.50, arguably the most formidable record in all of track and field. Carter broke an 11-year-old record by *nine feet*! He went on to a sensational college career at SMU, winning the NCAA championship and a silver medal at the 1984 Olympic Games. He then opted for a career in professional football, becoming an All-Pro defensive lineman for the NFL Champion San Francisco 49ers.

ALL-TIME HIGH SCHOOL 70-FOOTERS			
	12	*16*	*16–12*
1. Michael Carter '79	81-3.5	71-4.75	–9-10.75
2. Brent Noon '90	76-2	70-5.75	–5-8.25
3. Arnold Campbell '84	74-10.5	64-3	–10-7.5
4. Charles Moye '87	72-8	57-1	–15-7
5. Sam Walker '68	72-3.25	66-9.5	–5-5.75
6. Jesse Stuart '70	71-11i	68-11.5i	–2-11.5
7. Roger Roesler '96	71-2	61-6.25	–11-7.75
8. Kevin Bookout '02	71-1.5	(too early still)	
9. Doug Lane '68	70-11	66-11.25	–3-11.75
10. Dennis Black '91	70-7	68-10	–1-9
11. Ron Semkiw '72	70-1.75	70-0.5	–0-1.25

ALL-TIME U.S. TOP 11			
	16	*12*	*16–12*
1. Randy Barnes '90	75-10.25	66-9.5	+9-0.75
2. Brian Oldfield '75	75-0	58-10	+16-2
3. John Brenner '87	73-10.75	64-5.5	+9-5.25
4. Adam Nelson '02	73-10.25	63-2.25	+10-8
5. Kevin Toth '02	72-9.75	58-11	+13-10.75
6. George Woods '74	72-3i	60-11	+11-4
6. Dave Laut '82	72-3	65-9	+6-6
6. John Godina '99	72-3	64-1.25	+8-1.75
9. Gregg Trafalis '92	72-1.5	57-0	+15-1.5
10. Terry Albritton '76	71-8.5	67-9	+3-11.5
11. Andy Bloom '00	71-7.25	64-2.5	+7-4.74

Table 15.1: All-time high school and U.S. shot put performers as of August 2002.

(a) Compute the sample coefficient of determination, the proportion of variation "explained" by simple linear regression.

(b) Let $\alpha = 0.05$. Do these data provide convincing evidence that knowing a sister's height (x) helps one predict her brother's height (y)?

(c) Construct a 0.90-level confidence interval for the slope of the population regression line for predicting y from x.

(d) Suppose that you are planning to conduct a more comprehensive study of sibling heights. Your goal is to better estimate the slope of the population regression line for predicting y from x. If you want to construct a 0.95-level confidence interval of length 0.1, then how many sister-brother pairs should you plan to observe?

6. Table 15.2 displays height (x, measured in centimeters) and weight (y, measured in kilograms) measurements on $n = 30$ 11-year-old girls in Bradford, England.[6]

(a) Construct a scatter diagram of these data. Does it seem reasonable to assume that the sample was drawn from a bivariate normal distribution? Why or why not?

(b) Consider the population of all 11-year-old girls in Bradford. Assuming bivariate normality, estimate the proportion of such girls who weigh between 37 and 42 kilograms.

(c) Consider the population of all 11-year-old girls in Bradford whose height is $x = 150$ centimeters. Assuming bivariate normality, estimate the proportion of such girls who weigh between 37 and 42 kilograms.

7. Assume that the sample in Table 15.2 was drawn from a bivariate normal distribution.

(a) Compute the sample coefficient of determination, the proportion of variation "explained" by simple linear regression.

(b) Construct an analyis of variance table for testing $H_0 : \beta_1 = 0$ versus $H_1 : \beta_1 \neq 0$. Do these data provide convincing evidence that knowing x helps one predict y? (Assume a significance level of $\alpha = 0.05$.)

[6]The Open University (1983). *MDST242 Statistics in Society, Unit C3: Is my child normal?*, Figure 3.12. These data appear as Data Set 96 in *A Handbook of Small Data Sets*.

x	y	x	y	x	y	x	y	x	y	x	y
135	26	133	31	146	33	149	34	153	55	141	32
154	50	164	47	139	32	146	37	131	25	149	46
149	44	147	36	137	31	152	47	143	36	140	33
146	35	143	42	141	28	148	32	136	28	149	32
154	36	141	29	151	48	137	34	155	36	135	30

Table 15.2: Height (x, measured in centimeters) and weight (y, measured in kilograms) measurements on $n = 30$ 11-year-old girls in Bradford, England.

(c) Construct a 0.95-level confidence interval for the slope of the population regression line for predicting y from x.

8. A class of 35 students took two midterm tests. Jack missed the first test and Jill missed the second test. The 33 students who took both tests scored an average of 75 points on the first test, with a standard deviation of 10 points, and an average of 64 points on the second test, with a standard deviation of 12 points. The scatter diagram of their scores is roughly ellipsoidal, with a correlation coefficient of $r = 0.5$.

   Because Jack and Jill each missed one of the tests, their professor needs to guess how each would have performed on the missing test in order to compute their semester grades.

   (a) Jill scored 80 points on Test 1. She suggests that her missing score on Test 2 be replaced with her score on Test 1, 80 points. What do you think of this suggestion? What score would you advise the professor to assign?

   (b) Jack scored 76 points on Test 2, precisely one standard deviation above the Test 2 mean. He suggests that his missing score on Test 1 be replaced with a score of 85 points, precisely one standard deviation above the Test 1 mean. What do you think of this suggestion? What score would you advise the professor to assign?

9. Consider the simulated data in Table 15.3, in which 5 y-values are displayed for each x-value. Assume that $y_{i1}, \ldots, y_{i5}$ were drawn from a univariate normal distribution with unknown mean μ_i and unknown variance σ^2. Given x, we would like to predict y.

(a) Construct a scatter diagram for these data. Is simple linear regression appropriate? Why or why not?

(b) Construct an analysis of variance table for these data. Can we reject the null hypothesis that the μ_i lie on a straight line?

x	y_1	y_2	y_3	y_4	y_5
4	4.72	1.95	1.86	13.29	5.08
8	22.64	8.73	17.91	25.51	17.03
12	31.84	30.23	23.25	34.37	32.86
16	25.59	27.27	31.31	28.11	26.99
20	35.98	25.99	16.10	34.53	46.93
24	48.36	42.62	46.15	47.49	36.87
28	34.03	45.27	33.58	50.03	38.42
32	35.24	41.47	33.25	61.90	39.42
36	34.66	52.02	60.75	49.16	51.94
40	50.93	51.11	46.63	57.18	47.34
44	60.73	57.90	65.39	58.99	50.07
48	52.31	53.93	53.60	54.68	38.77
52	42.12	38.00	45.20	39.16	52.07
56	35.41	34.41	34.04	37.08	22.46
60	26.26	28.98	32.43	42.07	23.39

Table 15.3: Simulated data for testing the linearity of the regression function.

10. Table 15.4 displays several measurements on the tensile strength of cement (y, measured in kg/cm^2) for each of $k = 5$ different curing times (x, measured in days).[7] We will use these data to predict the tensile strength of concrete after 14 days of curing.

    (a) Construct a scatter diagram of the (x, y) pairs. Is simple linear regression appropriate? Why or why not?

    (b) Construct an analysis of variance table for the (x, y) pairs. Can we reject the null hypothesis that the regression function is linear?

    (c) A. Hald[8] suggested regressing log tensile strength on the re-

[7] These data appear as Data Set 8 in *A Handbook of Small Data Sets.*

[8] A. Hald (1952). *Statistical Theory with Engineering Applications.* John Wiley & Sons, New York, p. 45.

ciprocal of curing time. Construct a scatter diagram of the $(1/x, \log(y))$ pairs. Does simple linear regression seem appropriate? Why or why not?

(d) Construct an analysis of variance table for the $(1/x, \log(y))$ pairs. Can we reject the null hypothesis that the regression function is linear?

(e) Predict the tensile strength of cement after 14 days of curing.

x	y_1	y_2	y_3	y_4	y_5
1	13.0	13.3	11.8		
2	21.9	24.5	24.7		
3	29.8	28.0	24.1	24.2	26.2
7	32.4	30.4	34.5	33.1	35.7
28	41.8	42.6	40.3	35.7	37.3

Table 15.4: Tensile strength of cement (y, measured in kg/cm^2) for each of $k = 5$ different curing times (x, measured in days).

11. Table 15.5 displays two head measurements (in millimeters) for each of the first two adult sons in $n = 25$ families.[9] For each head, we will compute two variables:

```
> size <- length+breadth
> shape <- length-breadth
```

(a) Construct two scatter diagrams, one for head size and one for head shape. For each scatter diagram, let x denote the measurement on the first son and let y denote the measurement on the second son. Is it plausible that the sample of head sizes was drawn from a bivariate normal distribution? What about the sample of head shapes?

(b) Consider head size. Investigate the relation between first son head size and second son head size. Can we reject the null hypothesis that these variables are uncorrelated? Of the variation in second son head size, what proportion is explained by variation in first son head size?

[9] G. P. Frets (1921). Heredity of head form in man. *Genetica*, 3:193–384. These data appear as Data Set 111 in *A Handbook of Small Data Sets*.

(c) Consider head shape. Investigate the relation between first son head shape and second son head shape. Can we reject the null hypothesis that these variables are uncorrelated? Of the variation in second son head shape, what proportion is explained by variation in first son head shape?

(d) In another family from the same era, the first adult son's head had a length of 195 millimeters and a breadth of 160 millimeters. Use this information to guess the size of the second adult son's head.

First Son		Second Son	
Length	Breadth	Length	Breadth
191	155	179	145
195	149	201	152
181	148	185	149
183	153	188	149
176	144	171	142
208	157	192	152
189	150	190	149
197	159	189	152
188	152	197	159
192	150	187	151
179	158	186	148
183	147	174	147
174	150	185	152
190	159	195	157
188	151	187	158
163	137	161	130
195	155	183	158
186	153	173	148
181	145	182	146
175	140	165	137
192	154	185	152
174	143	178	147
176	139	176	143
197	167	200	158
190	163	187	150

Table 15.5: Head measurements (mm) on first two adult sons in $n = 25$ families.

Chapter 16

Simulation-Based Inference

Most of the inferential procedures that we have studied were devised before the advent of computers. Student's 1-sample t-test was introduced in 1908! Before computers, statistical inference relied on procedures for which (1) the quantities that depend on the data can be computed by hand, and (2) the sampling distributions needed to compute critical values are sufficiently tractable that the critical values can be tabled for widespread use. For decades, data analysis required tedious calculation and access to tables of quantiles for the standard normal and various chi-squared, t, and F distributions. Although the enormous practical value of procedures such as t-tests, ANOVA, and simple linear regression is beyond dispute, one reason that these procedures were central to 20th-century statistics was the simple fact that it was possible for a user to perform the necessary calculations. As computing power increased, the practice of statistics began to change. First desktop calculators, then computers, were used to ease the computational burden of the same familiar procedures. Eventually, computationally intensive procedures were introduced that would have been unthinkable in the days of Student and Fisher.

In a 1979 article titled "Computers and the Theory of Statistics: Thinking the Unthinkable," Bradley Efron wrote:

> The "unthinkable" mentioned in the title is simply the thought that one might be willing to perform 500,000 numerical operations in the analysis of 16 data points. Or one might be willing to perform a billion operations to analyze 500 numbers. Such statemennts would have seemed insane thirty years ago, when a slow and noisy fifty pound desk calculator which added, subtracted, multiplied, and divided was the most sophisticated computational aid available to most scientists

> Most of the statistical theory in common use was developed under the constraint of slow and expensive computation. Now computation is fast and cheap. It is not surprising that new theory is being developed, which takes advantage of the high-speed computer.[1]

Three decades later, computation is astoundingly faster and cheaper than it was in 1979, and the unthinkable has become common.

Let us consider some of the ways in which, through `R`, this book has relied on the computer:

1. We have used `R` as a calculator, to perform elementary arithmetic operations and to compute such quantities as logarithms.

2. Instead of tables, we have used `R` functions such as `pnorm` and `qnorm` to determine significance probabilities and critical values.

3. We have used `R` functions such as `sample` and `rnorm` to generate pseudorandom samples, thereby allowing us to explore the behavior of various statistics that do not have analytically tractable sampling distributions. For example, in Section 7.3.2 we used `rnorm` to generate pseudorandom normal samples in order to develop a sense of how normal probability plots behave in the case of a normal population.

The third item exemplifies the use of simulation in statistical inference.

Simulation is a fundamental tool in modern statistical inference, and a thorough treatment of simulation-based inference is a subject for other books and courses. We conclude with two examples of how statisticians use simulation to address problems that might otherwise be intractable.

16.1 Termite Foraging Revisited

In Sections 10.3, 11.2, and 14.3 we used simulation to approximate significance probabilities in situations where test statistics did not have familiar sampling distributions. This technique can be used in a wide variety of situations. To illustrate, we return to the termite foraging experiment described in Section 1.1.3. The probability model for random foraging, that "any previously unattacked roll was equally likely to be the next roll attacked," is an unconventional null hypothesis that cannot be tested by any of the procedures described in previous chapters. However, it is easy to simulate

---

[1] B. Efron (1979). Computers and the theory of statistics: thinking the unthinkable, *SIAM Review*, 21(4):460–480.

random foraging behavior on the computer. If we can develop a suitable quantitative measure of foraging behavior, then we can use simulation to decide whether or not the behavior observed in Figure 1.1 is consistent with random foraging.

The rolls in Plot 20 were positioned at regular intervals of 1.5 meters. For simplicity, we measure the Euclidean distance (d) between two rolls in units of 1.5 meters. Thus, in Figure 1.1, the distance between the two intially attacked rolls is $d = 2$ and the distance between the rolls labelled 1 and 3 is

$$d = \sqrt{3^2 + 1^2} = \sqrt{10} \doteq 3.16.$$

If *H. aureus* forages systematically, then we would expect to find that newly attacked rolls are close to previously attacked rolls. For example, the roll labelled 1 in Figure 1.1 is adjacent ($d = 1$) to a roll that had been previously attacked. If this pattern persisted throughout the attack sequence, then we might be inclined to conclude that *H. aureus* forages systematically. However, for us to reject the possibility that such a pattern is a chance occurrence, we must be convinced that the probability of such an occurrence is small. We need (1) a test statistic that quantifies how systematic we consider an attack sequence to be, and (2) a way of computing a significance probability.

Let C_0 denote the set of initially attacked rolls and let J denote the number of subsequently attacked rolls. For Plot 20, there are two rolls in C_0 and $J = 13$. We will index the J subsequently attacked rolls by the order in which they were attacked: $r_1, \ldots, r_J$. Because the rolls were not monitored continuously, we may not be able to determine a unique order of attack. In Figure 1.1 there are two rolls labelled 2, one in the second column and one in the fourth column. Either of these rolls might be assigned r_2, the other r_3. When different orderings are consistent with the observed data, it may be necessary to analyze each possible ordering. Fortunately, different orderings often produce identical results.

Given a unique order of attack, let

$$C_j = C_{j-1} \bigcup \{r_j\}$$

for $j = 1, \ldots, J$. The set C_j is the union of roll r_j with all of the rolls that were attacked before r_j. We define $d(r_j, C_{j-1})$, the distance of the individual roll r_j from the set of rolls C_{j-1}, to be the distance between r_j and the nearest roll in C_{j-1}. In Figure 1.1, r_1 is the roll labelled 1 and

$$d(r_1, C_0) = \min\{1, 3\} = 1.$$

If foraging is systematic, then each $d(r_j, C_{j-1})$ should be small. We sum the squared distances, forming the test statistic

$$T_J(r_1, \ldots, r_J; C_0) = \sum_{j=1}^{J} [d(r_j, C_{j-1})]^2 .$$

Small values of T_J are evidence of systematic foraging.

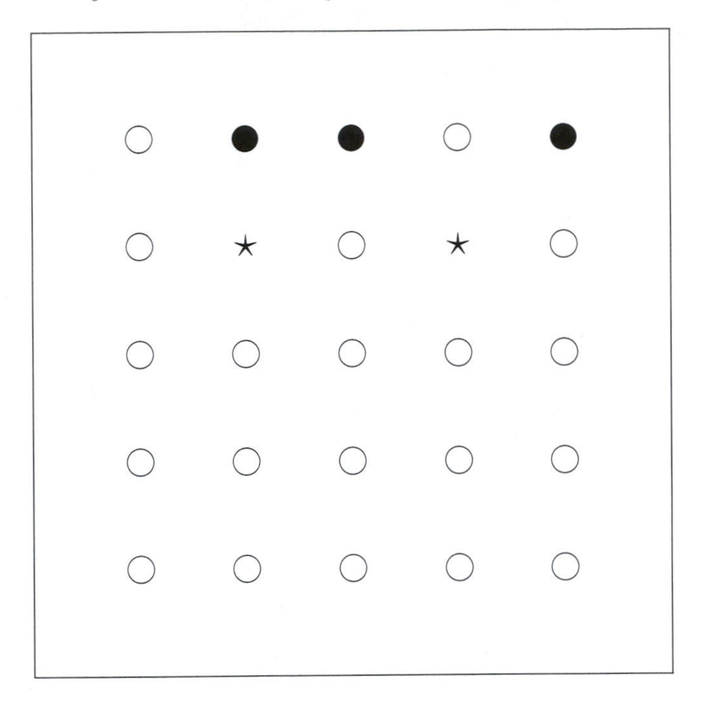

Figure 16.1: Plot 20, Week 2. $T_3 = 1 + (1 + 2) = 4$.

It is obvious that

$$T_1 = T_1(r_1; C_0) = [d(r_1, C_0)]^2 = 1^2 = 1.$$

To compute T_3, consider Figure 16.1, in which the rolls in C_1 are denoted by $\bullet$ and rolls r_2 and r_3 are denoted by $\star$. Regardless of which $\star$ is declared r_2 and which is declared r_3, we see that the squared distance of the left $\star$ from previously attacked rolls is 1 and the squared distance of the right $\star$ from previously attacked rolls is 2. Thus, for either order,

$$T_3 = T_3(r_1, r_2, r_3; C_0) = T_1 + (1 + 2) = 1 + 3 = 4.$$

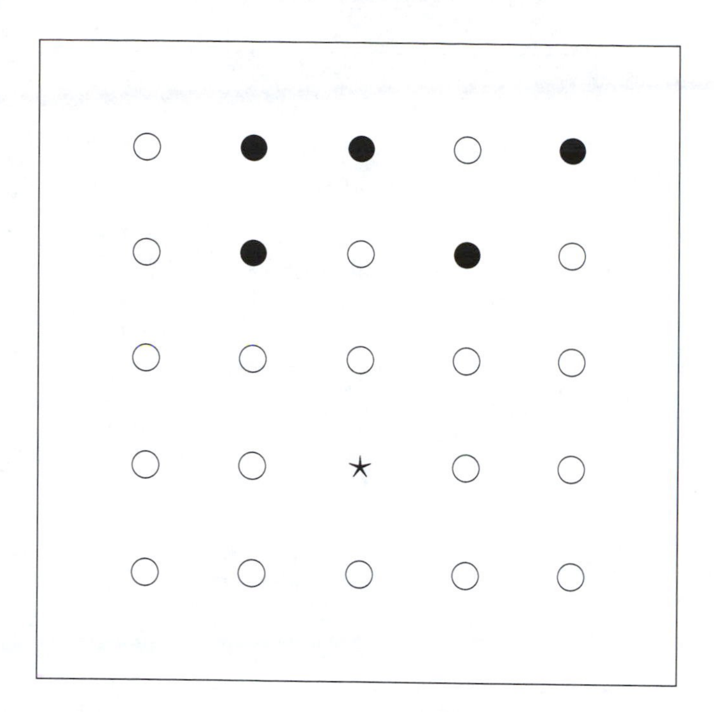

Figure 16.2: Plot 20, Week 3. $T_4 = 4 + 5 = 9$.

Calculating T_4 is easy, as only one new roll was attacked in Week 3. In Figure 16.2, the rolls in C_3 are denoted by $\bullet$ and roll r_4 is denoted by $\star$. The squared distance of $\star$ from the previously attacked rolls is 5 and

$$T_4 = T_4 (r_1, \ldots, r_4; C_0) = T_3 + 5 = 4 + 5 = 9.$$

Calculating T_{11} is potentially difficult, as seven new rolls were attacked in Week 4. In Figure 16.3, the rolls in C_4 are denoted by $\bullet$ and rolls $r_5, \ldots, r_{11}$ are denoted by $\star$. We see that each $\star$ is adjacent to a $\bullet$; hence, regardless of how we order the new rolls,

$$T_{11} = T_{11} (r_1, \ldots, r_{11}; C_0) = T_4 + 7 \cdot 1 = 9 + 7 = 16.$$

Finally, two new rolls were attacked in Week 5. In Figure 16.4, the rolls in C_{11} are denoted by $\bullet$ and rolls r_{12} and r_{13} are denoted by $\star$. Again, it does not matter which $\star$ is r_{12} and which is r_{13}:

$$T_{13} = T_{13} (r_1, \ldots, r_{13}; C_0) = T_{11} + (1 + 4) = 16 + 5 = 21.$$

Despite our inability to determine a unique attack order, the value of our test statistic is unambiguously $T_{13} = 21$. Noting that the smallest possible

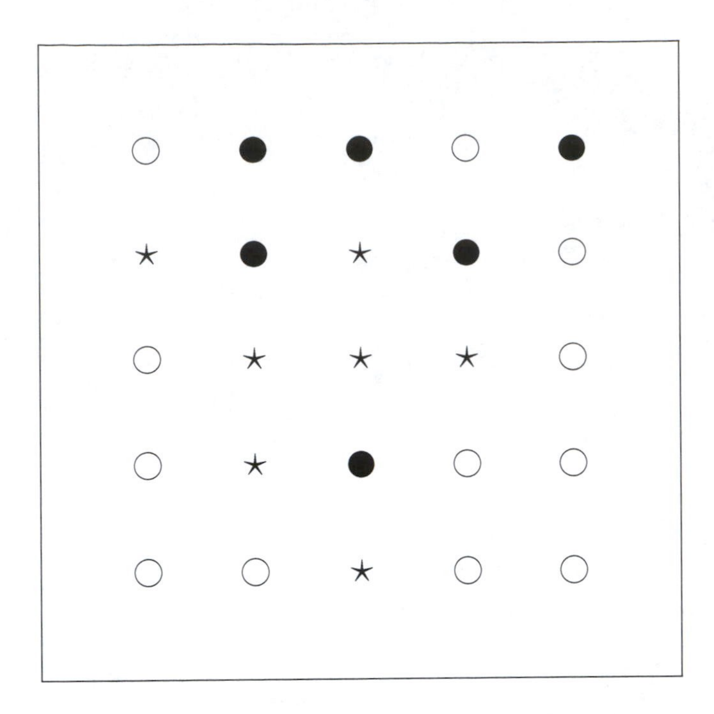

Figure 16.3: Plot 20, Week 4. $T_{11} = 9 + 7 \cdot 1 = 16$.

value of T_{13} is 13, the value of $T_{13} = 21$ seems rather small. But is it small enough to provide compelling evidence against the null hypothesis of random foraging? To answer that question, Jones et al. resorted to simulation. Given C_0 and $J = 13$, they simulated one million random attack sequences, computed T_{13} for each simulated attack sequence, tabulated the computed values, and calculated the proportion of simulated attack sequences for which $T_{13} \leq 21$. The results, displayed in Table 16.1, give an estimated significance probability of $48501/1000000 \doteq 0.0485$.[2]

The simulations reported in 1987 by Jones et al. were written in the C programming language and executed on an IBM PC/XT with 1 MB of RAM and a 10 MB hard drive. The CPU times varied by plot, according to C_0 and J. Some plots required nearly 20 hours. Today's computers are much more powerful, allowing us to replicate the original simulations in R. An R function that does so, `forage`, is described in Section R.3.5.

[2]Most of the other plots produced estimated significance probabilities greater than 0.05. On balance, Jones et al. were not able to conclude that *H. aureus* forages systematically.

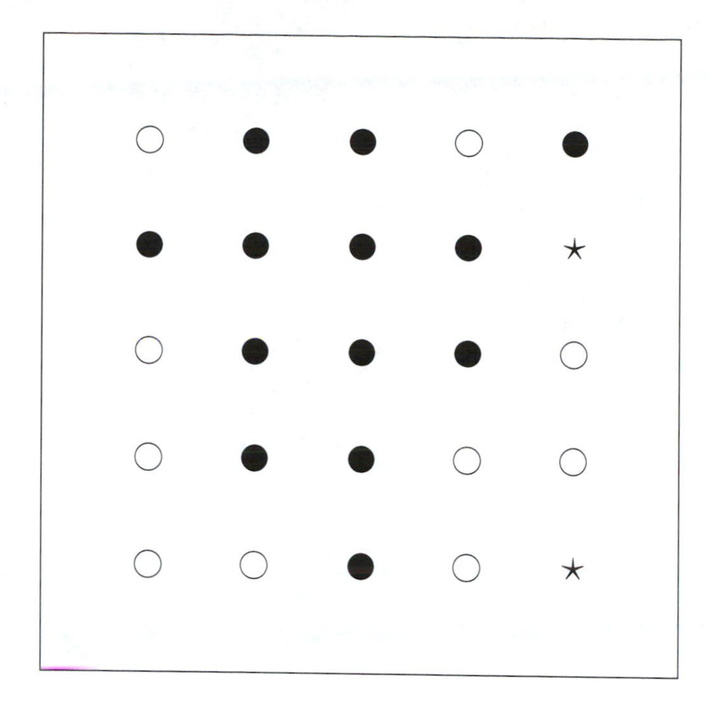

Figure 16.4: Plot 20, Week 5. $T_{13} = 16 + (1 + 4) = 21$.

16.2 The Bootstrap

One of the most useful tools in modern statistical inference is the *bootstrap*, introduced by Bradley Efron in 1979.[3] The bootstrap combines two key concepts: simulation and the plug-in principle. We illustrate the bootstrap with an example.[4]

In Section 14.2.2 we assumed that (X_i, Y_i) has a bivariate normal distribution and described a procedure for computing an approximate confidence interval for ρ, the population correlation coefficient. Recall that Fisher's z-transformation is

$$\zeta = \frac{1}{2} \log \left(\frac{1+\rho}{1-\rho} \right)$$

and that

$$\hat{\zeta} = \frac{1}{2} \log \left(\frac{1+\hat{\rho}}{1-\hat{\rho}} \right) \approx \text{Normal} \left(\zeta, \frac{1}{n-3} \right).$$

[3]B. Efron (1979). Bootstrap methods: another look at the jackknife, *Annals of Statistics*, 7(1):1–26.

[4]For a thorough but accessible exposition, see B. Efron and R. J. Tibshirani (1993), *An Introduction to the Boostrap*, Chapman & Hall/CRC, Boca Raton, FL.

T_{13}	Frequency	Cumulative Frequency
13	14	14
14	93	107
15	345	452
16	791	1243
17	2060	3303
18	4510	7813
19	7950	15763
20	12603	28366
21	20135	48501

Table 16.1: Frequency of selected values of T_{13} occurring in one million randomly generated attack sequences. The set of initial rolls, C_0, was taken to be the ● rolls in Figure 1.1.

Hence,

$$z \pm \frac{q_z}{\sqrt{n-3}}$$

is an approximate $(1-\alpha)$-level confidence interval for ζ, where

$$z = \frac{1}{2} \log \left(\frac{1+r}{1-r} \right)$$

and r is the sample correlation coefficient. After constructing a confidence interval for ζ, one applies an inverse transformation to recover a confidence interval for ρ.

Knowing that $\operatorname{Var} \hat{\zeta} \approx 1/(n-3)$ is essential to the construction. We can check this fact by simulation. The following R commands generate 10,000 pseudorandom samples of size $n = 28$ from a bivariate normal population with parameters $(\mu_x, \mu_y, \sigma_x^2, \sigma_y^2, \rho) = (0, 0, 4, 1, 0.5)$. The value of z is computed for each of the 10,000 simulated samples and the sample variance of the 10,000 values of z estimates $\operatorname{Var} \hat{\zeta}$. The resulting estimate, 0.03978227, is slightly smaller than the normal approximation, $1/(n-3) = 0.05$.

```
> pop <- c(0,0,4,1,.5)
> z <- 0
> for (b in 1:10000) {
+   Data.boot <- binorm.sample(pop,28)
```

```
+   pe <- binorm.estimate(Data.boot)
+   z.boot <- log((1+pe[5])/(1-pe[5]))/2
+   z <- c(z,z.boot)
+   }
> var(z[-1])
[1] 0.03978227
```

Notice that the quality of our estimate of Var $\hat{\zeta}$ depends on the number of pseudorandom samples. The more samples we generate, the more accurately we can estimate Var $\hat{\zeta}$.

Now suppose that (X_i, Y_i) does not have a bivariate normal distribution. The population correlation coefficient, ρ, is still defined and conveys meaningful information about the association between X_i and Y_i. We can attempt the same construction of a confidence interval for ρ, but it is not necessarily the case that Var $\hat{\zeta} \approx 1/(n-3)$.

How to proceed? If we knew how to sample from the population, then we could proceed as above, by simulation. For example, suppose that (X_i, Y_i) has a uniform distribution on a triangular region with vertices $(0,0)$, $(0,1000)$, and $(1000,1000)$. We can sample from this distribution by first drawing x_i from Uniform$(0,1000)$, then drawing y_i from Uniform$(0,x_i)$. The following R commands repeat our simulation procedure for estimating Var $\hat{\zeta}$:

```
> z <- 0
> for (b in 1:10000) {
+   x <- runif(28,max=1000)
+   y <- runif(28,max=x)
+   Data.boot <- cbind(x,y)
+   pe <- binorm.estimate(Data.boot)
+   z.boot <- log((1+pe[5])/(1-pe[5]))/2
+   z <- c(z,z.boot)
+   }
> var(z[-1])
[1] 0.04101275
```

Again, we see that Var $\hat{\zeta}$ is slightly smaller than the normal approximation, $1/(n-3) = 0.05$.

In practice, we don't know the population distribution. Thinking of Var $\hat{\zeta}$ as a population attribute, we apply the plug-in principle. Given a sample, the plug-in principle tells us to estimate an attribute of a population by computing the corresponding attribute of the empirical distribution associated with the sample. For some attributes, this is easy. Recall, for example, that the plug-in estimate of the population mean is just the sample mean. In contrast, the value of the attribute $\operatorname{Var}\hat{\zeta}$ for an empirical distribution is not easily computed. However, it is easy to generate pseudorandom samples from a discrete distribution; hence, we can use simulation to estimate the value of $\operatorname{Var}\hat{\zeta}$ for the empirical distribution, just as we previously used simulation to estimate the value of $\operatorname{Var}\hat{\zeta}$ for the population distribution. What we obtain is an approximation of the plug-in estimate.

The (nonparametric) bootstrap uses simulation to approximate the value of an intractable plug-in estimate. Here's how it works:

1. Let $\{(x_1, y_1), \ldots, (x_n, y_n)\}$ denote a sample from an unknown bivariate distribution P. Construct $\hat{P}_n$, the discrete probability distribution defined by assigning probability $1/n$ to each (x_i, y_i).

2. For $b = 1, \ldots, B$:

   (a) Generate a *bootstrap sample* by drawing n values from $\hat{P}_n$.

   (b) Compute $z(b)$, the value of z for bootstrap sample b.

3. Estimate $\operatorname{Var}\hat{\zeta}$ by computing the sample variance of $\{z(1), \ldots, z(B)\}$.

To illustrate, suppose that we observe the following sample, which we store in the 28×2 data matrix Data.

316	300	215	125	776	268	299	213
877	821	441	92	71	35	432	70
868	700	668	287	141	66	119	100
143	12	460	245	823	168	130	16
18	13	264	25	206	102	703	536
651	23	937	259	97	69	810	51
596	415	369	52	957	845	981	320

The bootstrap estimate of $\operatorname{Var}\hat{\zeta}$ was then computed by the following R commands:

```
> z <- 0
> for (b in 1:10000) {
+   i <- sample(x=1:28,size=28,replace=TRUE)
+   Data.boot <- Data[i,2]
+   pe <- binorm.estimate(Data.boot)
+   z.boot <- log((1+pe[5])/(1-pe[5]))/2
+   z <- c(z,z.boot)
+   }
```

```
> var(z[-1])
[1] 0.02722019
```

The resulting bootstrap estimate, 0.02722019, approximates the plug-in estimate. It underestimates the true value of $\text{Var}\,\hat{\zeta}$, which we know (by simulation) is approximately 0.041. We might improve the approximation of the plug-in estimate by generating more than $B = 10,000$ bootstrap samples, but the only way to improve the plug-in estimate itself is to draw more than $n = 28$ values of (X_i, Y_i) from the population.

We conclude on a cautionary note. One reason that Efron chose the name *bootstrap* was "to convey the self-help nature of the bootstrap algorithm." It was a clever choice, but the name can easily mislead naive users into thinking that the bootstrap does more than is actually possible, that somehow it magically creates something from nothing. *The bootstrap uses simulation to approximate intractable plug-in estimates*, and a plug-in estimate is only as good as the sample used to construct the empirical distribution. Although the bootstrap is a tool of extraordinary value, it does *not* create something from nothing.

16.3 Case Study: Adventure Racing

The June 2008 floods in the midwestern United States caused damage exceeding $6 billion. In Indiana, the worst flooding occurred on June 7. The floods caused Governor Mitch Daniels to declare a state of emergency in 23 counties and forced race directors to cancel the 2008 Planet Adventure (PA) Sprint Adventure Race (AR) in Cunot.[5] The race was rescheduled and took place on October 25 without incident.

Adventure racing is a team sport that combines several disciplines, typically including orienteering, paddling, mountain biking, and climbing.[6] ARs can last for days, but sprints typically last about six hours. The 2008 PA Sprint AR contained two loops, each starting and finishing at a transition area (TA) next to the Cunot Community Center. The first loop comprised three legs. First, the teams carried their packs and paddling gear from the TA to Cagle Mills Lake and paddled canoes to Lieber State Recreation Area. Second, the teams located six checkpoints on an orienteering course

[5] Perhaps there is some irony in the fact that a natural disaster forced cancellation of an *adventure* race. However, driving from Bloomington to Cunot and back was quite an adventure!

[6] The sport rose to prominence with Mark Burnett's *Eco-Challenge* races. Burnett went on to produce *Survivor*. Martin Dugard coined the phrase *adventure race*.

in Lieber SRA. Third, the teams paddled back across Cagle Mills Lake and returned to the TA. Results from the first loop are displayed in Table 16.2.

Type	Leg 1	Leg 2	Leg 3	Type	Leg 1	Leg 2	Leg 3
Male	21	80	24	Male	23	91	27
Coed	23	90	25	Male	26	83	29
Male	27	121	28	Male	26	94	30
Coed	26	77	29	Coed	26	100	31
Coed	26	76	29	Male	28	101	30
Coed	27	92	29	Coed	26	100	33
Male	26	102	30	Female	29	102	30
Coed	26	81	30	Male	31	73	29
Coed	27	92	29	Male	30	91	31
Male	27	131	32	Coed	31	99	31
Male	27	103	33	Male	30	144	32
Coed	29	154	31	Male	35	143	27
Male	28	88	32	Male	28	91	37
Female	30	145	34	Coed	31	163	35
Male	28	123	36	Male	28	147	39
Coed	33	83	33	Coed	36	163	35
Male	32	160	37	Coed	36	170	37
				Coed	42	48	48

Table 16.2: Loop 1 times (in minutes) for 35 teams in the 2008 PA Sprint AR. Three-person teams are displayed on the left; two-person teams are displayed on the right. Legs 1 and 3 involve paddling a canoe; Leg 2 involves orienteering. Two teams that missed an orienteering checkpoint are omitted.

Two obvious factors that affect orienteering time are fitness and navigational skill. Because one team member typically navigates and each team member runs (or jogs or walks) individually, a three-person team should not enjoy a competitive advantage over a two-person team.[7]

The paddle legs in this AR required scant navigational skill. Two obvious factors that affect paddling time are fitness and paddling technique. In contrast to orienteering, a team works together to propel one canoe. Thus, it is plausible that a three-person team might enjoy a competitive advantage over a two-person team. We proceed to investigate this possibility.

[7]It is possible to tether a slow runner to a fast runner, but tethers were infrequently used in this AR.

Before proceeding, we should concede that it is difficult to construe these data as a random sample. Teams do not compete independently. They are aware of each other, strive to keep pace, and sometimes follow each other when navigating. Nevertheless, the methods that we have developed can help to reveal patterns in these data. It is when one attempts to draw conclusions from these patterns that one must be cautious.

Summing the times for Legs 1 and 2, the average times for 17 three-person and 18 two-person teams are as follows:

	Orienteer	Paddle
3-person	105.8	57.9
2-person	117.8	62.9

On average, the three-person teams orienteered 12 minutes faster and paddled 5 minutes faster than the two-person teams. The superiority of the three-person teams is not surprising, as elite racers tend to compete on three-person teams.

The mean times do not provide evidence of a differential advantage in paddling. To look for one, we begin by adjusting each team's paddling time in light of its general AR prowess. The only quantity available for measuring general prowess is orienteering time, so we regress paddling time (y) on orienteering time (x). Figure 16.5 displays a scatter diagram of the (x, y) pairs. Although these data do not resemble a sample from a bivariate normal distribution, the fit of the sample regression line is impressive ($r^2 \doteq 0.48$).

The adjusted paddling times are just the residuals from regressing y on x. The 17 three-person teams have a mean residual of -1.509880 minutes and the 18 two-person teams have a mean residual of 1.425998 minutes. The difference is not large, but it is the case that the three-person teams tended to paddle faster than predicted by orienteering time and the two-person teams tended to paddle slower.[8]

Is the difference statistically significant? Here we must be cautious, as the statistical assumptions that we are accustomed to making are suspect. Nevertheless, suppose that the residuals constitute a random sample from a univariate probability distribution, which we estimate by the empirical distribution of the residuals. Suppose that we draw random samples of 17 hypothetical three-person teams and 18 hypothetical two-person teams from this distribution. What is the probability that the sample mean of the latter will exceed the sample mean of the former by at least $1.425998 - (-1.509880) = 2.935877$ minutes? Simulation suggests that the

[8]Because the residuals must sum to zero, these statements are equivalent.

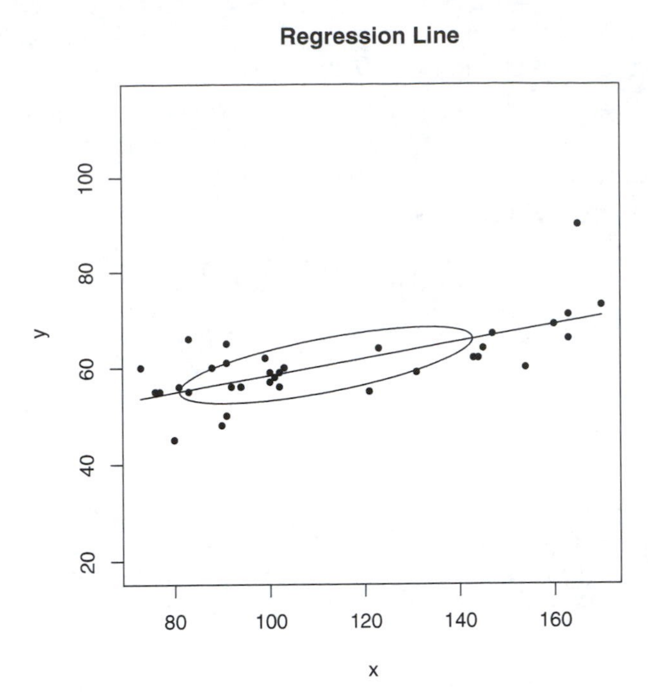

Figure 16.5: Scatter diagram of orienteering time (x) and total paddling time (y) for $n = 35$ teams in the 2008 PA Sprint AR. The sample concentration ellipse and the sample regression line are superimposed.

probability is about 6 percent, not quite small enough to claim an effect at the conventional sigificance level of $\alpha = 0.05$.

Notice that we have included coed, male, and female teams in our analysis. We might attempt to refine our analysis by considering each type of team separately. This is not possible for female teams, as only two female teams competed.[9] Repeating the entire analysis for male teams, the 8 three-man teams had a mean residual of -1.182086 minutes and the 10 two-man

[9]Using residuals from the full analysis, *Not Your Average Moms*, a two-women team, had an adjusted paddle time of 0.274 minutes and *Carpe Diem*, a three-women team, had an adjusted paddle time of -2.337 minutes. These quantities might be construed to mean that having a third team member gave *Carpe Diem* a competitive advantage on the paddling legs. Certainly, *Not Your Average Moms* could have used a third team member while *carrying* their canoe!

teams had a mean residual of 0.9456687 minutes. Simulation produced an estimated significance probability of about 16 percent. Repeating the entire analysis for coed teams, the 8 three-person teams had a mean residual of −1.306190 minutes and the 7 two-person teams had a mean residual of 1.492788 minutes. Simulation produced an estimated significance probability of about 20 percent.

Our results are provocative. In each case, the three-person teams had faster adjusted paddle times than the two-person teams. However, the numbers of teams involved are too small to produce definitive conclusions. As is so often the case, the statistician is obliged to request more data. Adventure racers are not likely to perform a controlled experiment, but I have no doubt that they will be delighted to generate more observational data!

16.4 Exercises

1. For the termite foraging experiment described in Section 1.1.3, suppose that a plot was initially attacked at its center roll and that $J = 2$ rolls were subsequently attacked.

   (a) How many attack sequences, $\{r_1, r_2\}$, are possible?

   (b) How many attack sequences result in $T_2 = 2$?

   (c) Compute $P(T_2 \leq 2)$ under the null hypothesis of random foraging.

2. Repeat the preceding exercise, but suppose that the plot was initially attacked at a corner roll instead of the center roll.

3. Figure 16.6 displays the data from Plot 21 in the termite foraging experiment described in Section 1.1.3.

   (a) Compute the value of T_8, the test statistic described in Section 16.1. Although the data displayed in Figure 16.6 do not determine a unique order of attack, $\{r_1, \ldots, r_8\}$, they do determine a unique value of T_8.

   (b) Use the R function `forage`, described in Section R.3.5, to estimate the probability that random foraging would produce a value of T_8 as small or smaller than the value observed in Plot 21. Does Plot 21 provide compelling evidence that termites forage systematically?

Figure 16.6: Order of *H. aureus* attack in Plot 21.

4. Figure 16.7 displays the data from Plot 29 in the termite foraging experiment described in Section 1.1.3.

   (a) The data displayed in Figure 16.7 do not determine a unique order of attack, $\{r_1, \ldots, r_{12}\}$. Furthermore, different orders lead to different values of T_{12}. Compute the smallest and the largest values of T_{12}, $t_{\min}$ and $t_{\max}$, that are consistent with the data displayed in Figure 16.7.

   (b) Use the R function `forage` to estimate significance probabilities corresponding to $t_{\min}$ and $t_{\max}$. Does Plot 29 provide compelling evidence that termites forage systematically?

5. Bootstrap samples are drawn from the empirical distribution. Thus, if the observed sample is $\vec{x} = \{x_1, \ldots, x_n\}$, then each bootstrap sample is generated by drawing n times *with replacement* from the set $\{x_1, \ldots, x_n\}$.

   Suppose that $n = 10$ and that $\{x_1, \ldots, x_{10}\}$ represent 10 distinct numbers (no ties).

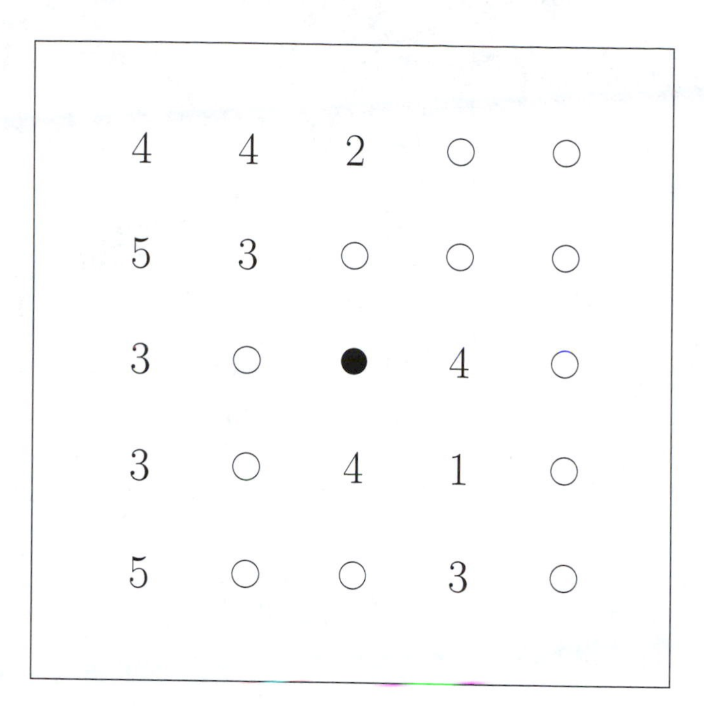

Figure 16.7: Order of *H. aureus* attack in Plot 29.

(a) How many ordered sequences of 10 numbers can be drawn from $\{x_1, \ldots, x_{10}\}$?

(b) How many ordered sequences can be drawn that contain 10 distinct values (no ties)?

(c) What is the probability that a bootstrap sample will contain 10 distinct values?

(d) How many ordered sequences can be drawn that contain a single value (one value repeated 10 times)?

(e) What is the probability that a bootstrap sample will contain a single value?

6. From the 28×2 data matrix reported in Section 16.2 we obtained a bootstrap estimate of $\text{Var}\,\hat{\zeta}$. Use this estimate (0.02722019) and a confidence coefficient of $1 - \alpha = 0.95$ to construct a confidence interval for ρ, the population correlation coefficient.

7. Let θ denote the population median and let $\hat{\theta}$ denote the plug-in estimator of θ, i.e., the sample median. The following random sample

was drawn from a continuous distribution:

0.240	4.360	0.613	0.638	1.985
3.051	2.022	1.875	4.519	2.303
0.759	2.768	1.359	0.444	0.984
4.492	1.252	0.794	3.163	0.759
4.406	2.850	0.969	0.842	1.591

(a) Compute the value of $\hat{\theta}$ for the above sample.

(b) Use the technique described in Section 10.2.3 to construct a confidence interval for θ that has a confidence level of approximately 0.90. What is the exact confidence level of this confidence interval?

(c) Compute $\hat{v}$, the bootstrap estimate of $\operatorname{Var} \hat{\theta}$.

(d) Construct the confidence interval $\hat{\theta} \pm 1.645 \cdot \sqrt{\hat{v}}$. The choice of the multiplier $1.645 \doteq$ `qnorm`(0.95) is intended to give a confidence interval that has a confidence level of approximately 0.90. What is the exact confidence level of this confidence interval?

8. In Section 16.3, we estimated significance probabilities by simulation. To do so, we performed one simple linear regression on the observed data, then resampled from the empirical distribution of the residuals. Another approach would be to resample from the entire bivariate distribution (as in Section 16.2), performing a simple linear regression for each bootstrap sample. Try this approach. Do your results differ substantially from the results in Section 16.3?

Appendix R

A Statistical Programming Language

R.1 Introduction

R.1.1 What Is R?

In the 1970s, researchers at AT&T Bell Laboratories developed `S`, a high-level statistical programming language that became popular with academic statisticians. Bell Labs subsequently licensed `S` to a company that added a variety of capabilities, creating the commercial product `S-Plus`. `R` is yet another implementation of `S`. The R Project for Statistical Computing is an ongoing effort by a group of statisticians to extend and improve `R`.

`R` is free, open-source software that can be downloaded in compiled or source code form. It runs on a variety of UNIX platforms and similar systems (including FreeBSD and Linux), Windows, and MacOS. The primary web site for information about `R` is:

`http://www.r-project.org/`

R.1.2 Why Use R?

This question encompasses several issues. First, there is the question of what role statistical software is to play in an introductory statistics course. Such courses may use software in different ways. Once upon a time, many instructors (myself included) avoided using software in the first semester. The rationale for doing so was that one should begin one's study of statistics by focusing on basic concepts and learn what the computer is doing before one uses the computer to do it. Unfortunately, this approach condemns one to analyzing fairly trivial data sets, and even then calculating by hand and/or calculator quickly becomes extremely tedious. As a result, this approach has fallen from favor.

At the other end of the spectrum, many introductory statistics courses use statistics packages like Minitab, SPSS, or SAS to analyze data. Such packages are extremely useful and every statistician should have some familiarity with at least one such package. However, if one begins to rely on such packages too quickly, the package may be viewed as a black box and the student may never really learn what that black box is doing.

There are many different ways to introduce the subject of statistics, and no one way is best for all students. This book is intended for students who want to *understand* what is going on inside the black box of procedures available in so many statistics packages. This intention determines the use that we shall make of the computer. We will strive for an intermediate approach, in which the computer is used to relieve the tedium of calculation, but in which the student is obliged to tell the computer what intermediate steps need to be performed in order to obtain the desired output. Such an approach requires a high-level, interactive programming language. Several such languages are available, but `S-Plus` and `R` have achieved the greatest popularity within the statistics community. Acquiring some familiarity with `S-Plus` and/or `R` will benefit students who continue to study statistics and/or analyze data in the future.

Why `R` instead of `S-Plus`? For most of the examples in this book, `R` and `S-Plus` are interchangeable—the same commands work for both. But `R` has two compelling advantages. First, `R` is available for certain operating systems for which `S-Plus` is not, e.g., MacOS. Second, `R` is free! As a result, students who begin using `R` in this course can be confident that they will always have access to `R`.

R.1.3 Installing R

To efficiently download software, documentation, etc., you should use a nearby CRAN (Comprehensive R Archive Network) mirror site, e.g., Statlib at Carnegie Mellon University:

`http://lib.stat.cmu.edu/R/CRAN/`

Most students will want to install `R` in compiled form by downloading executable binary files. On-line documentation and several manuals are included, although you may find it easier to get started using the examples provided in this book.

R.1.4 Learning about R

`R` is far too complicated to learn in one (or even several) lessons. I doubt that any one person—including the `R` developers—knows everything about `R`! But don't be intimidated: *the best way to learn* `R` *is to just start using* `R`. And, the best time to use `R` is when you're trying to accomplish a specific task. Try to learn bits and pieces of `R` as they're introduced in the text and/or as you develop an interest in a specific capability.

Of course, it's hard to learn anything without documentation. The material in this book, both the examples scattered throughout various chapters to illustrate various statistical methods and the tutorial material in this appendix, is a good way to get started. *I strongly suggest that you read the tutorial material in this appendix in front of a computer and that you try the various commands as you read about them.* Once you know the name of one `R` function, you can learn more about it and discover related functions using various utilities included in your `R` installation. If you're using the Windows version of `R` then you can start by exploring the `Help` menu in `RGui`, which will lead you to manuals, search utilities, and web pages. I tend to use `R functions (text)` for help on specific functions.

R.2 Using R

`R` is an interpreted language, designed to be used interactively. The user is prompted to issue a command as follows:

```
>
```

The cursor-up key allows the user to recall previous commands.

Except for a few standard arithmetic operations, `R` accomplishes things by executing various functions. For example, to exit `R` one executes the `quit` function:

```
> q()
```

When you quit, `R` will inquire if you want to "Save workspace image?" If you answer `yes` (`y`), then all of the objects in your current workspace, e.g., any data sets and functions that you created, will be saved and restored the next time that you start `R`.

R.2.1 Vectors

R can store and manipulate a variety of data objects, the most basic of which is a vector. In R a vector is an ordered list of numbers, i.e., a list of numbers with a designated first element, second element, etc. Vectors can be created in various ways. In each of the following examples, the created vector is assigned the name x.

Note that R has a large number of built-in functions. Assigning their names to user-created objects will mask the built-in functions. For this reason certain simple names, e.g., c and t, should be avoided.

Example R.1 To enter a list of numbers from the keyboard, use the concatenate function:

```
> x <- c(20,5,15,18,5,13,1)
```

Notice that this can be done recursively, e.g.,

```
> x <- c(20,5,15)
> x <- c(x,18,5)
> x <- c(x,13,1)
```

To display the vector, type its name:

```
> x
```

Just typing

```
> c(20,5,15,18,5,13,1)
```

causes R to display the vector without saving it for future use.

Example R.2 To read a list of numbers from an ascii text file, say data.txt, use the scan function. In most situations, you will need to specify the complete path of data.txt. How one does this depends on which operating system your computer uses.

For example, suppose that you are using the Windows version of R and data.txt resides in the directory c:\Stat320. Then the following command will read the contents of data.txt into the vector x:

```
> x <- scan("c:\\Stat320\\data.txt")
```

Notice that the single slashes in a Windows path name must be entered as double slashes in R.

Example R.3 Several functions are useful for creating sequences of numbers, e.g.,

```
> x <- seq(from=1,to=15,by=2)
> x <- rep(1,times=10)
```

Consecutive integers are especially easy, e.g.,

```
> x <- 11:20
```

Example R.4 R has a variety of functions for generating pseudorandom samples.[1]

To draw 10 numbers from a uniform distribution on $(0, \pi)$:

```
> x <- runif(10,min=0,max=pi)
```

To draw 20 numbers from a normal distribution with mean 5 and standard deviation 1.5:

```
> x <- rnorm(20,mean=5,sd=1.5)
```

To simulate rolling a fair die 30 times:

```
> die <- 1:6
> x <- sample(x=die,size=30,replace=TRUE)
```

A subset of a vector can be identified by a vector of index values. For example, to extract the 2nd, 3rd, and 5th elements of the vector x, one might type:

```
> k <- c(2,3,5)
> x[k]
```

To extract the other elements, just type:

```
> x[-k]
```

One may wish to rearrange the elements, e.g.,

```
> y <- sort(x)
```

The preceding command is equivalent to

```
> y <- x[order(x)]
```

[1] The precise meanings of the phrases that follow are explained in Chapters 3–5.

R.2.2 R Is a Calculator!

R provides a variety of arithmetical operations and mathematical functions. These operations/functions have been vectorized, i.e., they work on entire vectors, not just individual numbers. Several examples follow.

First, let's create two vectors:

```
> x <- 10:20
> y <- seq(from=1.8,to=2.2,length=length(x))
```

Now, each of the following is a valid R command:

```
> x+100
> x-20
> x*10
> x/10
> x^2
> sqrt(x)
> exp(x)
> log(x)
> x+y
> x-y
> x*y
> x/y
> x^y
```

R.2.3 Some Statistics Functions

R provides hundreds of functions that perform or facilitate a variety of statistical analyses. Most R functions are not used in this book. (You may enjoy discovering and using some of them on your own initiative.) Tables R.1 and R.2 list some of the R functions that are used.

R.2.4 Matrices

On certain occasions it is useful to organize vectors of numbers into rectangular arrays (matrices). Suppose that x and y are vectors of equal length, say 10. Then the R command

```
> M <- cbind(x,y)
```

creates a matrix with 10 rows and 2 columns, the first column containing the elements of x and the second column containing the elements of y. The

Function	Distribution	Section
pgeom	Geometric	4.2
phyper	Hypergeometric	4.2
pbinom	Binomial	4.4
punif	Uniform	5.3
pnorm	Normal	5.4
pchisq	Chi-Squared	5.5
pt	Student's t	5.5
pf	F	5.5

Table R.1: Some `R` functions that evaluate the cumulative distribution function (cdf) for various families of probability distributions. The prefix `p` designates a cdf function; the remainder of the function name specifies the distribution. For the analogous quantile functions, use the prefix `q`, e.g., `qnorm`. To evaluate the analogous probability mass function (pmf) or probability density function (pdf), use the prefix `d`, e.g., `dnorm`. To generate a pseudorandom sample, use the prefix `r`, e.g., `rnorm`.

`transpose` function transposes `M` into a matrix `N` with 2 rows and 10 columns, the first row containing the elements of `x` and the second row containing the elements of `y`:

```
> N <- t(M)
```

Alternatively, we could create `N` directly, as follows:

```
> N <- rbind(x,y)
```

The `matrix` function organizes the elements of a vector into a matrix of specified dimensions. By default, `matrix` fills by column, not row. Compare the results of the following commands:

```
> matrix(1:12,nrow=4,ncol=3)

> matrix(1:12,byrow=TRUE,nrow=4,ncol=3)
```

The `matrix` function provides yet another way of creating `M` and `N`:

```
> M <- matrix(c(x,y),ncol=2)
> N <- matrix(c(x,y),byrow=TRUE,nrow=2)
```

Function	Used to Compute/Display
`sum`	sample sum
`mean`	sample mean
`median`	sample median
`var`	sample variance
`quantile`	sample quantile(s)
`summary`	several useful quantities
`plot.ecdf`	empirical cdf
`boxplot`	box plot(s)
`qqnorm`	normal probability plot
`plot, density`	kernel estimate of pdf

Table R.2: Some `R` functions that compute or display useful information about one or more univariate samples. See Chapter 7.

The `matrix` function is especially useful when used in conjunction with the `scan` function. For example, Chapters 14 and 15 describe procedures for analyzing $n \times 2$ data matrices. If such a matrix is stored in the text file `c:\Stat320\data.txt`, then the following commands recreate the matrix in R:

```
> v <- scan("c:\\Stat320\\data.txt")
> M <- matrix(v,byrow=TRUE,ncol=2)
```

If one desires, one can then extract vectors, `x` and `y`, that correspond to the first and second columns of `M`:

```
> x <- M[,1]
> y <- M[,2]
```

One reason why it may be useful to organize numbers into matrices is that one can then use the `apply` function to perform the same operation on each row or column of a matrix. For example, the command

```
> apply(M,1,mean)
```

computes the means of each row of `M` and the command

```
> apply(M,2,mean)
```

computes the means of each column of `M`.

R.2.5 Creating New Functions

The full power of R emerges when one writes one's own functions. To illustrate, I've written a short function named **Edist** that computes the Euclidean distance between two vectors. When I type **Edist**, R displays the function:

```
> Edist
function(u,v){
return(sqrt(sum((u-v)^2)))
}
>
```

Edist has two arguments, u and v, which it interprets as vectors of equal length. **Edist** computes the vector of differences, squares each difference, sums the squares, then takes the square root of the sum to obtain the distance. Finally, it returns the computed distance. I could have written **Edist** as a sequence of intermediate steps, but there's no need to do so.

I might have created **Edist** in any of the following ways:

Example R.5

```
> Edist <- function(u,v){ return(sqrt(sum((u-v)^2))) }
>
```

Example R.6

```
> Edist <- function(u,v){
+ return(sqrt(sum((u-v)^2)))
+ }
>
```

Notice that R recognizes that the command creating **Edist** is not complete and provides continuation prompts (+) until it is.

Examples R.5 and R.6 are useful for very short functions, but not for anything complicated. Be warned: if you mistype and R cannot interpret what you did type, then R ignores the command and you have to retype it. Using the cursor-up key to recall what you typed may help, but for anything complicated it is best to create a permanent file that you can edit. This can be done within R or outside of R.

Example R.7 To create moderately complicated functions in R, use the edit function. For example, I might start by typing

```
> Edist <- function(u,v){u-v}
```

This creates an R object called Edist, but not the Edist that we want—this Edist returns the vector of differences.[2] So, I use edit to modify Edist.[3] This process is initiated with the command

```
> Edist <- edit(Edist)
```

After making and saving the desired changes to Edist, I close the editor, thereby returning control to R. R checks the edited version of Edist: if R can interpret the edited version, then R replaces the previous version with the edited version; if R cannot interpret the edited version, e.g., because of typographical errors, then R issues an error message and retains the previous version. Fortunately, R also retains a temporary version of whatever modifications I attempted to make, so I have another chance at getting it right. To access the temporary version, I type

```
> Edist <- edit()
```

Note that I should *not* retype

```
> Edist <- edit(Edist)
```

as this command returns to the original unedited version and discards whatever changes I attempted to make.

Example R.8 Objects created in R can be lost, e.g., if one forgets to save one's workspace image when one quits R. For this reason, I prefer to create my R functions outside of R. To accomplish this, I first use a text editor to create an ascii text file that contains whatever R commands I want to execute, e.g., the command that creates Edist. For example, I might use the Windows Notepad editor to create an ascii text file that contains the following:

```
Edist <- function(u,v)
{
return(sqrt(sum((u-v)^2)))
}
```

[2]Using the return function is good practice, but often unnecessary. An R function will automatically return the last quantity that it computes.

[3]Each installation of R has a default editor. For the Windows operating system, the default editor is the Windows Notepad editor.

Let's suppose that I call this file `myRfcns.txt` and save it in the directory `c:\Stat320`. Then, I can start `R` and use the `source` function to execute the commands in `myRfcns.txt`:

```
> source("c:\\Stat320\\myRfcns.txt")
```

To check that I succeeded in creating `Edist`, I can produce a list of all the objects in my workspace by typing

```
> objects()
```

If I want to change `Edist`, then I simply edit `myRfcns.txt` and `source` the edited file into `R`.

Creating new `R` functions that perform repetitive tasks, e.g., calculations needed for homework exercises, can save you hours of time! Two simple examples of such functions are the `iqr` function, described in Section 7.3, and the `iqrsd` function, described in Section 7.5. The remainder of Appendix R describes several other `R` functions that I created for use with this book.

R.3 Functions That Accompany This Book

The researchers who created `S` envisioned it as a high-level programming language, a language that would contain enough building blocks that users could easily create functions that serve their special needs. The concluding section of Appendix R documents 20 `R` functions that accompany this book. These functions are reproduced for easy examination, but you should *not* retype them! Download the ascii text files that contain these functions from the web page for this book, then `source` the contents into your `R` workspace. For example, suppose that you have a Windows operating system and that you save `binorm.R` in the directory `c:\Stat320`. Then the following command instructs `R` to execute the commands in `binorm.R` that create the six `binorm` functions:

```
> source("c:\\Stat320\\binorm.R")
```

Notice that the functions that follow contain comments that describe the functions' inputs and outputs. The `#` symbol is used to insert comments, as `R` ignores lines that begin with `#`.

R.3.1 Inferences about a Center of Symmetry

Procedures for drawing inferences about a center of symmetry are discussed in Section 10.3. These procedures are based on the Wilcoxon signed rank test. The R functions `W1.p.sim` and `W1.p.norm` provide two ways to approximate a significance probability in the absence of ties; `W1.p.ties` approximates a significance probability in the presence of ties. The functions `W1.walsh` and `W1.ci` provide point estimates and confidence intervals. To use these functions, download the ascii text file `symmetric.R` from the web page for this book, then `source` its contents into your `R` workspace. For convenience, the commands in `symmetric.R` are reproduced here.

The functions `W1.p.sim`, `W1.p.ties`, and `W1.ci` rely on simulation. Each contains an argument, `draws`, that specifies the number of samples generated by simulation. The default value of `draws` is 1000, but the user can increase `draws` to obtain more accurate estimates, as in Examples 10.10 and 10.11.

```
W1.p.sim <- function(n,tplus,draws=1000) {
#
# This function approximates a significance probability
#   for the Wilcoxon signed rank test by simulation in
#   the absence of ties.
# n is the sample size, tplus is the observed value
#   of the test statistic, and draws is the number
#   of values drawn from the sampling distribution
#   under the null hypothesis.  The default value of
#   draws is 1000.
#
m <- n*(n+1)/4
d <- abs(tplus-m)
c1 <- m-d
c2 <- m+d
tot <- 0
for (i in 1:draws) {
  signs <- rbinom(n,size=1,prob=.5)
  Tplus <- sum(which(signs > .5))
  if (Tplus <= c1 || Tplus >= c2) {tot <- tot+1}
  }
return(tot/draws)
}
```

```
W1.p.norm <- function(n,tplus) {
#
# This function approximates a significance probability for
#   the Wilcoxon signed rank test by normal approximation
#   in the absence of ties.
# n is the sample size and tplus is the observed value
#   of the test statistic.
#
m <- n*(n+1)/4
sd <- sqrt(m*(2*n+1)/6)
d <- abs(tplus-m)
z <- (.5-d)/sd
return(2*pnorm(z))
}
```

```
W1.p.ties <- function(x,theta0,draws=1000) {
#
# This function estimates a significance probability
#   for the Wilcoxon signed rank test by simulation in
#   the presence of ties.
# x is the sample, theta0 is the hypothesized center
#   of symmetry, and draws is the number of draws
#   from the sampling distribution under the null
#   hypothesis.  The default value of draws is 1000.
#
n <- length(x)
m <- n*(n+1)/4
x <- x-theta0
d <- sort(abs(x))
dif <- d-c(0,d[-n])
eps <- min(dif[dif>0])/3
tot <- 0
for (i in 1:draws) {
  y <- x+runif(n,min=-eps,max=eps)
  pos <- which(y>0)
  d <- abs(y)
  r <- rank(d)
  tplus <- sum(r[pos])
```

```
  d <- abs(tplus-m)
  c1 <- m-d
  c2 <- m+d
  signs <- rbinom(n,size=1,prob=.5)
  Tplus <- sum(which(signs > .5))
  if (Tplus <= c1 || Tplus >= c2) {tot <- tot+1}
  }
return(tot/draws)
}


W1.walsh <- function(x) {
#
# This function computes the median of the Walsh averages
# for the sample x.
#
x <- matrix(x,ncol=1)
e <- matrix(1,nrow=nrow(x),ncol=1)
Sums <- x %*% t(e) + e %*% t(x)
return(median(Sums[upper.tri(Sums,diag=TRUE)])/2)
}


W1.ci <- function(x,alpha,draws=1000) {
#
#  This function constructs several possible confidence
#    intervals for a center of symmetry, each with an
#    approximated probability of coverage.
#  x is the sample, 1-alpha is the desired probability
#    of coverage, draws is the number of simulations
#    used in W1.p.sim.
#
n <- length(x)
q <- qnorm(1-alpha/2)
m <- n*(n+1)/2
k <- round(0.5+m/2-q*sqrt(m*(2*n+1)/12))
k <- max(c(k,3))
k <- (k-2):(k+2)
x <- matrix(x,ncol=1)
e <- matrix(1,nrow=n,ncol=1)
```

```
w <- x %*% t(e) + e %*% t(x)
w <- w[lower.tri(w,diag=TRUE)]/2
w <- sort(w)
m <- m-1
CI <- matrix(0,nrow=5,ncol=4)
dimnames(CI)[[2]] <- c("k","Lower","Upper","Coverage")
CI[,1] <- k
for (i in 1:5) {
  CI[i,2] <- w[k[i]]
  CI[i,3] <- w[m-k[i]]
  CI[i,4] <- 1-W1.p.sim(n,k[i]-1,draws)
  }
return(CI)
}
```

R.3.2 Inferences about a Shift Parameter

Procedures for drawing inferences about a general shift parameter are discussed in Section 11.2. These procedures are based on the Wilcoxon rank sum test. The R functions `W2.p.sim` and `W2.p.norm` provide two ways to approximate a significance probability in the absence of ties; `W2.p.ties` approximates a significance probability in the presence of ties. The functions `W2.hl` and `W2.ci` provide point estimates and confidence intervals. To use these functions, download the ascii text file `shift.R` from the web page for this book, then `source` its contents into your R workspace. For convenience, the commands in `shift.R` are reproduced here.

The functions `W2.p.sim`, `W2.p.ties`, and `W2.ci` rely on simulation. Each contains an argument, `draws`, that specifies the number of samples generated by simulation. The default value of `draws` is 1000, but the user can increase `draws` to obtain more accurate estimates, as in Examples 11.7 and 11.8.

```
W2.p.sim <- function(n1,n2,tx,draws=1000) {
#
# This function approximates a significance probability
#   for the Wilcoxon rank sum test by simulation in the
#   absence of ties.
# n1 & n2 are the sample sizes, tx is the observed value
#   of the test statistic, and draws is the number
#   of values drawn from the sampling distribution
```

```
#   under the null hypothesis.  The default value of
#   draws is 1000.
#
N <- n1+n2
m <- n1*(N+1)/2
d <- abs(tx-m)
c1 <- m-d
c2 <- m+d
tot <- 0
for (i in 1:draws) {
  subset <- sample(1:N,n1)
  Tx <- sum(subset)
  if (Tx <= c1 || Tx >= c2) {tot <- tot+1}
  }
return(tot/draws)
}


W2.p.norm <- function(n1,n2,tx) {
#
# This function approximates a significance probability
#   for the Wilcoxon rank sum test by normal approximation
#   in the absence of ties.
# n1 & n2 are the sample sizes and tx is the observed value
#   of the test statistic.
#
N <- n1+n2
m <- n1*(N+1)/2
sd <- sqrt(n1*n2*(N+1)/12)
d <- abs(tx-m)
z <-(.5-d)/sd
return(2*pnorm(z))
}


W2.p.ties <- function(x,y,Delta0=0,draws=1000) {
#
# This function approximates a significance probability
#   for the Wilcoxon rank sum test by simulation in
#   the presence of ties.
```

```
# x & y are the samples, Delta0 is the hypothesized
#   shift, and draws is the number of draws from the
#   sampling distribution under the null hypothesis.
#   The default value of draws is 1000.
#
n1 <- length(x)
n2 <- length(y)
N <- n1+n2
m <- n1*(N+1)/2
x <- x-Delta0
d <- sort(c(x,y))
dif <- d-c(0,d[-N])
eps <- min(dif[dif>0])/3
tot <- 0
for (i in 1:draws) {
  z <- c(x,y)+runif(N,min=-eps,max=eps)
  r <- rank(z)
  tx <- sum(r[1:n1])
  d <- abs(tx-m)
  c1 <- m-d
  c2 <- m+d
  subset <- sample(1:N,n1)
  Tx <- sum(subset)
  if (Tx <= c1 || Tx >= c2) {tot <- tot+1}
  }
return(tot/draws)
}


W2.hl <- function(x,y) {
#
# This function computes the Hodges-Lehmann estimate
# of the shift parameter from the samples x & y.
#
x <- matrix(x,ncol=1)
y <- matrix(y,nrow=1)
X <- x %*% matrix(1,nrow=1,ncol=ncol(y))
Y <- matrix(1,nrow=nrow(x),ncol=1) %*% y
return(median(X-Y))
}
```

```
W2.ci <- function(x,y,alpha,draws=1000) {
#
#  This function constructs several possible confidence
#    intervals for a shift parameter, each with an
#    approximated probability of coverage.
#  x & y are the samples, 1-alpha is the desired
#    probability of coverage, draws is the number of
#    simulations used in W1.p.sim.
#
n1 <- length(x)
n2 <- length(y)
q <- qnorm(1-alpha/2)
m <- n1*n2
k <- round(0.5+m/2-q*sqrt(m*(n1+n2+1)/12))
k <- max(c(k,3))
k <- (k-2):(k+2)
x <- matrix(x,ncol=1)
y <- matrix(y,nrow=1)
X <- x %*% matrix(1,nrow=1,ncol=ncol(y))
Y <- matrix(1,nrow=nrow(x),ncol=1) %*% y
d <- sort(as.vector(X-Y))
m <- m+1
CI <- matrix(0,nrow=5,ncol=4)
dimnames(CI)[[2]] <- c("k","Lower","Upper","Coverage")
CI[,1] <- k
mTx <- n1*(n1+1)/2
for (i in 1:5) {
  CI[i,2] <- d[k[i]]
  CI[i,3] <- d[m-k[i]]
  CI[i,4] <- 1-W2.p.sim(n1,n2,mTx+k[i]-1,draws)
  }
return(CI)
}
```

R.3.3 Inferences about Monotonic Association

The concept of monotonic association is discussed in Section 14.3. Two equivalent measures of monotonic association are considered, the probability of concordance (κ) and Kendall's $\tau = 2\kappa - 1$. The R functions kappa.distinct, kappa, and kappa.p.sim can be used to draw inferences about κ. To use these functions, download the ascii text file mono.R from the web page for this book, then source its contents into your R workspace. For convenience, the commands in mono.R are reproduced at the end of this section.

The function kappa.distinct accompanies Example 14.8 and performs the calculations described therein. It is not intended for general use, as it assumes that both the x-values and the y-values are distinct. Instead, use kappa to compute $\hat{\kappa}$.

Use kappa.p.sim to test the null hypothesis $H_0 : \kappa = 0.5$ versus $H_1 : \kappa \neq 0.5$. This function uses simulation to estimate a significance probability. The argument draws specifies the number of samples generated under the null hypothesis. It has a default value of 1000, but the user can increase draws to obtain more accurate estimates of **p**, as in Example 14.9.

```
kappa.distinct <- function(X) {
#
# This function computes the sample probability of concordance
# in the case of no ties.  X is an n-by-2 data matrix.
#
n <- nrow(X)
x <- X[,1]
y <- X[,2]
z <- y[order(x)]
concord <- 0
for (i in 2:n) {
  concord <- concord + sum(as.numeric(z[i:n]>z[i-1]))
  }
return(concord*2/(n^2-n))
}
```

```
kappa <- function(X) {
#
# This function computes the sample probability of concordance
# in the presence of ties.  X is an n-by-2 data matrix.
#
n <- nrow(X)
```

```
m <- (n^2-n)/2
concord <- 0
discord <- 0
for (i in 2:n) {
  x <- X[i-1,]
  Y <- matrix(X[i:n,],ncol=2)
  p <- apply(t(Y)-x,2,prod)
  p <- sign(p)
  con <- sum(p>0)
  dis <- sum(p<0)
  concord <- concord+con
  discord <- discord+dis
  }
uninform <- m-concord-discord
return((concord+uninform/2)/m)
}


kappa.p.sim <- function(X,draws=1000) {
#
# This function estimates a signficance probability
#   for testing the null hypothesis that the population
#   probability of concordance is 0.5.
# X is an n-by-2 data matrix.
# The number of simulated samples generated under H0
#   is draws, with a default value of draws=1000.
#
n <- nrow(X)
m <- (n^2-n)/2
dif <- abs(kappa(X)-.5)
tot <- 0
for (j in 1:draws) {
  z <- sample(x=1:n,size=n,replace=FALSE)
  con <- 0
  for (i in 2:n) {
    con <- con + sum(as.numeric(z[i:n]>z[i-1]))
    }
  if (abs(con/m-.5) >= dif) {tot <- tot+1}
  }
return(tot/draws)
```

```
}
```

R.3.4 Exploring Bivariate Normal Data

Chapters 14 and 15 use the following six `R` functions to explore the structure of bivariate data:

Function	Input	Output
`binorm.ellipse`	parameter vector	graphics
`binorm.sample`	parameter vector	data matrix
`binorm.estimate`	data matrix	parameter vector
`binorm.scatter`	data matrix	graphics
`binorm.regress`	data matrix	graphics
`binorm.resid`	data matrix	data matrix

To use these functions, download the ascii text file `binorm.R` from the web page for this book, then `source` its contents into your `R` workspace. For convenience, the content of `binorm.R` is reproduced at the end of this section.

The binorm functions were designed to be compatible with each other. The functions `binorm.ellipse` and `binorm.sample` require the user to input a vector of five parameters that specify a unique bivariate normal distribution. These parameters must be listed in the following order: mean of x, mean of y, variance of x, variance of y, and correlation coefficient. The functions `binorm.estimate`, `binorm.scatter`, `binorm.regress`, and `binorm.resid` require the user to input a data matrix with two columns. The first column must correspond to x, the second to y. The data matrices returned by `binorm.sample` and `binorm.resid` are suitable for input to `binorm.estimate`, `binorm.scatter`, `binorm.regress`, and `binorm.resid`. The vector of estimates returned by `binorm.estimate` is suitable for input to `binorm.ellipse` and `binorm.sample`.

```
binorm.ellipse <- function(pop) {
#
#  This function plots the concentration ellipse of a
#  bivariate normal distribution.  The 5 parameters are
#  specified in the vector pop in the following order:
#  mean of X, mean of Y, variance of X, variance of Y,
#  correlation of (X,Y).  Example: pop <- c(0,0,1,4,.5)
#
ndots <- 628
```

```
m <- matrix(pop[1:2], nrow = 2)
off <- pop[5] * sqrt(pop[3] * pop[4])
C <- matrix(c(pop[3], off, off, pop[4]), nrow = 2)
E <- eigen(C,symmetric=TRUE)
a <- 0:ndots/100
X <- cbind(cos(a), sin(a))
X <- X %*% diag(sqrt(E$values)) %*% t(E$vectors)
X <- X + matrix(rep(1, ndots + 1), ncol = 1) %*% t(m)
xmin <- min(X[, 1])
xmax <- max(X[, 1])
ymin <- min(X[, 2])
ymax <- max(X[, 2])
dif <- max(xmax - xmin, ymax - ymin)
xlim <- c(m[1] - dif, m[1] + dif)
ylim <- c(m[2] - dif, m[2] + dif)
par(pty = "s")
plot(X,type="l",xlab="x",ylab="y",xlim=xlim,ylim=ylim)
title("Concentration Ellipse")
}


binorm.sample <- function(pop,n) {
#
#  This function returns a sample of n observations drawn
#  from a bivariate normal distribution.  The 5 parameters
#  are specified in the vector pop in the following order:
#  mean of X, mean of Y, variance of X, variance of Y,
#  correlation of (X,Y).  Example: pop <- c(0,0,1,4,.5)
#  The sample is returned as an n-by-2 data matrix,
#  each row of which is an observed value of (X,Y).
#
m <- matrix(pop[1:2], nrow = 2)
off <- pop[5] * sqrt(pop[3] * pop[4])
C <- matrix(c(pop[3], off, off, pop[4]), nrow = 2)
E <- eigen(C,symmetric=TRUE)
Data <- matrix(rnorm(2 * n), nrow = n)
Data <- Data %*% diag(sqrt(E$values)) %*% t(E$vectors)
return(Data + matrix(rep(1, n), nrow = n) %*% t(m))
}
```

```
binorm.estimate <- function(Data) {
#
#  This function estimates bivariate normal parameters.
#  Each row of the n-by-2 matrix Data contains a single
#  observation of (X,Y).  The function returns a vector
#  of 5 estimates: mean of X, mean of Y, variance of X,
#  variance of Y, correlation of (X,Y).
#
n <- nrow(Data)
m <- c(sum(Data[, 1]), sum(Data[, 2]))/n
v <- c(var(Data[, 1]), var(Data[, 2]))
z1 <- (Data[, 1] - m[1])/sqrt(v[1])
z2 <- (Data[, 2] - m[2])/sqrt(v[2])
r <- sum(z1 * z2)/(n - 1)
return(c(m,v,r))
}
```

```
binorm.scatter <- function(Data) {
#
#  This function plots a scatter diagram of the data
#  contained in the n-by-2 data matrix Data.  It also
#  superimposes the sample concentration ellipse.
#
ndots <- 628
xmin <- min(Data[, 1])
xmax <- max(Data[, 1])
xmid <- (xmin + xmax)/2
ymin <- min(Data[, 2])
ymax <- max(Data[, 2])
ymid <- (ymin + ymax)/2
dif <- max(xmax - xmin, ymax - ymin)/2
xlim <- c(xmid - dif, xmid + dif)
ylim <- c(ymid - dif, ymid + dif)
par(pty = "s")
plot(Data,xlab="x",ylab="y",xlim=xlim,ylim=ylim,
     pch=".",cex=2)
#
#  Value of cex sets size of plotting symbol.
```

```
#
title("Scatter Diagram")
v <- binorm.estimate(Data)
m <- matrix(v[1:2], nrow = 2)
off <- v[5] * sqrt(v[3] * v[4])
C <- matrix(c(v[3], off, off, v[4]), nrow = 2)
E <- eigen(C,symmetric=TRUE)
a <- 1:ndots/100
Y <- cbind(cos(a), sin(a))
Y <- Y %*% diag(sqrt(E$values)) %*% t(E$vectors)
@
Y <- Y + matrix(rep(1, ndots), nrow = ndots) %*% t(m)
lines(Y)
}


binorm.regress <- function(Data) {
#
#  This function plots a scatter diagram of the data
#  contained in the n-by-2 data matrix Data.  It also
#  superimposes the sample concentration ellipse and
#  the regression line.
#
ndots <- 628
xmin <- min(Data[, 1])
xmax <- max(Data[, 1])
xmid <- (xmin + xmax)/2
ymin <- min(Data[, 2])
ymax <- max(Data[, 2])
ymid <- (ymin + ymax)/2
dif <- max(xmax - xmin, ymax - ymin)/2
xlim <- c(xmid - dif, xmid + dif)
ylim <- c(ymid - dif, ymid + dif)
par(pty = "s")
plot(Data,xlab="x",ylab="y",xlim=xlim,ylim=ylim,
     pch=".",cex=2)
#
#  Value of cex sets size of plotting symbol.
#
title("Regression Line")
```

```
v <- binorm.estimate(Data)
m <- matrix(v[1:2], nrow = 2)
off <- v[5] * sqrt(v[3] * v[4])
C <- matrix(c(v[3], off, off, v[4]), nrow = 2)
E <- eigen(C,symmetric=TRUE)
a <- 0:ndots/100
Y <- cbind(cos(a), sin(a))
Y <- Y %*% diag(sqrt(E$values)) %*% t(E$vectors)
Y <- Y + matrix(rep(1, ndots + 1), ncol = 1) %*% t(m)
lines(Y)
x <- xlim[1] + (2 * dif * (0:ndots))/ndots
slope <- v[5] * sqrt(v[4]/v[3])
y <- v[2] + slope * (x - v[1])
Y <- cbind(x, y)
Y <- Y[Y[, 2] < ymax, ]
Y <- Y[Y[, 2] > ymin, ]
lines(Y)
}


binorm.resid <- function(Data) {
#
#  This function computes residuals from simple linear
#    regression.  The n-by-2 matrix Data contains x in
#    the first column and y in the second column.
#  It returns an n-by-2 data matrix that contains x in
#    the first column and the residual errors in the
#    second column.
#
v <- binorm.estimate(Data)
slope <- v[5] * sqrt(v[4]/v[3])
yhat <- v[2] + slope * (Data[,1] - v[1])
resid <- Data[,2] - yhat
return(cbind(Data[,1],resid))
}
```

R.3.5 Simulating Random Termite Foraging

Sections 1.1.3 and 16.1 describe a study of termite foraging behavior. The test statistic, T, assumes a small value when each subsequently attacked roll is near a previously attacked roll. Thus, small values of T are evidence against a null hypothesis of random foraging, under which each unattacked roll is equally likely to be attacked next. To compute a significance probability for a particular plot, e.g., Plot 20 depicted in Figure 1.1, we require the probability distribution of T. This discrete distribution cannot be calculated by the methods of Chapter 4; instead, we resort to computer simulation.

Dana Ranschaert and I created an `R` function, `forage`, that approximates the pmf of T by simulation. To use `forage`, download the ascii text file `termites.R` from the web page for this book, then `source` its contents into your `R` workspace. For convenience, the commands in `termites.R` are reproduced at the end of this section.

To use `forage`, one must specify four arguments:

1. `initial`, a vector that contains the numbers of the initially attacked rolls. The 5×5 rolls are numbered as follows:

1	2	3	4	5
6	7	8	9	10
11	12	13	14	15
16	17	18	19	20
21	22	23	24	25

   In Figure 1.1, the vector of initially attacked rolls is `c(3,5)`.

2. `nsubsequent`, the number of subsequently attacked rolls. In Figure 1.1, there are 13 such rolls.

3. `nsim`, the number of simulated foraging histories. In the original study, each plot was simulated 1 million times.

4. `maxT`, the largest value of T to be tabulated.

For example, the command

```
> pmf20 <- forage(c(3,5),13,10000,30)
```

computes a matrix with $30-13+1$ rows and 2 columns. The first column of `pmf20` contains values of T, from 13 to 30. Corresponding to each value of T, the corresponding number in the second column of `pmf20` tabulates how many of the 10000 simulated foraging histories produced that value of T.

```
forage <- function(initial,nsubsequent,nsim,maxT) {
#
# This function simulates nsim termite foraging histories.
# initial is the vector of initially attacked rolls;
# nsim is the number of subsequently attacked rolls.
# The function returns a matrix in which the first column
# contains values of the test statistic T (from nsubsequent
# to maxT) and the second column contains the corresponding
# number of histories that produced that value of T.
#
v <- rep(1:5,5)
w <- rep(1:5,rep(5,5))
D <- cbind(v,w)
D <- (diag(25) - matrix(1/25,25,25)) %*% D
D <- D %*% t(D)
v <- diag(D)
H <- diag(v) %*% matrix(1,25,25)
D <- H+t(H)-2*D
D[D<0] <- 0
H <- matrix(100,25,25)
for (rowi in 2:25)
  for (colj in 1:(rowi-1)) {
    H[rowi,colj] <- 0
    }
v <- 1:length(initial)
w <- 1:(length(initial)+nsubsequent)
pmf <- rep(0,maxT)
for (isim in 1:nsim){
  rolls <- c(initial, sample(x=(1:25)[-initial],
             size=nsubsequent, replace=FALSE))
  D0 <- D[rolls,rolls] + H[w,w]
  distance <- apply(D0,1,min)
  total <- round(sum(distance[-v]))
  if (total < maxT+0.5) {
    pmf[total] <- pmf[total]+1
    }
return(cbind(nsubsequent:maxT,pmf[-(1:(nsubsequent-1))]))
}
```

Index